AF815662

ORPHAN

HOLD
DO NOT SHIP

H1 – 000000001148

ISBN 9780412005411

Producers and Scroungers Strategics of Exploitation and P...

CLIENT Alternate Sourcing

STATION TPP Receiving Station 02

HELD 2026 – 10 – 30 12:03

SHELF

If its order arrives, scan this label at a receiving station.

PRODUCERS AND SCROUNGERS: STRATEGIES OF EXPLOITATION AND PARASITISM

PRODUCERS AND SCROUNGERS

STRATEGIES OF EXPLOITATION AND PARASITISM

EDITED BY C.J. BARNARD

CROOM HELM
London & Sydney

CHAPMAN & HALL
New York

Croom Helm Ltd, Provident House, Burrell Row,
Beckenham, Kent BR3 1AT
Croom Helm Australia Pty Ltd, First Floor,
139 King Street, Sydney, NSW 2001, Australia

British Library Cataloguing in Publication Data

Producers and scroungers.
1. Parasites
I. Barnard, C.J.
574,5'249 QL757

ISBN 0-7099-0690-0 (Croom Helm)

First Published in the U.S.A. by
Chapman & Hall
in association with Methuen, Inc.
733 Third Ave, New York, NY 10017

Library of Congress Cataloging in Publication Data
Main entry under title:

Producers and scroungers.

Includes indexes.
1. Host-parasite relationships. 2. Parasitism.
I. Barnard, C.J. (Christopher J.)
QL757.P77 1984 591'.05249 84-1885
ISBN 0-412-00541-7 (Chapman & Hall)

Typeset by Columns of Reading
Printed and bound in Great Britain

CONTENTS

List of Contributors vii

Preface 1

1. When Cheats may Prosper *C.J. Barnard* 6

2. Host-Parasite Interrelationships: Strategies of Coexistence and Coevolution *C.R. Kennedy* 34

3. Kleptobiotic Interactions in Invertebrates *Fritz Vollrath* 61

4. The Evolution of Food-scrounging Strategies Within and Between Species *C.J. Barnard* 95

5. The Producer/Scrounger Model and its Relevance to Sexuality *G.A. Parker* 127

6. Sneaky Breeders *Anthony Arak* 154

7. Brood Parasitism Within Species *Malte Andersson* 195

8. Exploitation in Co-operative Societies: Models of Fitness Biasing in Co-operative Breeders *Sandra L. Vehrencamp* 229

9. Models of Producer/Scrounger Relationships Between and Within Species *Richard Sibly* 267

Author Index 288

Subject Index 294

LIST OF CONTRIBUTORS

MALTE ANDERSSON Department of Zoology, University of Gothenburg, PO Box 25029, S-400 31 Gothenburg, Sweden.

ANTHONY ARAK Department of Zoology, University of Cambridge, Downing Street, Cambridge CB2 3EJ, UK.

C.J. BARNARD Department of Zoology, University of Nottingham, University Park, Nottingham NG7 2RD, UK.

C.R. KENNEDY Department of Biological Sciences, The University, Exeter EX4 4PS, UK.

G.A. PARKER Department of Zoology, University of Liverpool, Brownlow Street, Liverpool L69 3BX, UK.

RICHARD SIBLY Department of Zoology, University of Reading, Whiteknights, Reading RG6 2AJ, UK.

SANDRA L. VEHRENCAMP Department of Biology, University of California, San Diego, La Jolla, California 92093, USA.

FRITZ VOLLRATH Smithsonian Tropical Research Institute, Panama, and Department of Zoology, South Parks Road, Oxford OX1 3PS, UK.

PREFACE

Many associations between organisms, both intra- and interspecific, can be usefully regarded as 'producer/scrounger' (P/S) relationships. One or more individuals or species (scroungers) within the association in some way use the behavioural or physiological investment of others (producers) to reduce their costs of obtaining a limited resource. Examples are legion: many parasites take up more or less permanent residence in or on a host individual and use the host's ingestive, digestive or circulatory processes to reduce their own feeding costs; pirate (kleptoparasitic) individuals or species exploit the foraging investment of 'host' individuals/species by stealing procured food; inter- and intraspecific brood parasites exploit the nest-building behaviour and parental care of other species/individuals; non-displaying 'satellite' males may usurp matings (kleptogamy) from higher ranking males whose display effort has attracted females to a courtship gathering, and so on.

Scroungers appear to reduce the costs of exploiting a resource by letting producers invest the necessary time and energy in foraging, building, incubating, displaying, defending, evolving anti-predator adaptations, etc., and then usurping the results of their efforts. The utility of scrounging, however, is conditional on a number of factors including the availability of producers, the number of scroungers exploiting them, the cost of scrounging in terms of producer avoidance, defence and/or retaliation, the value of the limited resource and the scope for alternative strategies of resource exploitation.

This book brings together theoretical and empirical studies of P/S relationships in a wide variety of contexts and species. It focuses in particular on the circumstances and associations in which scrounging is likely to be profitable and examines the population and community processes which encourage and limit the spread of scrounger strategies. Because of the way natural selection works, we should expect exploitative rather than mutualistic or co-operative relationships to underlie associations between organisms. The book investigates outwardly co-operative relationships as well as those which are clearly exploitative and shows how apparently mutualistic associations may be based on subtle forms of exploitation.

The major emphasis of the book is on the evolutionary relationship between producers and scroungers, particularly strategies of adaptation and counteradaptation by each party to the impact of the other. As well as including empirical studies which highlight such counteradaptive arms races, theoretical models are developed which explore evolutionary, developmental, opportunistic and motivational possibilities of different P/S relationships.

Barnard begins in Chapter 1 by distinguishing two forms of scrounger strategy: 'usurpation' and 'scrounging by deception'. 'Usurpation' refers to the theft of resources procured by another organism and Barnard points out how usurpation may evolve into manipulation whereby other organisms are influenced into procuring resources or making them available at a time and place suitable to the scrounger. A simple conceptual model is then developed which views interactions between organisms as detector/detectee relationships and shows how these provide the driving force for the evolution of different 'scrounging by deception' strategies. Deception here refers to the mimicking of other organisms (models) which enhances the fitness of the mimic in terms of, for instance, access to limited resources or reduced risk of predation, but usually decreases the fitness of the model. Inter- and intraspecific 'scrounging by deception' strategies and their long-term evolutionary consequences are discussed.

In Chapter 2, Kennedy discusses the coevolution of host/parasite relationships. As he says, parasites are 'the best scroungers in the business' in the sense of their usually complete dependence on the host and the arms races between parasites and host defence systems have had complex consequences for the evolution of both parties. In particular, Kennedy examines the effects on the dispersion of parasites through their host populations, resource partitioning between parasites and strategies of coexistence. These are discussed in the context of the energetic benefits derived from the parasitic way of life and the costs to parasites of physiological and other host defence responses.

The book then moves on to discuss the evolution of resource-stealing strategies in free-living animals. Vollrath begins in Chapter 3 by distinguishing three forms of 'kleptobiotic' interactions in relation to food: piracy (kleptoparasitism) (the usurpation of food items by force), pilfering (usurpation by stealth) and peculation (usurpation by deception) and laying down ground rules for the use of his terms. Piracy, pilfering and peculation are viewed as stages along a continuous scale of competitive interactions between predators and are illustrated by examples from invertebrate relationships. A detailed study of web-building spiders shows how a complex kleptoparasitic relationship may have evolved and how the efficacy of kleptoparasitism improves with experience. While Vollrath starts by considering food-stealing, he extends his discussion at the end of the chapter to the usurpation of parental care, a topic which is also discussed by Parker, Arak, Andersson and Vehrencamp in later chapters.

Barnard continues in Chapter 4 by considering the selection pressures which have favoured the evolution of food-scrounging strategies within and between species and the kinds of counteradaptative strategies which have evolved in food-scrounging arms races. The chapter deals mainly with kleptoparasitic relationships and discussion centres on strategies of evasion, retaliation and compensation/toleration by host lineages and the adoption of alternative food procurement or competition restriction strategies by kleptoparasites. The chapter finishes by considering how races might end and the factors which may bias the outcome in favour of one lineage or the other.

In Chapter 5, Parker examines the evolution of phenotype-limited producer and scrounger strategies. In these cases the strategy an individual adopts is conditional on its phenotypic characteristics. For instance, it may pay individuals to search for their own resources (play producer) if they are large and able to out-compete smaller individuals, but usurp already acquired items (play scrounger) if they are small and unable to gain access to resources. The evolution of phenotype-limited P/S strategies is examined under a variety of conditions. Parker then moves on to introduce anisogamous sex as fundamentally a P/S relationship. He sees females as producers in the sense that they typically invest heavily in provisioning their gametes and caring for the zygote, and males as scroungers in that their contribution to zygote survival is typically virtually zero. The scrounger analogy depends on the fact that females might be expected to benefit if males invested more in the zygote.

While Parker deals with intersexual scrounging, Arak in Chapter 6 discusses scrounger strategies within the sexes. Arak points out that, wherever individuals, usually males, invest time or energy in attracting mates or caring for offspring, other individuals will be found exploiting this investment. For simplicity, these intrasexual scroungers are referred to as 'sneaky breeders'. Arak discusses the conflict of interest that inevitably arises between males as a result of 'sneaky' strategies in the population and the sometimes complex antisneak and counterantisneak adaptations that have evolved. Of particular interest is the evolution of apparently co-operative associations between producer and scrounger males. As well as discussing the consequences of scrounger strategies within sexes, Arak also considers their effects on the 'battle of the sexes' and mate selection as individuals are selected to distinguish 'sneaks' from better quality mates.

Exploitation within sexual relationships is taken further in Chapter 7 when Andersson discusses intraspecific brood parasitism. In a surprisingly large number of species, particularly birds, some females usurp the parental care of conspecific females. In birds this takes the form of laying eggs in another female's nest and allowing the host female to incubate the eggs and raise the young. Andersson reviews the evidence for brood parasitism in altricial and precocial species and raises the interesting possibility, backed up by a simple model, that some cases of brood parasitism have evolved through kin selection. Since brood parasitism is likely to affect the survival of the host's offspring. Andersson considers the various strategies open to the host in taking account of egg dumping by parasites.

Co-operative societies in animals have been the subject of much interest among ethologists, evolutionists and population geneticists over the past 20 years. While several feasible mechanisms have been proposed for their evolution, recent studies have shown that outward co-operation may conceal subtle exploitative strategies. In Chapter 8, Vehrencamp reviews the evidence for exploitation within apparently co-operative societies. Where certain individuals are more successful at, for instance, finding food or raising offspring, an opportunity exists for them to exploit the benefits of group living to their own advantage. Less successful individuals will be selected to resist exploitation and a conflict of interest arises. Depending on the extent to which individuals can capitalise on their potential to exploit, different degrees of what Vehrencamp calls 'fitness biasing' may occur within groups. Vehrencamp develops an extensive model of fitness biasing in what appear to be co-operatively breeding species and supports the model with evidence from bird, mammal and insect societies. An important consideration taken into account in the model is the effect of relatedness between group members on the direction and degree of fitness biasing expected.

In the final chapter, Sibly pulls together some of the ideas and examples discussed in the book and explores a series of modelling techniques for investigating P/S relationships. After considering simple frequency-dependent models, Sibly introduces a more complex graphical model which takes into account both frequency and density-dependent factors. The model represents the payoff to producers and scroungers in populations of different P/S composition as two intersecting surfaces. The central concept of the model is the equal payoff line (EPOL) which is the line of intersection of the payoff surfaces that populations should track in terms of their P/S composition at evolutionary equilibrium. Sibly extends the concept of the EPOL to traditional models of species population dynamics, considering the conditions under which populations become stable, and to the epidemiology of infectious disease. He then considers a means of estimating payoff surfaces which was originally devised by de Wit as a method for analysing two-species interactions. Finally, Sibly suggests various ways in which his models could be applied in field and laboratory studies.

The most immediately striking feature of the discussions presented in this volume is the extensive range of topics they cover within the definition of producer/scrounger relationships. The evolution of the sexes is discussed in the same breath as that of intraspecific assessment cues, brood parasitism and parasite population and community dynamics. That we are able to do this is indicative of the enormous advances made by recent trends in ethological and population ecological research. The focusing of attention on plausible adaptive explanations based on the quantitative principles of population genetics enables us to ask functional questions which probe beyond the superficial appearance of relationships between organisms. This is nowhere better illustrated than in the reviews of apparent co-operative breeding and intraspecific brood parasitism by Vehrencamp and Andersson. By transcending phylogenetic and ecological boundaries and supporting ideas with predictive models and well-worked examples, this book aims to consolidate some of the developments in evolutionary thinking over the past few years and to stimulate new directions of research in the areas on which it touches.

C.J. Barnard

1 WHEN CHEATS MAY PROSPER

C.J. Barnard

The outstanding success of recent evolutionary models of behaviour and population ecology has been due largely to the development of the concept of evolutionarily stable strategies (ESSs) (Maynard Smith, 1974). ESS models assume that strategies for survival in organisms have a genetic basis and are the result of natural selection choosing between alleles for alternative strategies over evolutionary time. Ultimately, the allele which is likely to go to fixation in a population is the one coding for a strategy (the ESS) which cannot be bettered by any feasible alternative strategy in the sense of increasing individual reproductive potential and therefore allele fitness (see Dawkins, 1980, for an excellent review). In many cases, the ESS is not a simple, single strategy, but may be a complex mixture of responses each played with a different probability or conditional for its performance on certain environmental conditions or strategies played by other organisms. Despite the complex ecology of most natural populations, it has been possible to quantify pure, mixed and conditional ESSs both in the field and in the laboratory (Parker, 1974; Davies and Halliday, 1979; Milinski, 1979). The aim of this introductory chapter, however, is not to review the progress of ESS theory, but to examine the evolution of certain types of alternative strategy which provide the driving force of natural selection and determine the form the ESS will take.

The chapter takes as its starting point the fact that most organisms depend for their existence on limited natural resources, like nutrients and water, and that these resources usually require the investment of time, energy and predation risk for their exploitation. The relative contributions of time, energy and risk costs and the nutrient or other benefit of the resource to the organism's probability of survival and successful reproduction is what determines the strength of action of natural selection on the organism's resource harvesting strategy. Any strategy which reduces costs relative to benefits will be favoured and organisms will evolve towards optimal foraging strategies (where 'foraging' is used in the broad sense of seeking any resource, e.g. food, females, nesting sites, shelter). The most direct way in which optimisation proceeds is by honing the organism's perceptual, sampling and resource handling abilities so that harvesting becomes more efficient. However, the fact that selection favours efficient foragers means that one organism's searching and other investment can be made use of as part of another's survival strategy. In other words, the investment of one organism (*producer*) provides an adaptive short-cut for evolution in another (*scrounger*) and a mixed (producer/scrounger) instead of pure (producer only) strategy may now become the ESS. The purpose of this introduction is to review the ways in which adaptive short-cuts arise and the types of scrounger strategies which are evolutionary responses to them.

Scrounger strategies can evolve in one of two ways. Firstly, and most simply, they may be strategies of usurpation. Producer individuals invest time and effort in procuring a resource which is then usurped by scrounger individuals. Parasitism, piracy and satellite mating strategies are examples of scrounging by usurpation. Secondly, what I shall call *cost-increasing models* may arise in the environment, convergence in characteristics to which improves the converging organism's reproductive potential while reducing that of the model. Batesian mimics are good examples although, as we shall see later, mimetic relationships are only one consequence of this far more wide-reaching evolutionary process.

Strategies of Usurpation

The investment of time and energy by an organism in procuring a resource leaves it open to exploitation by non- or low-investing usurpers. Unless there is some strong form of retaliation by the producer, usurpation is likely to have an immediate and large selective advantage because it sidesteps the major costs of resource procurement. Once producer strategies have evolved, usurping scrounger strategies are inevitable if appropriate mutations arise, even though many may not ultimately reach an equilibrium in the population or community. Because usurpation involves capitalising on another organism's investment, usurpation strategies are essentially parasitic.

Parasitic Strategies

The term 'parasite' is normally restricted to those organisms like tapeworms and fleas which have evolved dependence on a host organism for survival. Indeed, as Kennedy points out in the next chapter, parasites are supreme scroungers in the sense implied above. They not only capitalise on the nutrients and energy made available by the host's foraging efforts, they may also use the host for transport, dispersal and, in the case of endothermic hosts, the maintenance of a thermally-regulated environment. Parasites, however, are only the extreme end of a range of scrounger strategies which can be described as parasitic. Organisms can capitalise on the resource procurement of others without becoming physiologically dependent on them. Kleptoparasitism (food-stealing or piracy), for instance, has evolved on an opportunistic basis both within and between species (Brockmann and Barnard, 1979; Barnard, Ch. 4 this volume; Vollrath, Ch. 3 this volume). Kleptoparasites may subsist almost entirely by stealing items procured by host individuals (Källander, 1977; Vollrath, 1979) or kleptoparasitism may be part of a wider range of feeding strategies (Brockmann and Barnard, 1979). In both cases, however, the impact of kleptoparasitism on host feeding efficiency may be sufficient to cause a counteradaptive arms race between the two parties with consequences for both population and community dynamics.

Food is only one resource which requires time and energy to procure. For sexually reproducing species, mates are another. Just as kleptoparasites capitalise on the food procurement of hosts, 'satellite' and 'sneaky' breeders capitalise on the mate attraction abilities of other individuals (Clutton-Brock *et al.*, 1979; Howard, 1978; Arak, Ch. 6 this volume). In bullfrogs (*Rana catesbeiana*), for instance, large males call to attract females while small males position themselves near to callers and intercept females. In ruff (*Philomachus pugnax*), a lekking bird species, certain males with characteristically dark head plumage defend breeding territories to which females are attracted. Some non-resident males, however, intrude on to defended territories (and, interestingly, are sometimes tolerated) and may mate with attracted females while the resident is engaged in territorial defence (Hogan-Warburg, 1966). These satellite males have paler head plumage than residents and are easily distinguishable. In red deer (*Cervus elephus*), young stags sometimes steal hinds (kleptogamy) from the harems of older stags while the latter are driving off rivals (Clutton-Brock *et al.*, 1979), but become harem defenders, and therefore producers, themselves as adults.

Sometimes the usurped resource is a shelter or a nesting site. Female golden digger wasps (*Sphex ichneumoneus*), for instance, lay their eggs in underground burrows which they provision with katydids (Orthoptera: Tettigoniidae). Brockmann *et al.* (1979) discovered that females obtained burrows in one of two ways. They either dug a burrow themselves and provisioned it, or they entered and provisioned a burrow dug by another female. Digging a burrow is expensive, taking on average 100 minutes to complete. Entering an already completed burrow is therefore clearly advantageous. A careful quantitative analysis of the costs and benefits of digging and entering showed that in natural populations, both strategies fared equally well in terms of their expected reproductive potential (Brockmann *et al.*, 1979). In the wasps, therefore, producer and scrounger strategies appeared to exist in evolutionary equilibrium.

Usurpation of nests or other brood care sites may extend to brood parasitism where the full parental investment of the host is exploited. Familiar interspecific brood parasites are, of course, the cuckoos (Cuculidae) and cowbirds (Icteridae) which lay their eggs in the nests of other species and allow the host parent to raise the chick at the expense of its own offspring. However, brood parasitism also occurs extensively *within* species, although it is harder to detect. Intraspecific brood parasitism may be analogous to the interspecific cuckoo strategy in that it involves laying eggs in the nests of other breeding females (see Andersson, Ch. 7 this volume) or it may be a subtle strategy for ensuring over-representation of an individual's own offspring in a communally reproducing system (see Vehrencamp, Ch. 8 this volume). By a variety of strategies, therefore, the parental care of conspecifics can be exploited to reduce the cost of breeding.

The producer/scrounger relationships discussed above are, at first sight, examples of scrounger individuals usurping resources procured or created by producer individuals. However, as Brockmann *et al.* (1979) pointed out in their analysis of digger wasp populations, it is the relative success of the genes coding for producer or scrounger *strategies* which is important rather than the reproductive success of particular individuals (see also Dawkins, 1982). Thinking in terms of strategies rather than individuals leads to an interesting view of anisogamous sexual reproduction as a producer/scounger relationship. In anisogamously reproducing species, males can be seen as being in a sense parasitic on females because their small, low-investment gametes (sperm) exploit the well-provisioned, high-investment gametes (eggs) of females (Parker *et al.*, 1972). Much of the evolution of sexual behaviour, morphology and physiology is predictable from the assumption of male exploitation (see Arak, Ch. 6, and Parker Ch. 5 this volume). An important point, however, which is not usually made clear, is that the costs of egg and sperm production should be measured in terms of replacement cost rather than size. The size difference is likely to be offset by the large number of sperm produced per ejaculate. Because females produce fewer gametes during their lifetime than males, however, each gamete wasted represents a more serious loss. The fact that females produce fewer gametes may indeed be due to the greater material investment in each, but it is the increased value of eggs in terms of their limited number on which sexual selection acts.

Manipulation

While selection is likely to favour the exploitation of procured or created resources, it may go a step further. Relying on a producer individual to make a resource available is risky. Producers may not be present or perhaps cannot be located at the time the resource is needed. One way round the problem is for scroungers to *manipulate* producers into making the resource available at the appropriate time. The concept of manipulation in evolutionary terms is recent and few direct studies have yet been made. However, it is clear that several parasites rely on manipulating the behaviour of at least one of their hosts in order to complete their life cycle successfully. Several good examples are cited by Holmes and Bethel (1972).

Holmes and Bethel point out that many parasites which have intermediate hosts in their life cycle alter the behaviour of the intermediate host to increase its vulnerability to predation by the definitive host. Intermediate host behaviour appears to be altered in a number of ways which can be interpreted as manipulation by the parasite.

Reduced Activity. Perhaps the simplest way in which hosts can be made more vulnerable to predators is by reducing their activity. Several examples are known among helminths using fish as intermediate hosts. The plerocercoid of *Ligula intestinalis*, for instance, is found in the coelom of several fish species and appears to make them sluggish and float in areas where they are more likely to be spotted by avian predators. Dence (1958) reported that infected shiners (*Notropis cornutus*) were slower moving and less gregarious than uninfected fish and tended to inhabit the shallow, warmer inshore water where piscivorous birds fed. Similarly, Holmes and Bethel (1972) report infected spottail shiners (*N. hudsonius*) and yellow perch (*Perca flavescens*) lagging behind schools and swimming close the surface. In sticklebacks (*Gasterosteus aculeatus*), fish infected with the plerocercoids of *Schistocephalus solidus* not only tend to swim near the water surface but are more easily caught by hand (Baer, cited in Holmes and Bethel, 1972).

Samples of fish taken by avian predators support the suggestion that these changes in behaviour render the intermediate host more vulnerable to predation. Van Dobben (1952), for instance, found that 30 per cent of the roach (*Rutilus rutilus*) in the diet of cormorants (*Phalocrocorax carbo*) were infected with *Ligula* compared with only about 6.5 per cent in the general population. Infected fish also seem to be more vulnerable to dip netting and therefore possibly to avian predators. Hopkins (cited in Holmes and Bethel, 1972) collected a sample of sheepshead minnows (*Cyprinodon variegatus*) that was 100 per cent infected with the metacercariae of *Ascocotyle*. A later sample collected with a seine net was only 10 per cent infected.

Changes in host activity in relation to parasite burden have also been recorded under controlled laboratory conditions. Rau (1983a) recorded the open-field behaviour of laboratory mice infected with larvae of the aphasmid nematode *Trichinella spiralis*. Ambulatory and exploratory activity declined significantly as infections progressed and the onset of these behavioural changes coincided with the expected time of larval release by adult worms. Levels of activity were also negatively correlated with the initial dose of worms administered to the mouse. Again, the reduction in activity may render the host more vulnerable to predation.

Altered Responses. While reduced activity in the intermediate host may make parasite transmission more likely, it may be the result of gross pathological damage to the host, for instance in terms of changed blood chemistry or hormonal output (Arme and Owen, 1968). Similarly disorientation and increased conspicuousness (Holmes and Bethel, 1972), both common changes in intermediate hosts as a result of infection, result from either damage to the central nervous system or to gross morphological changes. Of more interest from an evolutionary viewpoint are changes in host behaviour which are reflected in altered host responses to particular stimuli. These are not due simply to gross damage, but to more subtle interference by the parasite.

Some of the best-known examples of this kind of behavioural modification come from trematode infections in ants. Carney (1969) discovered several behavioural peculiarities in carpenter ants (*Camponotus* spp) carrying metacercariae of *Brachylecithum mosquensis*. The metacercariae were situated in or around the supraoesophageal ganglion and the tissues of the gaster. Infected ants typically wandered in a slow circling fashion on open rock surfaces, often remaining motionless for several hours, and did not respond to changes in light intensity produced by shading. Ninety-one per cent of infected worker ants were found in the open compared with only 9 per cent in the adjacent wooded areas where most of the population occurred. Graham (1966) described similar changes in responsiveness to light in *Tribolium confusum* beetles infected with the cysticercoid of the chicken tapeworm *Raillietina cesticillus*. Infected beetles showed reduced photophobic tendencies compared with uninfected beetles and did not seek concealment.

One of the most detailed studies of altered host responses is that of the freshwater shrimp *Gammarus lacustris* infected with cystacanths of the acanthocephalan *Polymorphus paradoxus* (Holmes and Bethel, 1972). Following reports that gammarids infected with *P. paradoxus* were often attached to, or closely associated with, floating material and were difficult to dislodge (a response which contrasted sharply with that of benthic uninfected individuals), Holmes and Bethel carried out a series of choice tests. When provided with a choice between light and dark zones in an aquarium, uninfected *Gammarus* remained in the dark zone almost all the time and showed no preference for the upper part of the lighted zone. Infected *Gammarus*, however, spent over 80 per cent of their time in the lighted zone and 70 per cent of that in the upper part of the zone clinging to floatage provided by the experimenters. Further experiments varying the direction of the light source indicated that the behaviour of the infected animals was due to a change in phototaxis (from negative to positive) rather than geotaxis. Holmes and Bethel also tested for differences in the tendency to cling to different types of floatage. They found that the only *Gammarus* to cling actively to floatage were those infected with *P. paradoxus*. Most of these preferred to cling to rough reeds rather than sticks, *Potamogeton* or other types of floatage. When dislodged, infected *Gammarus* skimmed across the water surface and clung to the first suitable object encountered.

In predator tests, Holmes and Bethel allowed mallard (*Anas platyrhynchos*) to forage for infected and uninfected *Gammarus* in a tank. In all tests mallard were attracted to floatage and fed around it before diving or tipping. When the floatage was disturbed, some *Gammarus* skimmed away but most remained clinging. Ducks fed at first on the clinging *Gammarus* and then on those which skimmed away, returning repeatedly to search the floatage. Infected *Gammarus* even clung to the birds' feathers so that some were ingested during preening. The result of the different responses shown by infected and uninfected *Gammarus* was that a disproportionately high number of infected individuals (four times that of uninfected individuals) were eaten by the mallard. As Holmes and Bethel (1972) point out, the altered responses of infected *Gammarus* are also likely to make them more vulnerable to muskrats and beavers, the other definitive hosts of *P. paradoxus*.

Some interesting observations by Rau (1983b) suggest that parasites may alter the social responses of their intermediate hosts. Working with laboratory mice infected with larvae of *Trichinella spiralis*, Rau found that infection influenced the establishment and maintenance of dominance relationships between males. Infected animals became subordinate to uninfected individuals both when tested in an unfamiliar test arena and when tested in their home cage where they were normally dominant. Rau (1983b) suggests that this will make infected animals more likely to be evicted from resident groups and therefore become more vulnerable to predation. Butler (1980) found that 97 per cent of house mice emigrating from colonies were subordinate individuals. Numerous other correlations between an individual's social status and its parasite burden have been reported (see references cited in Rau, 1983b).

Dawkins (1978, 1982) sees manipulation operating in a much wider context. His concept of the extended phenotype envisages the phenotypic effect of genes reaching beyond the body in which the genes reside to influence the actions of other bodies in ways which directly enhance the genes' own fitness. Outwardly, therefore, individuals appear to be manipulated into improving the survivorship and reproductive potential of other individuals. As Dawkins and Krebs (1978) point out, this may well be the underlying function of many acts of communication which appear on the surface to be co-operative acts aimed at maximising the accurate transfer of information between individuals. By exploiting the signalling channels of other individuals, con- or heterospecific scroungers may be able to manipulate their investment of time and energy into the production of exploitable resources. In the next section, we shall see one way in which manipulative scrounging may evolve.

Cost-increasing Models and Scrounging by Deception

So far, I have discussed scrounging only in the sense of usurping some limited resource. However, the costs of survival can be reduced in a less direct way which depends on what I shall call detector/detectee relationships.

We can envisage organisms as *detectees* which are impinged upon by a number of *detectors* trying to distinguish them from features of the environment (*background*) that are irrelevant or deleterious to the detector's goal-directed behaviour. Successful discrimination by a detector enhances the reproductive potential of the detector and decreases that of the detectee. The terms 'detector', 'detectee' and 'background' refer to any discriminatory relationship which influences the fitness of the organisms concerned. A detector may be, for instance, a predator, a host immune system or an individual of a mate-choosing sex and the corresponding detectee a potential prey individual, an invading parasite and an individual of the candidate sex (see Table 1.1). Before showing how this is applicable to scrounging, I must first develop a simple general model of the evolution of detector/detectee relationships.

Table 1.1: Some detector/detectee relationships with apparent detection cues and background features (see text). 1: from discussion in Edmunds (1974). 2: from Rohwer (1977). 3: from Falls, J.B. and Brooks, R.J. (1975). Individual recognition by song in white-throated sparrows II. Effects of location. *Can. J. Zool. 53*: 1412-20. 4: from Leuthold, W. (1966). Variations in territorial behavior of Uganda kob *Adenota kob thomasi* (Neumann 1896). *Behaviour 27*: 215-58. 5: from discussion in Bloom, B.R. (1979). Games parasites play: how parasites evade immune surveillance. *Nature 279*: 21-6.

	Detector	Detectee	Detection cue	'Background'
1	Marmoset (predator)	Stick insect (prey)	Insect leg position	Physical habitat
2	High-status Harris sparrow (assessor)	Low-status bird (competitor)	Bib size	Bib size of high-status birds
3	Territorial white-crowned sparrow (assessor)	Neighbour (rival)	Song type + position on territory boundary	All song types + positions
4	♀ Uganda kob (choosing sex)	♂ Uganda kob (candidate sex)	TG territory possession by male	All males present
5	Host antibodies (detector)	Schistosome (parasite)	Schistosome antigen	Host antigen

Detection Cues

Since we are dealing with discriminatory tasks which influence directly the reproductive potential of detector and detectee parties, we can envisage selection acting on detectors to improve their ability to distinguish detectees from background and on detectees to reduce their risk of detection.

More specifically, because detectees are part of a detector's environment, detectors will be selected to use those features of the environment which result in the most efficient discrimination. This will be a function of the reproductive benefit accruing from successful discriminations devalued by the costs of the time taken in the discrimination process, the penalty of making a mistake, the cost of evolving the appropriate pattern-processing machinery, and so on. The features which are eventually used may be only part of the detectee, the whole detectee or more than the detectee (for instance, the use of shadows in conjunction with body morphology by predators detecting cryptic prey (Edmunds, 1974). Such features can be referred to as *detection cues*. A detection cue is therefore the subset of all possible environmental cues which maximises a given detector's discrimination efficiency for a given detectee. There is clear evidence that specific detection cues are used (Table 1.1), but, as yet, none that those cues are the most efficient that could be used in each case.

At the same time as the detector is evolving to maximise the efficiency with which it discriminates the detectee, the detectee will evolve to minimise its probability of discovery. It can do this only by reducing the *conspicuousness* of the detection cue. From Table 1.1 and the above discussion it is clear that we are dealing with two fundamentally different types of discrimination. In the first, involving e.g. predator/prey and host/parasite relationships, the detectee seeks to avoid being distinguished from an inedible or otherwise irrelevant background. In the second, involving e.g. mate choice and assessment between conspecifics during a contest, the detectee seeks to avoid being distinguished from some preferred population criterion (e.g. individuals with a particular form of secondary sexual characteristic or weaponry). While in the first case selection is acting on detectees to be *overlooked* by detectors and in the second it is acting on them to be *noticed* (seen as a preferred mate type or a high-ranking contestant), the evolutionary process underlying the modification of detection cues is the same. We are now in a position to formulate the model.

Detectors and Detectees: a General Model

Consider detectee I sought by detector D using detection cues D_I. For I to avoid detection by D, D_I must converge in characteristics to some feature of what D considers background. More specifically, since we assume we are dealing with naturally selected convergence, D_I should converge to that background feature E, a unit increase in similarity to which most increases the detector's discrimination costs (see above).

'Similarity' may therefore be the result of convergence in any of a number of characteristics, e.g. tone, colour distribution, mode of performance of a particular behaviour, an olfactory characteristic. The perception of similarity is seen as a multi-dimensional feature-matching process, as envisaged by Tversky (1977) and Krumhansl (1978). A unit increase in similarity is that degree of convergence in D_I to a background feature which decreases *D*'s reproductive potential by a given amount (e.g. by causing it to overlook prey or not to recognise a low-status mimic of high-status competitive ability).

There is an infinite number of ways in which the conspicuousness of D_I could possibly vary relative to *E*. We can therefore envisage a hyperspace (*total cue space*) surrounding *E* which contains an infinite variety of possible D_I cue states (Figure 1.1). However, also surrounding *E* and some distance from it, is a conspicuousness threshold beyond which D_I is deleteriously conspicuous to *D*; that is, the probability of detection by *D* is so high that *I* is unlikely to reproduce (e.g. because it is unlikely to survive or to attract a mate). This threshold can therefore be represented as a boundary enclosing the set of evolutionarily possible cue states for D_I within total cue space. I shall refer to this enclosed set as *safe cue space* (SCS). For convenience we can represent the SCS as a circle with centre *E* (Figure 1.1).

At first sight, the best cue state for *I* would seem to be *E*, since then *I* matches *D*'s idea of background. If D_I is displaced from *E* to state P (Figure 1.1) for any reason, selection will favour convergence towards *E* along the vector PE. PE therefore represents the route of maximum convergence rate. While selection will tend to modify D_I towards *E*, it is unlikely that D_I will ever reach *E*. The reason is easy to see. If evolution in *D* was somehow halted, D_I would converge to *I*'s optimal distance D_{Iopti} from *E*. D_{Iopti} reflects a trade-off between the benefits of reduced conspicuousness to *D* and the costs of convergence (e.g. the costs of producing pigmentation, modifying morphology and restricting mobility). Conversely, if evolution in *I* was halted, then D_I would reach the optimal state D_{Ioptd} for *D* which is maximum possible conspicuousness. Since evolution proceeds simultaneously in the two parties, the actual equilibrium state for D_I, D_I', will be a compromise between these two optima. The exact position of D_I' within total cue space will depend on the relative intensity of selection acting on *I* and *D*. In predator/prey systems, for instance, D_I' is likely to be nearer D_{Iopti} than D_{Ioptd} because selection will probably be acting more intensely on anti-predator adaptations in *I* (Dawkins and Krebs, 1979). D_I, however, will be an unstable equilibrium because the arms race between I and D is a continuing one. There will therefore be a continual change in D_I by I and counteracting change in the criteria for E by D.

Figure 1.1: Convergence of detection cue D_I to background feature E so that the equilibrium cue state D_I' lies within the detector's SCS (represented by the circle). See text for explanation of the symbols

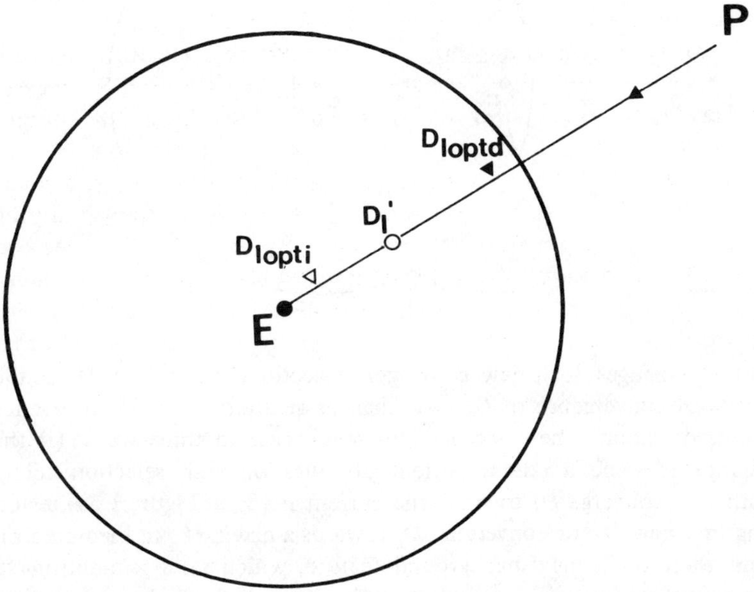

Assuming that D persists in seeking I and does not shift its attention to alternative detectees, selection may act in two ways:

(a) D_I continues to converge along the current vector (Figure 1.2). Here, movement of D_I' towards D_{Iopti} causes D to change criteria for E so that the distance a (the compromise equilibrium distance of D_I from E) is restored and the SCS moves through total cue space. We can therefore, for instance, envisage successive stages in the evolution of crypsis as proceeding through the modification of ground colour intensity (E_O in Figure 1.2), marking pattern (E_1), orientation behaviour (E_2) and so on. A 'ratchet' mechanism will operate because back-mutation of D_I away from E will be selected against while forward mutation will be selected for. All D can do to counter convergent change in D_I therefore is to refine its current definition of E.

Figure 1.2: As the detector/detectee arms race progresses, the criterion for E changes ($E_{0...2}$) and with it the equilibrium detection cue state D_I'. Dashed circle boundaries represent changes in the boundary of the SCS

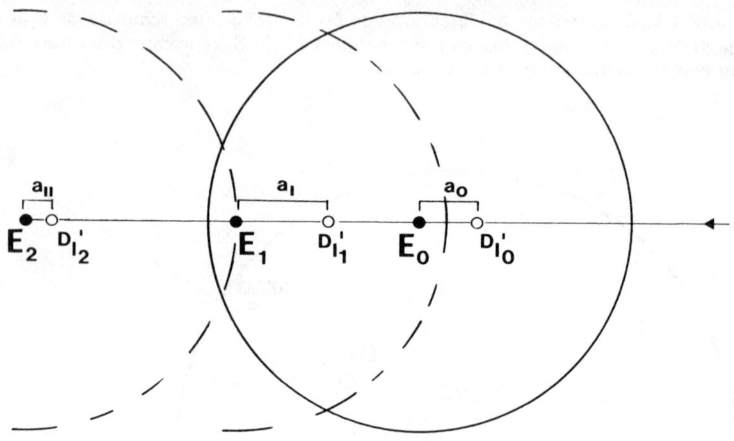

(b) D_I changes to a new convergence vector (Figure 1.3). Here, the mode of convergence of D_I to E changes qualitatively and convergence proceeds along a new vector. This may occur in three ways: (i) the change may be a random effect of mutation with selection acting either to converge D_I towards the current E (E_o in Figure 1.3a) resulting in a new D_I' or converging D_I towards a new E (E_1 in Figure 1.3a), and therefore a new background feature, which results in the maximum rate of increase in similarity to background from the new starting point. For instance, convergence during the evolution of a cryptic appearance may have been proceeding along a colour intensity vector, but a mutation event now means it is better to alter shape. Changes in vector may be caused by development in the organism. Increase in body size with age is one factor which may compromise the efficacy of one cryptic pattern and cause a shift to another. The larvae of several lepidopterans, e.g. papillionids and sphingids, shift from one cryptic colouration to another as they grow and their size becomes inappropriate to the original background (Cott, 1940; Edmunds, 1974). (ii) The cost of convergence along the current vector may outweigh the benefit in increased reproductive potential for I or there may be structural or physiological constraints preventing further modification of the same character. In this case, D_I reaches D_{Imax} (Figure 1.3b) and detector-specific change in the detection cue ceases. Selection will now favour

Figure 1.3: (a) The convergence vector for a given detection cue may switch to a new *E* as a result of random mutation (mu). Open circle represents detection cue state after mutation, the solid arrow represents the convergence vector to a new *E* (E_1) and the open arrow represents convergence to the previous *E* (E_0). Which of the vectors is chosen depends on the relative rate of increase in similarity to *E* which they produce. Dashed circle boundary as in Figure 1.2. (b) Detection cue (D_l) state ceases to converge to E_0 because it reaches D_{max} at which point the costs of further convergence outweigh the benefits of reduced detectability. Mutations shifting D_l to points where new convergence vectors can be followed will be selected for. Symbols as in (a). (c) If a CIM (see text) arises in the environment, selection may favour a switch in convergence vectors to *E* for the CIM (E_m). Symbols as in (b)

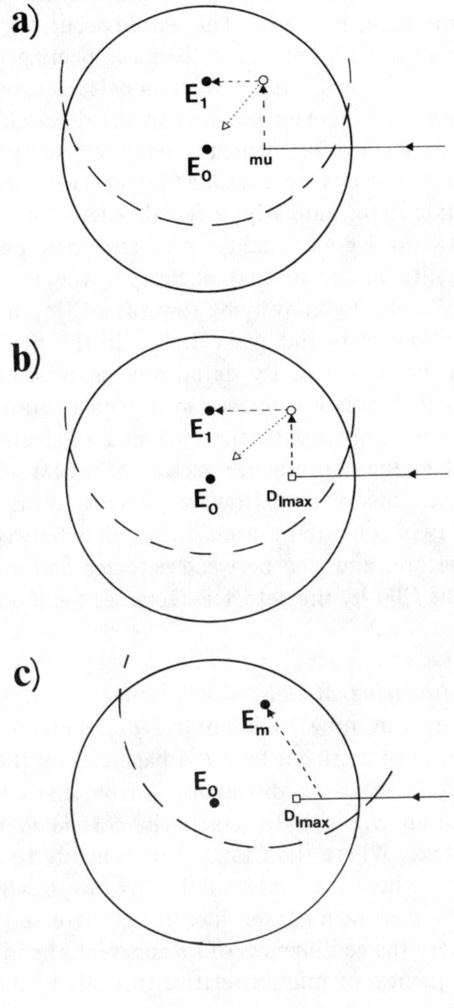

any mutation which changes D_I to a new convergence vector. (iii) Change may occur as the result of a *model* arising in the environment which acts as E (E_m in Figure 1.3c). The model may act as E because convergence towards it by I substantially increases D's sampling costs (as might happen if D was a predator of I and I converged in appearance towards a poisonous, distasteful or even simply unprofitable species (model), or, if D was a conspecific assessor and I converged in characteristics towards a conspecific higher in competitive ability than D), therefore causing it to overlook I. Alternatively, the model may act as E because D seeks the model and being mistaken for the model increases I's reproductive potential. This would occur, for instance, where D was a female selectively mating with males bearing particular characteristics (model). We can refer to such models as *cost-increasing models* (CIMs) because they increase the cost to the detector of making a mistake. Taking prey (detectee) which is being selected to resemble a toxic species (model) increases a predator's (detector) chances of accidentally ingesting a toxic item. Similarly, a female (detector) attempting to discriminate between a good quality, preferred male phenotype (model) and a low-quality mimic of that phenotype (detectee) risks rejecting a good-quality mate. By converging towards a CIM, therefore, detectees increase detection costs and make it less likely, at least temporarily, that they will be detected. By definition, an increase in similarity to a CIM results in a greater increase in detectee reproductive potential than increases in similarity to the current E. Selection will therefore favour a shift to the convergence vector for a CIM at any point along a current convergence vector. However, a shift towards a CIM is likely to result in two concurrent arms races, one between detector and detectee as before, and one between detectee and model (where convergence to the CIM by the detectee decreases the model's reproductive potential).

Arms Races Involving Models. Convergence of D_I towards E_m proceeds as before but now to a point D_{Icrit} (Figure 1.4). At D_{Icrit}, the benefit to D of distinguishing I is balanced by the cost of making a mistake and D ceases to distinguish I from the CIM. The distance b of D_{Icrit} from E_m will be positively correlated with the cost of making a mistake. Where the CIM is, say, a highly toxic 'prey' species, b will be large, where it is only mildly noxious, b will be small. However, unless the cost of mistaken identity is extremely high, it is likely to pay D to test the genuineness of I's apparent phenotype by *probing*. When the frequency of mimics relative to CIMs in the environment is

Figure 1.4: Convergence of D_I towards E for a CIM proceeds to state D_{Icrit} at which point the benefit to the detector of distinguishing the detectee (I) is balanced by the cost of mistaking the CIM for I. Selection may then favour convergence towards other CIMs or environmental features. See text for explanation of b

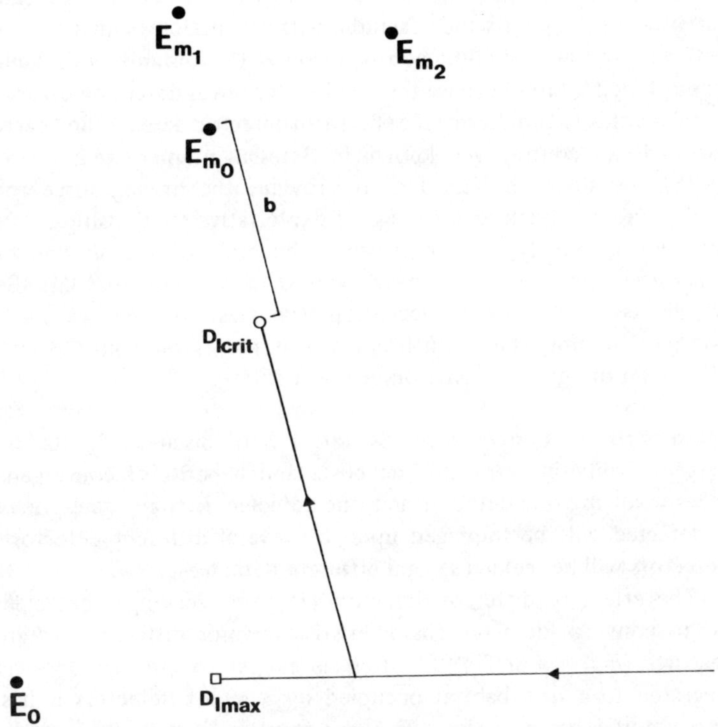

low, probing is likely to be in response to mistaken identity and D will pay the appropriate cost. The frequency of probing will therefore be low. Because challenges by detectors are rare, detectee mimics will increase in frequency. As they do, however, the cost of probing to detectors decreases because more probes are responses to correct identification and the frequency of probing will also increase. Since the cost of convergence for detectees is now high, two things may now happen: the frequency of detectee mimics in the population may decrease as a result of an increased rate of detection or detectees may converge further towards the CIM and continue the detector/detectee arms race. Increases in the frequency of mimic detectees, however, will have an additional effect: more detectee samples may be needed by naive detectors before they learn or are selected to associate a convergent phenotype with the fitness-influencing properties of a CIM and the cost of sampling errors to the CIM will increase. This results in a second arms race. Selection will favour detectors which distinguish reliably appropriate detectees (because reliable discrimination is likely to be cheaper than probing) and add in to their assessment mechanism any variation in the CIM which facilitates its distinction from a convergent detectee and will simultaneously favour CIM mutants with such distinguishing features because they will suffer fewer detection errors.

While this twofold arms race between detector/detectee and detectee/model is, of course, well known in Batesian mimicry (e.g. Sheppard, 1975), we shall see later that it provides the driving force for the evolution of a much wider range of exploitative relationships. Batesian mimicry is merely one interspecific example of the *scrounging by deception* strategies which have evolved in response to CIMs. Before we discuss scrounging by deception strategies, however, we must first consider an important constraint which may limit convergence to a given CIM or any other environmental feature.

Constraints on Convergence. So far, I have discussed limits to convergence only in terms of the costs and benefits of convergence in races involving one detector and one detectee. In many cases, however, a detectee will be impinged upon by several different detectors and detectors will be seeking several different detectees.

The effect of different detectors ($D_{1...n}$) seeking a detectee will be to converge detectee characteristics towards different background features ($E_{1...n}$ or $Em_{1...n}$). It is easiest to envisage I as having migrated to a new habitat occupied by a set of detectors it has not encountered before. Also, I's characteristics lie initially outside the SCS for each detector (Figure 1.5a). The distances *a, b, c* in Figure 1.5a represent the selection pressure on I to converge to D_I' for each of three detectors, say a predator, a mate-choosing sex and a conspecific assessor of competitive ability (background features $E_{1...3}$). The greater the distance, the stronger the selection pressure. Since I must end up within the SCS for each detector, we can represent convergence to $D_I'{}_{1...3}$ as a migration of SCSs towards I (Figure 1.5b). The proportion of the zone of SCS overlap which is made up by a sub-space of a given detector's SCS is inversely related to the constraint imposed on I's characteristics by that detector. Clearly, therefore, I is unlikely to attain D_I' for any given detector. Instead, detection cues will converge to a new equilibrium $D_I'{}_n$ which will reflect a com-

Figure 1.5: (a) A detectee (*I*) may be outside the SCS for detectors when first encountered. $E_{1...3}$ represents the background features to which *I* must converge to avoid detection by detectors 1. . .3. $D_{I'_{1...3}}$ represent the independent equilibrium cue states for the detection cues used by detectors 1. . .3 to distinguish *I*. Circles represent SCS boundaries for each detector. See text for explanation of a, b and c. (b) As the arms race between *I* and the three detectors proceeds, $D_{I_{1...3}}$ will converge to their respective *E*s. The eventual equilibrium cue state for each detection cue (small open circles) must lie within the SCS for each detector (represented here as the shaded zone of overlap) and is therefore likely to be different from the state represented in (a) (see text)

(a)

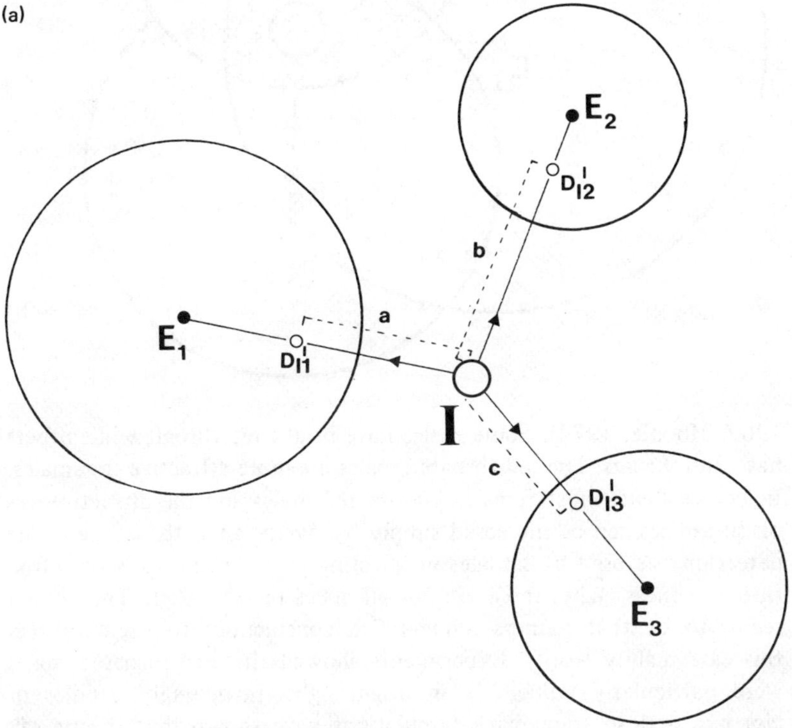

promise between the selection pressures exerted by the set of n (in this case three) detectors. However, the extent to which the constraint imposed by any given detector is reflected in its domination of *I*'s characteristics depends on how close to *E* for that detector the detector's sub-space lies.

A nice example of a trade-off in characteristics resulting from the simultaneous selection pressures of more than one detector comes from three-spined sticklebacks (*Gasterosteus aculeatus*). Male sticklebacks exist in different colour morphs during the breeding season (McPhail,

(b)

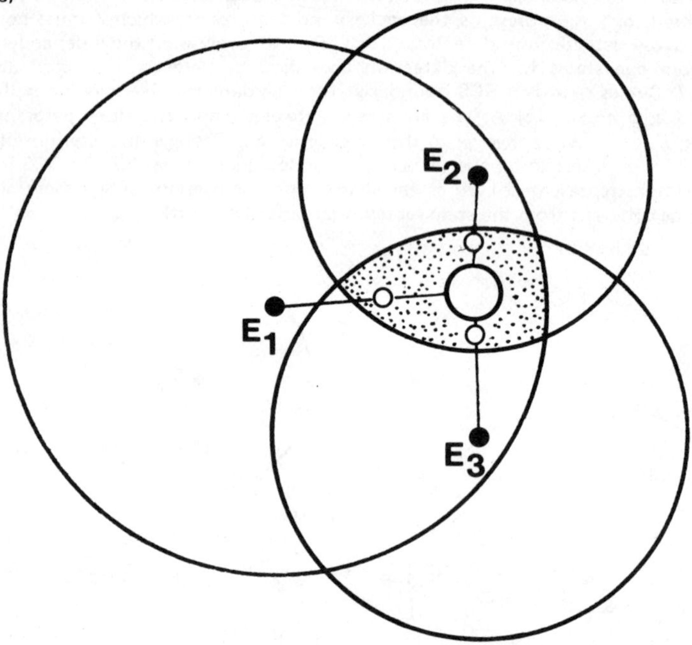

1969; Moodie, 1972). Some males have bright red throats while others have dull throats. The red-throated males are more attractive to females. In choice tests, most females choose red males and the attractiveness of dull males can be increased simply by dyeing their throats red. The detection cue used by females in selecting preferred males is therefore throat redness. Why, then, are not all males red-throated? The answer seems to be that redness renders fish conspicuous to predators (in this case mainly trout). Experiments showed that red-throated males were particularly vulnerable in bright light. Intriguingly, studies of North American stickleback populations have shown that the prevalence of red-throatedness depends on depth in the water. In deep water, most males are red-throated while in shallow water most are dull. In deep water, where light intensity is low, convergence of the red throat detection cue can proceed further towards the preferred criterion for mate-choosing females because counteracting selection by predators is reduced.

Just as detectees are likely to be impinged upon by more than one detector, so detectors are likely to be selected to detect more than one detectee. The consequences of this can be interpreted in terms of adaptation budgets (Dawkins and Krebs, 1979). We can envisage organisms as having limited metabolic resources (adaptation budget) with which to respond adaptively to selection pressures. Clearly, if only one selection pressure is operating, then a large proportion of the adaptation budget can be allocated in response. If, on the other hand, several selection pressures are acting jointly, only a small portion of the budget can be given over to counter each one. In a predator/prey arms race, for example, a generalist predator is unlikely to be as efficient at dealing with a given prey item as a specialist predator. In general, therefore, detection cues used by generalist detectors will converge less closely to E for each detectee. For detectee I and generalist detector D, D_I' is likely to lie closer to D_{Iopti} than it would for a specialist detector.

Scrounging by Deception

The model developed above shows how detector/detectee relationships are likely to be a wide-ranging influence on the evolution of organism characteristics. While they are the driving force behind well-known phenomena like crypsis and interspecific mimicry, they are also a major selective pressure favouring intraspecific scrounging. In this case, however, scroungers arise as mimics of intraspecific CIMs to which detectors attend and which influence detector fitness. Inter- and intraspecific CIMs provide adaptive short-cuts by presenting an opportunity for potential scroungers to capitalise on their detector fitness-influencing properties and reduce the costs of survival. Clearly, the range of contexts in which scrounging by deception could occur is immense and a full coverage cannot be given here. However, it is worth detailing some examples to illustrate both the circumstances in which scrounging by deception can arise and the possible consequences of arms races involving scroungers, detectors and models.

The Evolution of Assessment Cues

Barnard and Burk (1979) point out that scrounging by deception may explain the evolution of complex assessment cues and ultimately individual recognition within species populations. Barnard and Burk distinguish three different ways in which aggressive contests between conspecifics can result in stable dominance hierarchies within groups. Only one of these types of hierarchy, *assessment hierarchies*, however, is of interest here. Because of the time and injury costs which are likely to be incurred by fighting, contestants will be selected to remember features (assessment cues) of opponents which correlated with winning or losing previous fights and to avoid contests with individuals of higher competitive ability (status). Assessment cues become CIMs because low-status mimics converging in their characteristics towards them are likely to benefit through reduced interference and increased priority of access to limited resources. Assessment cue mimics are therefore scroungers because they exploit the resource-accruing properties of high competitive ability. An arms race develops between mimics (detectees) emulating high-status assessment cues and assessors (detectors) trying to distinguish them from genuinely high-status individuals. In this case, however, individuals play both roles simultaneously, being selected to emulate high status and detect emulation in others.

As the frequency of mimics increases in the population, so also does the frequency of probing by assessors, for the reasons discussed above. This increases both the cost of assessment for assessors (because more time and energy needs to be spent probing) and the cost of expressing status to genuinely high-status individuals (because the frequency of challenges is high). Barnard and Burk (1979) argue that, when the cost of assessment exceeds the benefit gained from distinguishing between opponents, assessors should develop alternative means of detecting mimics. Simultaneously, high-status models will be selected to develop distinguishing features to reduce the cost of expressing status. Barnard and Burk suggested that selection would favour elaboration of the current assessment cue (C_1) to C_{1max} (equivalent to D_{Imax} in Figure 1.3b) at which point a second, reference, cue would be added into the assessment mechanism to test the veracity of C_1. Depending on the relative severity of the assessor's cost of assessment and the model's cost of expressing status, reference cues may be added in before the current cue C_n reaches C_{nmax}. A similar arms race and elaboration process may occur with the reference cue and ultimately several reference cues may be incorporated into the assessment mechanism. It is interesting that, where assessment cues have been investigated in detail (Rohwer and Rohwer, 1977; Davies and Halliday, 1978), they have turned out to be multifactorial. The arms race may result in a highly complex (many reference cues) assessment cue being used which is functionally equivalent to individual recognition or in a less complex but metabolically expensive 'status-limited' cue (e.g. large antlers, deep vocalisations) which can be possessed only by truly high competitive ability individuals.

Social Mimicry

An intriguing feature of some mixed species social aggregations, particularly among birds, is the character convergence between associating species which Moynihan (1960) termed *social mimicry*. Moynihan (1960, 1981) interprets social mimicry as a signal economy mechanism. He argues that character convergence between associating species is favoured because it increases the economy of interspecific communication. The number of displays with which the sensory-perceptual systems of associating individuals must cope is reduced and presumably time that might be wasted in redundant display and interpretation can be used for other activities. Barnard (1982) disagrees with Moynihan's signal economy interpretation because it is difficult to envisage a plausible evolutionary mechanism for mutualistic interspecific communication (see also Dawkins and Krebs, 1978). Instead, Barnard argues that social mimicry reflects interspecific exploitation and manipulation. His argument is based on the assumption that association is an adaptation to exploiting limited resources and/or avoiding predation.

One of the many benefits that may accrue to animals joining a group is the use of local information about the type and whereabouts of limited resources. The local information advantage has been particularly well studied in single and mixed species feeding flocks of birds (Krebs *et al.*, 1972; Krebs, 1973; Rubenstein *et al.*, 1977; Barnard and Sibly, 1981). By copying the behaviour of visibly successful companions (in terms of e.g. choice of foraging area or diet), birds may increase their own feeding efficiency. Copying, however, may be costly to recipients (individuals which are copied) in terms of reduced feeding efficiency, especially where interactions lead to kleptoparasitism (see Vollrath, Ch. 3 this volume; Barnard, Ch. 4 this volume). In these cases, recipients may evolve various countermeasures, such as reducing the handling time (and therefore the availability) of items, omitting telltale orientation postures prior to capturing prey, avoiding or rejecting vulnerable prey or aggressive retaliation (Barnard, Ch. 4 this volume). Such countermeasures are likely to reduce the success rate and/or increase the interaction costs for the interactor (copier or kleptoparasite). Selection is therefore likely to favour interactors which find a way to reduce the likelihood of countermeasures being taken.

Barnard (1982) argues that one way this could be achieved is by character convergence between interactor and recipient. Studies have shown that interactions are more successful when the interactor is nearer to the recipient as the latter finds food (Barnard, 1978; Burger *et al.*, 1979; Caldwell, 1981). Close approach of recipients is likely to be facilitated if interactors are difficult to distinguish from recipients. Social mimicry might therefore be analogous to so-called aggressive mimicry (Wickler, 1968) where the mimic emulates a harmless or mutualistic individual in order to approach and successfully attack a victim. Of course, selection will favour increased discriminatory abilities in recipients and hence further improvement of deception by interactors. The resultant arms race will hone not only the convergence in external features but also the way these features are used. In short, interactors will converge in both signal form and signal function. Although the end result may be difficult to distinguish from convergence by signal economy, there is a clear distinction between the selective mechanisms underlying the exploitation and economy models.

Similarity, of course, can also be a consequence of shared ancestry, with selection acting against divergence rather than in favour of convergence. Caldwell (1981) cites an interesting possible example in New World herons (*Ciconiiformes*). She noticed that adult little blue herons (*Egretta caerulea*) feeding in mixed flocks caught significantly fewer fish per unit time than immature birds. When feeding solitarily, however, they did better than immatures. The difference appeared to be due to young herons foraging much nearer than adults to snowy egrets (*E. thula*) in mixed flocks and was maintained by egrets selectively attacking adult rather than immature herons. Intriguingly, immature herons are white, like snowy egrets, while adults are blue. Caldwell points out that snowy egrets are good at locating locally rich feeding areas and the young herons' white coloration allows them to integrate more successfully with egrets and enhance their feeding efficiency. Exploitation of egrets may therefore be one factor selecting for a delay in adopting blue plumage. The fact that egrets chase away adult birds suggests there is a cost to herons feeding close by.

Automimicry

Many Batesian mimetic relationships are based on character convergence between innocuous prey species and a toxic model. While some toxic species synthesise their own noxious chemicals, others like the monarch butterfly (*Danaus plexippus*) sequester them from their food. Monarch caterpillars feed on the milkweed plant (*Asclepias curassavica*) which secretes toxic alkaloids to discourage herbivores, store the alkaloids in their tissues and retain them when they metamorphose. Brower (1971) showed that adult toxicity was acquired during larval feeding by raising monarchs on cabbage. Blue jays fed cabbage-reared monarchs showed no ill-effects, while those fed milk-weed-reared butterflies vomited soon afterwards.

Brower pointed out that an interesting frequency-dependent scrounging strategy, based on toxin sequestering, appears to have evolved in monarch populations. Sequestering toxins may well cost the monarch caterpillar something in terms of reduced growth rate compared with non-sequestering larvae on harmless plants. Non-sequesterers may therefore be able to exploit sequesterers by gaining the benefits of protection from predators (predators cannot distinguish non-toxic from toxic conspecifics) while not paying the metabolic cost of sequestering. When the frequency of sequesterers in the population is high, therefore, some females would be expected to lay their eggs on non-toxic plant species. When it is low, most females would be expected to lay on toxic plants, otherwise predators might learn that monarch butterflies were palatable. Brower (1971) regards non-sequesterers as Batesian mimics of their toxic conspecifics and refers to them as *automimics*.

Transvestism

For sexually reproducing species, receptive mates constitute an important limited resource. Selection has therefore favoured a variety of adaptations in both sexes which facilitate the acquisition of one or more mates (Halliday, 1978; Wittenberger, 1981; Parker, Ch. 5 this volume; Arak, Ch. 6 this volume). As with the acquisition of any other limited resource, acquiring mates requires investment of time and energy. Acquirers are therefore open to exploitation by scroungers unless there are severe penalties or other limitations militating against it. In Chapters 5 and 6 Parker and Arak discuss a range of scrounger strategies based on mate exploitation. Most of these strategies are strategies of usurpation (see above). In some cases, however, deceptive strategies have evolved which depend on trans-sexual mimicry. One example is the 'cuckolder' male strategy in blue-gill sunfish (*Lepomis machrochirus*) (Gross and Charnov, 1980; Arak, Ch. 6 this volume). Gross and Charnov (1980) identified three male behavioural types in their study populations: (a) *parentals* which constructed and remained in nests, attracted females and were highly aggressive towards intruders; (b) *sneakers* which remained close to the substrate, showed rapid nest entry and exit and were non-aggressive; and (c) *satellites* which showed slow nest entry and were sometimes aggressive to fish of equal size. A particularly intriguing observation was that, while parental and sneaker males were light in body colour, satellites were dark with dark, vertical bars and, in both colour and movement, mimicked female fish.

It turns out that sneaker and satellite are two developmental stages of a cuckoldry strategy which exploits the parental investment of parental males. Males which become parentals delay maturation until they are about seven years old when they begin territorial defence, nest-building and attracting females to spawn. Cuckolders, on the other hand, first mature at two years and follow the developmental sequence sneaker → female-mimicking satellite to gain access by deception to spawning females. The sneaker/satellite strategy in *Lepomis* is a clear example of a parasitic mating strategy which depends for part of its development on visual deception.

An interesting 'transvestite' strategy has also been reported in plants. Gilbert (1975) studied the complex coevolved relationship between pollen-feeding heliconiid butterflies, their pollen source vines *Anguria* and *Gurania* and their larval food plant, *Passiflora*. While the male vines supply the heliconiids with pollen and nectar, they depend on the butterflies for pollen transfer to the sparsely distributed female plants. Gilbert was intrigued by the excessive production of pollen and nectar by male vines, but interpreted it as a sexually-selected strategy by individual males to maintain the attention of heliconiids so that their pollen was the most likely to be transferred to any female plant encountered by the butterflies. A major problem for female vines, however, is to encourage pollen-bearing heliconiids to visit them. Since heliconiids visit vines only to feed on pollen and nectar, there is no reason for them to visit females. Females appear to have got round the problem by mimicking male plants. Male mimicry has occurred in two ways. Firstly, the female inflorescence has become telescoped so that it appears more like the dense male inflorescence. Male flowers are concentrated at the end of a long peduncle whereas those of females usually occur in a more spaced out fashion at each node. Secondly, female plants have evolved pollen-like nodules on the stigma. Experiments have shown that heliconiids are attracted to objects (e.g. small glass beads) which are shaped like pollen grains and nodules are likely to result in butterflies depositing pollen on the receptive stigma. Females therefore cash in on the investment of male plants in nectar and excess pollen to maintain the visiting rate of foraging heliconiids.

The aim of this chapter has been to point out the ways in which scrounger strategies have evolved. While I have discussed scrounging under the two headings of usurpation and deception, it is clear that these are not strictly alternatives. Exploitative social mimicry, for instance, may contain elements of both. The chapter has used only a few examples to illustrate the main points. Nevertheless, it has shown that 'scrounging' embraces a wide range of evolutionary processes from parasitism to the evolution of intraspecific assessment cues and is a major evolutionary shaping force. The chapters which follow examine in detail some of the scrounger strategies which have so far been studied or modelled and discuss their evolutionary consequences.

Acknowledgements

I thank Dennis Lendrem and Des Thompson for reading through and criticising the manuscript. The model was prompted by discussion and correspondence, some years ago, with John Allen.

References

Arme, C. and Owen, R.W. (1968). Occurrence and pathology of *Ligula intestinalis*. Infections in British fishes. *J. Parasit. 54:* 272-80

Barnard, C.J. (1978). Aspects of winter flocking and food-fighting in the house sparrow (*Passer domesticus domesticus* L.). D. Phil. Thesis, University of Oxford

Barnard, C.J. (1982). Social mimicry and interspecific exploitation. *Am. Nat. 120*: 411-15

— and Burk, T. (1979). Dominance hierarchies and the evolution of 'individual recognition'. *J. theor. Biol. 81*: 65-74

— and Sibly, R.M. (1981). Producers and scroungers: a general model and its application to feeding flocks of house sparrows. *Anim. Behav. 29*: 543-50

Brockmann, H.J. and Barnard, C.J. (1979). Kleptoparasitism in birds. *Anim. Behav. 27*: 487-514

— Grafen, A. and Dawkins, R. (1979). Evolutionary stable nesting strategy in a digger wasp. *J. theor. Biol. 77*: 473-96

Brower, L.P. (1971). Prey coloration and predator behavior. In V. Dethier (ed.), *Topics in Animal Behavior, topics in the study of life*. The BIO source book, Part 6, New York, Harper and Row

Burger, J.D., Hahn, C. and Chase, J. (1979). Aggressive interactions in mixed species flocks of migrating shorebirds. *Anim. Behav. 27*: 459-69

Butler, R.G. (1980). Population size, social behaviour and dispersal in house mice: a quantitative investigation. *Anim. Behav. 28*: 78-85

Caldwell, G.S. (1981). Attraction to tropical mixed-species heron flocks: proximate mechanism and consequences. *Behav. Ecol. Sociobiol. 8*: 99-103

Carney, W.P. (1969). Behavioral and morphological changes in carpenter ants harboring dicrocoelid metacercaria. *Am. Midl. Nat. 82*: 605-11

Clutton-Brock, T., Albon, S.D. and Guinness, F. (1979). The logical stag: adaptive aspects of fighting in red deer (*Cervus elephus L.*). *Anim. Behav. 27*: 211-25

Cott, H.B. (1940). *Adaptive Coloration in Animals*. London, Methuen

Davies, N.B. and Halliday, T.R. (1978). Deep croaks and fighting assessment in toads (*Bufo bufo*). *Nature 274*: 683-5

—, and Halliday, T.R. (1979). Competitive mate searching in male common toads *(Bufo bufo)*. *Anim. Behav. 27*: 1253-67

Dawkins, R. (1978). Replicator selection and the extended phenotype. *Z. Tier-*

psychol. 47: 61-76
— (1980). Good strategy or evolutionarily stable strategy? In G.W. Barlow and J. Silverberg (eds.), *Sociobiology: beyond nature/nurture*: 331-67. Boulder, Colorado, Westview Press
— (1982). *The Extended Phenotype*. London, Freeman
— and Krebs, J.R. (1978). Animal signals: information or manipulation? In J.R. Krebs and N.B. Davies (eds.), *Behavioural Ecology: an evolutionary approach*: 283-309. Oxford, Blackwell
— and Krebs, J.R. (1979). Arms races between and within species. *Proc. Roy. Soc. Lond. B 205*: 489-511
Dence, W.A. (1958). Studies on *Ligula*-infected common shiners (*Notropis cornutus* Agassiz) in the Adirondaks. *J. Paras. 44*: 334-8
Dobben, W.H. van (1969). The food of the cormorant in the Netherlands. *Ardea 40*: 1-63
Edmunds, M. (1974). *Defence in Animals.* London, Longmans
Gilbert, L.E. (1975). Ecological consequences of a coevolved mutualism between butterflies and plants. In L.E. Gilbert and P.H. Raven (eds.), *Coevolution of Animals and Plants*: 210-40. University of Texas Press
Graham, G.L. (1966). The behavior of beetles, *Tribolium confusum*, parasitized by the larval stage of a chicken tapeworm *Raillietina cesticillus. Trans. Am. microsc. Soc. 85*: 163
Gross, M.R. and Charnov, E.L. (1980). Alternative male life history strategies in bluegill sunfish. *Proc. Nat. Acad. Sci. USA 77*: 6937-40
Halliday, T.R. (1978). Sexual selection and mate choice. In J.R. Krebs and N.B. Davies (eds.), *Behavioural Ecology: an evolutionary approach*: 180-213. Oxford, Blackwell
Hogan-Warburg, A.J. (1966). Social behaviour of the ruff (*Philomachus pugnax* L.) *Ardea 54*: 109-229
Holmes, J.C. and Bethel, W.M. (1972). Modification of intermediate host behaviour. In E.U. Canning and C.A. Wright (eds.), *Behavioural Aspects of Parasite Transmission*: 123-49. London, Academic Press
Howard, R.D. (1978). The evolution of mating strategies in bullfrogs, *Rana catesbiana. Evolution 32*: 850-71
Källander, H. (1977). Piracy by black-headed gulls on lapwings. *Bird Study 24*: 186-94
Krebs, J.R. (1973). Social learning and the significance of mixed-species flocks of chickadees, *Parus* spp. *Can. J. Zool. 51*: 1275-88
— , MacRoberts, M.H. and Cullen, J.M. (1972). Flocking and feeding in the great tit *Parus major* – an experimental study. *Ibis 114*: 507-30
Krumhansl, C.L. (1978). Concerning the applicability of geometric models to similarity data: the interrelationship between similarity and spatial density. *Psych. Rev. 85*: 445-63
Maynard Smith, J. (1974). The theory of games and the evolution of animal conflicts. *J. theor. Biol. 47*: 209-21
McPhail, J.D. (1969). Predation and the evolution of a stickleback, *Gasterosteus. J. Fish Res. Bd. Canada 26*: 3183-208
Milinski, M. (1979). An evolutionarily stable feeding strategy in sticklebacks. *Z. Tierpsychol. 51*: 36-40
Moodie, G.E.E. (1972). Predation, natural selection and adaptation in an unusual three-spined stickleback. *Heredity 28*: 155-67
Moynihan, M.H. (1960). Some adaptations which help to promote gregariousness. *Proc. 12th Int. Ornith. Conf. Helsinki 1958*: 523-41
— (1981). A coincidence of mimicries and other misleading coincidences. *Am. Nat. 117*: 372-8

Parker, G.A. (1974). The reproductive behaviour and the nature of sexual selection in *Scatophaga stercoraria* L. IX. Spatial distribution of fertilization rates and evolution of male search strategy within the reproductive area. *Evolution 28*: 93-108

—, Baker, R.R. and Smith, V.G.F. (1972). The origin and evolution of gamete dimorphism and the male-female phenomenon. *J. theor. Biol. 36*: 529-53

Rau, M.E. (1983a). The open-field behaviour of mice infected with *Trichinella spiralis. Parasitology 86*: 311-18

— (1983b). Establishment and maintenance of behavioural dominance in male mice infected with *Trichinella spiralis. Parasitology 86*: 319-22

Rohwer, S. and Rohwer, F.C. (1977). Status signalling in Harris sparrows: experimental deceptions achieved. *Anim. Behav. 26*: 1012-22

Rubenstein, D., Barnett, R., Ridgely, D. and Klopfer, P. (1977). Adaptive advantages of mixed-species flocks among seed-eating finches in Costa Rica. *Ibis 119*: 10-21

Sheppard, P.M. (1975). *Natural Selection and Heredity*. London, Hutchinson

Tversky, A. (1977). Features of similarity. *Psych. Rev. 84*: 327-52

Vollrath, F. (1979). Behaviour of the kleptoparasitic spider *Argyrodes elevatus* (Araneae, Theridiidae). *Anim. Behav. 27*: 515-21

Wickler, W. (1968). *Mimicry in Plants and Animals*. New York, McGraw-Hill

Wittenberger, J.F. (1981). *Animal Social Behavior*. Boston, Duxbury Press

2 HOST-PARASITE INTERRELATIONSHIPS: STRATEGIES OF COEXISTENCE AND COEVOLUTION

C.R. Kennedy

Although there is no general agreement on the definition of a parasite, there is a consensus that parasites are metabolically dependent on their hosts and obtain from them all or part of their nutritional requirements. Parasites can therefore be regarded as scroungers in the sense that the term is employed in this volume, and the host-parasite relationship as a producer-scrounger relationship, in that parasites exploit their host's resources by using the behavioural, physiological and energetic investment of the host (a producer) to reduce the cost of obtaining their own resources. They scrounge not only food but also shelter and many other requirements. The cost of this mode of life, however, is the difficulty of finding a host and the attendant heavy egg and larval mortality, and then, because the relationship is so intimate, of overcoming host defence mechanisms directed specifically against them.

Stable contemporary host-parasite relationships must therefore represent a state of equilibrium between parasite provocation and host defence measures, and must be the product of long-term selection and coevolution of host and parasite. Indeed, in complex life cycles, the parasite may have coevolved with up to three separate species of host. Because parasitism is so widespread and parasites probably affect every living organism at some point in its lifetime, parasites must potentially influence the evolution of virtually every species of free-living animal as well as the functioning of most populations and every ecosystem.

Parasitism, however, is not only a relationship between an individual parasite and host but also, as Crofton (1971a, b) has emphasised, a population relationship, in which one species population takes up its abode in another. It is therefore necessary to consider parasites at both the individual and population levels simultaneously, and to relate the one to the other. The way in which a parasite population is dispersed through that of its host has important consequences for the stability of the relationship (Anderson, 1978; Anderson and Gordon, 1982; Anderson and May, 1978). The fact that a particular parasite or a certain level of infection may be harmful or even lethal to an individual host has to be evaluated in the context of the proportion of hosts infected and carrying high parasite burdens and the proportion that may be genetically resistant to infection (Wakelin, 1978). It is also necessary to bear in mind that the population relationships between parasite and host are a consequence of selection acting on individual genotypes, and are not selected for *per se*.

In this brief account it is impossible to cover all aspects of the functioning and evolution of such a widespread mode of life. Instead, I shall concentrate on four selected aspects of obligate host-parasite relationships that will illustrate the role of parasites as scroungers and highlight the evolutionary processes involved in the relationship. These are: the dispersion of parasites throughout their host populations; the benefits of the association to the parasites as revealed by a study of parasite energetics; the coevolution and coexistence of parasites and hosts and the cost of the relationship to the parasite in terms of host defences and parasite counter measures; and the way in which a limited resource, the host, is partitioned between individuals of the same and different species of parasite. Although treated separately for convenience, the four aspects are clearly closely interrelated.

Dispersion of Parasites

The distribution of parasites in space is heterogeneous, both between and within host species. Some parasites are narrow specialists, capable of infecting only one species or even strain of host. These are at one end of a broad spectrum: at the other end are generalists capable of infecting several species of host, often phylogenetically related or ecologically similar (Kennedy, 1975). Of these, one or more species will be the principal or required host to which the parasite is particularly adapted and in which it will achieve its optimum growth and/or reproductive success, and the others will be auxiliary, additional or accidental hosts. A generalist parasite can therefore form different relationships with different species of host, and the status of each host species and the distribution and flow of parasites through them can change from one locality to another. Generalists are thus opportunists and exhibit different degrees of adaptation, both phenotypic and genotypic, to different species of host, whereas specialists retain their specificity in all localities.

The distribution of parasites within a host species population is almost invariably overdispersed (variance greater than mean) (Crofton,

1971a, b; Anderson, 1978; Anderson and Gordon, 1982), such that some host individuals may be uninfected whilst others carry heavy parasite burdens, and the greater proportion of parasites are located in only a small proportion of hosts (Figure 2.1). Such a dispersion pattern can generally be described by the negative binomial model. The essential fact is that parasites are very seldom, and then generally only for a short period, randomly distributed throughout their host population.

Figure 2.1: Dispersion patterns of parasites in host populations. a: Random dispersion of metacercariae of *Tylodelphys podicipina* in the initial stages of colonisation of a population of perch *Perca fluviatilis*. b: Overdispersion of adults of *Acanthocephalus clavula* in a population of eels *Anguilla anguilla*. c: Diagrammatic representation of a dispersion spectrum

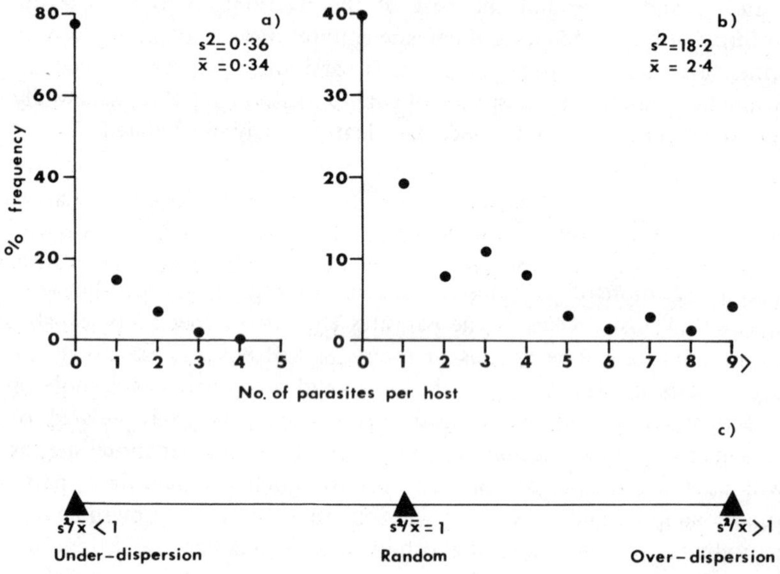

Source: Modified after Anderson and Gordon, 1982

Overdispersion can arise for a number of reasons, and is essentially due to heterogeneity in the probability of infection of a host. This can be caused in turn by heterogeneity in host behaviour, heterogeneity in effective immune responses (including those genetically based) within the host population, direct reproduction within a host and spatial heterogeneity in the distribution of infective stages. The degree of spatial clumping of infective stages influences the degree of overdispersion in the host population, but overdispersion will result even when the infective stages are arranged experimentally in a uniform pattern (Keymer and Anderson, 1979). Parasite frequency distributions, however, are not static. They are dynamic, and change in space and time under the influence of a number of factors that act to increase or decrease the degree of overdispersion (Anderson and Gordon, 1982).

The consequences of overdispersion are important. The greater the degree of overdispersion, the greater the probability of a parasite encountering a sexual partner. Furthermore, overdispersion minimises any deleterious effects of the parasite on the host population. In general, the higher the parasite density, the greater the degree of overdispersion and so the greater the extent to which the majority of parasites are confined to relatively few host individuals. Although this may result in the death of those particular individuals, the majority of hosts will only be lightly affected or will be completely unaffected. This has considerable implications for the regulation of parasite population numbers (Crofton, 1971a; Anderson and May, 1978), since a greater proportion of parasites than hosts will be removed by the death of a heavily infected host individual, and indeed regulatory processes may only operate in the few hosts with heavy parasite burdens (Anderson and Gordon, 1982; Kennedy, 1982). This has further evolutionary implications where, as is often the case, resistance to parasite infection is genetically based, since such mortality may effectively act as a selection pressure in favour of resistant individuals by killing off the heavily infected susceptible ones (Wakelin, 1978). It is probable that genetics, selection and density of host and parasite and overdispersion are interrelated and may interact in a complex and homeostatic manner in both population regulation and coevolution of both partners. As Pimentel (1961, 1968) has pointed out, parasite density may act as a selection pressure on hosts, resulting in changes in the genetics of the host population which will in turn affect the population density of the parasite.

Energetics of Parasites – The Benefits of Scrounging

For a parasite to complete its life cycle, several host species and conditions must overlap in time and space to create a resource patch for it (Price, 1980). The probability of colonising a new host individual is thus low, and attendant larval parasite mortality high. It is not surprising, therefore, to find that parasites have adopted the reproductive strategy of a successful coloniser, and tend towards the '*r*' extreme of the *r*/*K* selection continuum (Esch, Hazen and Aho, 1977). The associated high fecundity of parasites is generally held to be one of their most characteristic features (Crofton, 1971b; Kennedy, 1975), but they are only able to divert a large proportion of the energy they assimilate into growth and reproduction because they are able to minimise their energy expenditure on many other activities by letting the host accomplish these and so exploit the host's energy investment. This is achieved by several means:

1. By living in close proximity to, or surrounded by, a renewable food source. No energy is therefore required for the demanding activities associated with prey searching and capture.
2. By ingesting food in a fluid, readily assimilable or even partially predigested state. Energy costs in handling and preliminary breakdown of solid particulate foods are thus avoided or minimised.
3. By using host hormones and host products as stimuli for development rather than expending energy on synthesising their own.
4. By relying on overdispersion and site selection to achieve aggregation and increased probability of sexual contact, thus saving energy on mate seeking.
5. By reducing energetically expensive locomotory activity within a host to the minimum necessary for site selection, feeding and release of eggs.
6. By relying on host movements for dispersal in space and dissemination of reproductive products, rather than on their own activities. Host mobility allows parasite immobility and conservation of energy reserves (Kennedy, 1976).
7. By passing free-living periods as immobile, resting stages in which energy reserves are used solely for survival, or, if free-living stages are active, using the reserves for adaptive behaviour patterns that conserve energy and are directly related to the probability of locating a new host. A consequence of both strategies is that the energy reserves of each stage can be minimal.

As long, therefore, as the host is alive and its metabolism is not too disrupted, the relationship will always be profitable to the parasite as there will be a source of energy available, whether renewable or invested, that it can exploit at minimum cost to itself. There have in fact

Table 2.1: *Hymenolepis diminuta* in a rat: major pathways and estimated energy distribution over 168 days

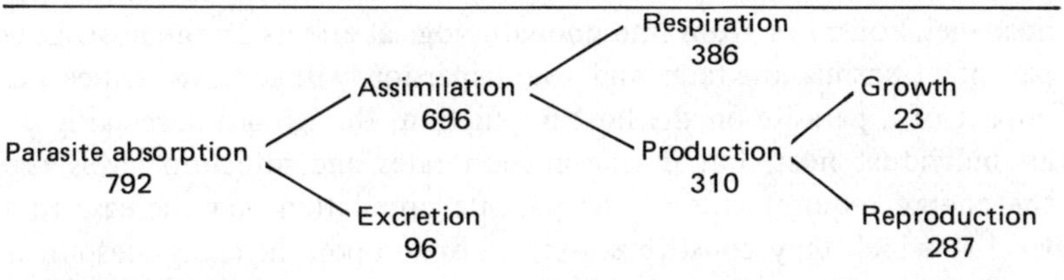

Source: after Bailey, 1975

been very few investigations conducted into the energy burden of parasites, but Bailey (1975) has shown that of the energy assimilated by the cestode *Hymenolepis diminuta* in rats, 45 per cent goes to production. Of this 93 per cent is used for reproduction and only 7 per cent for growth (Table 2.1), a proportion commensurate with the demands of high fecundity and *r* selection. The efficiency of the system was estimated at 0.8 per cent, which falls within the range reported for prey-predator systems. Compared to a predator, whose impact is variable and intermittent, the parasite levied a small but steady drain on the host which, on a population level, was considered to be of the same order as that of predation (Bailey, 1975). In this system the presence of from 1-16 parasites had no detectable effect on any of the rate processes measured in the rat, but in other systems the energy drain of a single parasite on an individual host may be much higher. Larvae of the cestode *Schistocephalus solidus* are energetically more efficient than their fish host, and over short periods of time parasite assimilation may exceed that of the host (Walkey and Meakins, 1970). This is achieved at the expense of host growth, fecundity and overall production.

Although some parasites appear to have no detectable effect upon host growth or production (Uznanski and Nickol, 1980), the majority exert effects of varying severity of which host mortality can be regarded as an extreme (Anderson, 1978). There are numerous well-documented examples of the deleterious effects of parasites on growth and survival rates of hosts, and Anderson (1978) believes that all parasites have a detrimental effect upon the natural intrinsic growth rate of their host population, and so on host production, even though these effects have seldom been quantified. In some cases the parasite may actually compete for host resources with the host (Anderson, 1978; Lanciani and May, 1982), but in others the loss of host production may be due not to the energy requirements of the parasite as such but to other effects resulting from direct parasite damage to host tissues, from toxic metabolic products of the parasite which interfere with host metabolism or from immunopathological effects on the host. Low parasite transmission rates and overdispersion will tend to reduce the impact of a parasite on the host population, though not necessarily on an individual host, but if transmission rates and infection levels rise the energy requirements of the parasite population may increase to a level at which they constitute a severe drain upon the energy resources of the host.

Coevolution and Coexistence of Host and Parasite

Coevolution

There appears to be no coherent body of theory for the evolutionary biology of parasites (Price, 1980), but there is general acceptance of the view that contemporary host-parasite systems are the product of variable periods of coevolution of host and parasite in the course of which each has exerted strong selection pressures on the other (Pimentel, 1961, 1968). The frequent occurrence of asexual reproduction and self-fertilisation can lead to single individuals founding a new colony, which may be genetically distinct, homogeneous and discrete and with little gene flow to other colonies. This will promote rapid race and strain formation, and evolutionary and speciation rates amongst parasites are believed to be high (Price, 1980). Contemporary observations support this view. Strain formation is widespread amongst parasites and specificity to the level of host strain-parasite strain is common (Kennedy, 1975; Wakelin, 1978). Coevolution can also occur very rapidly, as evidenced by the coevolution of *Fasciola hepatica* and the Australian snail *Lymnea tomentosa* within recent historical time (Boray, 1969). Schom *et al.* (1981) have also shown significant changes in the *Hymenolepis citelli–Tribolium confusum* association over a period of only two years (16 generations) in the laboratory and have provided convincing evidence that high selection pressures have been applied to the parasite by the host and vice versa. Pimentel and his co-workers (Pimentel, Nagel and Madden, 1963; Pimentel and Bellotti, 1976) have provided additional examples of the rapid coevolution of host and parasite in experimental systems. The relative ease and speed with which it is possible experimentally to select and breed resistant and susceptible strains of host (Figure 2.2) and infective strains of parasite (Wakelin,

1978) also indicates the potential for rapid coevolution in host-parasite relationships.

Figure 2.2: Variation in resistance and susceptibility of host strains to parasitic infection. The course of experimental infections with *Trichuris muris* in various inbred strains of mice

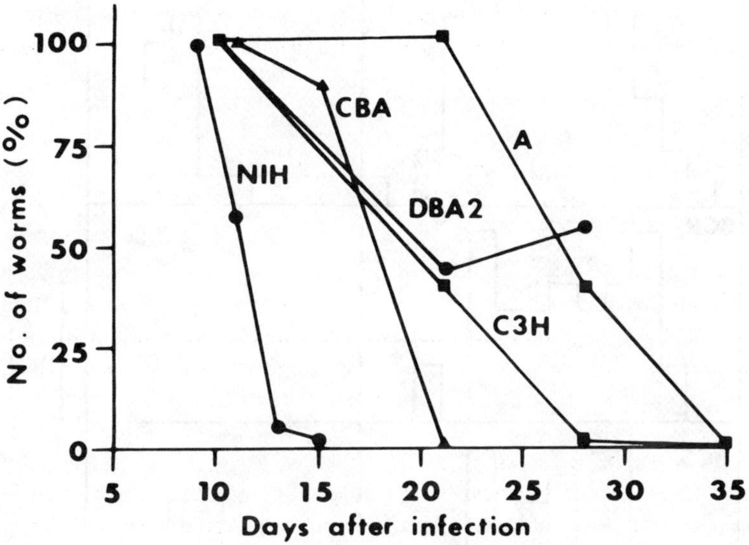

Source: reproduced from Wakelin (1978) by permission of the author and Academic Press Ltd.

It is in fact becoming increasingly clear that there is a genetic background to host defences, susceptibility and resistance (Figure 2.3) (Bradley, 1974; Wakelin, 1978). The capacity of hosts to respond immunologically to particular parasite antigens is often under direct genetic control, so that immune mediated resistance may be genetically determined and show variation with the genetic diversity that exists between strains and individuals of a host species (Wakelin, 1978). There is thus considerable intraspecific host variation in susceptibility and resistance to infection such that some strains and some members of a random-bred population may be completely refractory, others partially refractory and still others completely susceptible (Figures 2.2 and 2.3). Genetic control of host resistance and/or immunological responsiveness has been demonstrated in systems as diverse as *Plasmodium* in

Figure 2.3: The genetic control of host resistance and susceptibility. Liver parasite burdens (in Leishman-Donovan (LD) units), 14-16 days after infection with *Leishmania donovani*, in mice derived by crossing resistant (R) and susceptible (S) strains and by back-crossing (BC) F1 progeny to R and S parents

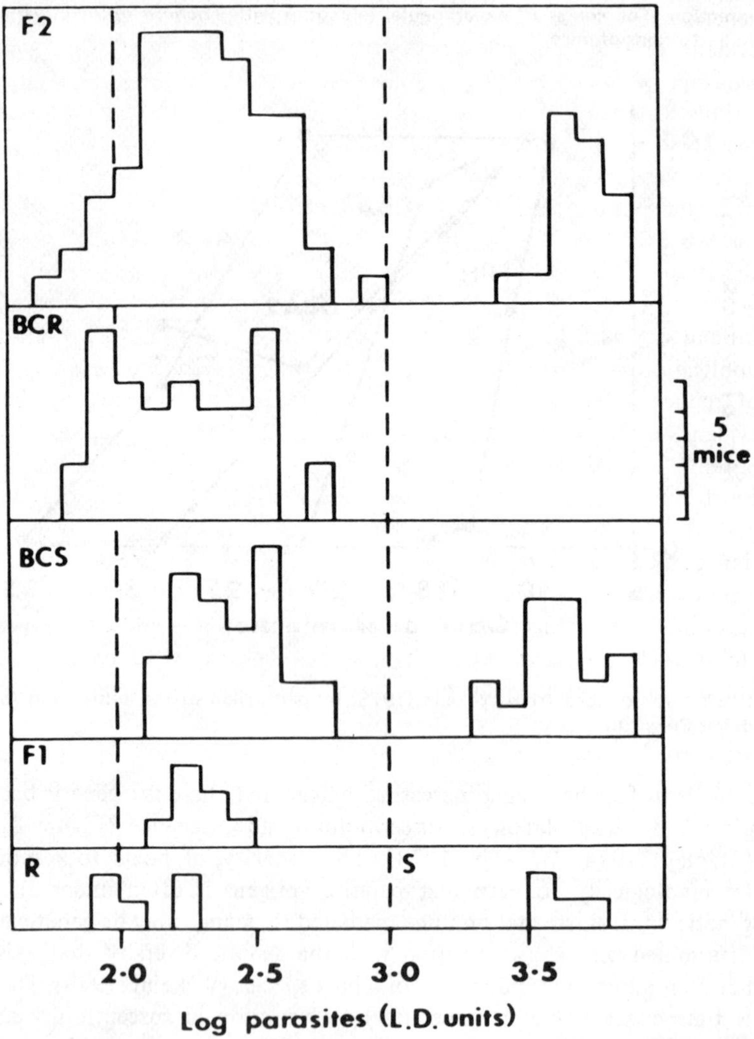

Source: from Bradley (1974) and Wakelin (1978) by permission of the authors and Academic Press Ltd.

mosquitoes and man, larval *Schistosoma* in snails and adult cestodes and nematodes in mammals and almost certainly will prove to be more widespread and perhaps even universal. In many relationships there is also evidence of genetic variability in the parasite population that controls invasiveness and virulence, such that establishment of an infection requires genetic compatibility in both members of the association. It has become evident (see Wakelin, 1978; Anderson and May, 1982) that some parasites can exert very strong selection pressures on their hosts' populations leading in some cases to the maintenance of polymorphisms in the host (as in sickle cell haemoglobin in humans), or, if the parasite is very virulent, to the elimination of susceptible genotypes. However, since the parasite can also respond to selection pressure exercised by the host, this may lead eventually to the evolution of less virulent parasite strains.

It is a frequently expressed axiom of parasitology that the successful and well-adapted parasite is one that does not harm its host, and the implication of this is that coevolution should proceed in the direction of reduced parasite 'harm'. At first sight this appears attractive and evident, and it may indeed be the best strategy where the reproductive potential of a parasite is invested in a long-lived adult, as in cestodes. Further consideration, however, reveals that this is not necessarily the best strategy for all parasites, since in many cases successful transmission of the parasite may be related to its pathogenicity. For example, transmission of many larval parasites to their definitive host can only be accomplished by ingestion of the intermediate hosts containing them and in such cases it will be of advantage to the parasite to harm or even kill the host if this will increase the probability of predation upon it (Barnard, Ch. 1 this volume). Similarly, transmission of many protozoans from one host to another is often more successful if the infected host is injured in some way. In such a case the best strategy for the parasite may be to reproduce as rapidly as possible before the current host is severely or terminally affected or before it can mount an effective immune response. There is thus no single 'most-suitable' coevolutionary pathway for parasites and hosts but many pathways are possible, the suitability in each case depending upon the relationship between parasite pathogenicity and transmission efficiency.

The relationship between parasite virulence, reproductive rate and transmission success has been examined in detail by Anderson and May (1982). They have demonstrated that transmission success is positively correlated with parasite virulence and reproductive success, and so they believe that the complicated interplay of virulence and transmission leaves room for many coevolutionary paths and many endpoints. They show by reference to examples and models that selection pressure on the host by pathogenic/virulent parasites can result in stable host polymorphisms, and that the possibility of co-existence and maintenance of two or more parasite strains depends on both the virulence and transmissibility of the parasite. If a parasite is too virulent, it may kill off too many hosts, thus destroying itself directly and reducing its reproductive rate and transmission success. This in turn may lead to the elimination of susceptible genotypes from the host population. On the other hand, if the parasite is not virulent enough, transmission rates and reproductive success will also decline. For many parasites, therefore, an intermediate level of virulence is best for transmission and is associated with maximum basic reproductive rate.

Anderson and May (1982) therefore argue that a well-balanced host-parasite association is not necessarily one in which the parasite does little harm, and in support of this argument they cite the facts that: (1) transmission efficiency is often positively linked to virulence and so reproductive rate; (2) parasitism is a far commoner type of relationship than the harmless relationships of amensalism and commensalism; (3) that the existence of genetic polymorphisms in hosts controlling their immune responses must be due to the continual selection pressure exerted by parasites; and (4) that the complexity and sophistication of host immune systems is hard to envisage if the parasite is not of major importance to host fitness. Indeed, if selection always favoured harmless parasites, one would not expect to find polymorphisms in host susceptibility, yet these are in fact probably the rule. They envisage that in many instances coevolution will proceed by selection pressure acting on the parasite in favour of intermediate pathogenicity or the maintenance of strains of intermediate pathogenicity, and on the host for resistance. Such selection pressure on the host will in turn exert pressure on the parasite. In other cases there may be selection pressure on the parasite for increased pathogenicity since it aids transmission, provided that the host is not killed prematurely and before the parasite has had time to reach the infective or reproductive stage. If, however, the parasite evolves to a completely harmless state, then numbers could build up in individual hosts to levels at which they constituted an unacceptable drain on the host's energy reserves. The interplay of such selection pressures is not yet understood, but well-balanced systems may be the exception in view of the temptation to cheat (Anderson, 1982). In any event, the parasite faces a constant battle to survive, and pathogenicity, because of its positive relationship with transmission and reproductive success, is beneficial to it. The ways in which it survives host defences can more easily be understood and appreciated by studying coexistence of host and parasite in contemporary relationships.

Coevolution, and coexistence, of hosts and parasites can also be considered in terms of an arms race, as discussed by Dawkins and Krebs (1979) in the context of predator and prey. Any adaptation by the parasite may change the selection pressure on the host, leading in turn to counter-adaptations by the latter. The evolution of such a host-parasite system may involve escalation of ever-more refined mutual counter-adaptations: the equivalent, in contemporary terminology, of an arms race. Such coevolution, however, can proceed along several different lines and have several possible outcomes: one species may cause extinction of the other, one may reach an optimum stable state thereby preventing the other from doing so, they may both reach a mutual local optimum, or there may be no stable end but continuous cycling. Many evolutionary changes result from such arms races, and they have pronounced effects upon the speed of evolution of each of the species involved in them.

Coexistence

On first encountering a potential host, a parasite will be recognised initially as 'non-self', and the host defence mechanisms, whether non-specific or specific humoral or cell-mediated, will be mobilised against it. It appears highly probable that all parasites elicit immune responses of a specific nature in their host, and many often also elicit non-specific responses. Parasites, by their nature, represent a complex mosaic of antigens; antigenic differences may occur at distinct life cycle stages of the parasite, but variation or modulation can occur at any given stage (Capron and Dessaint, 1982). Antigens may be surface-bound or circulating, and the complexity of the host immune response to them results in large measure from the diversity of expression of the parasite antigens themselves. Immune responses to metazoan parasites especially tend to develop slowly, and are seldom efficient at eliminating the parasite. The size of parasites, their antigenic complexity and their complex life cycles make them very difficult to control immunologically. The responses may, however, be effective in restraining resident parasites, modifying parasite characters, rendering the infected host resistant to reinfection, or even, rarely, producing sterile immunity in some host genotypes (Mitchell, 1982). For the host-parasite relationship to continue, therefore, the parasite must have evolved strategies or tactics to enable it to negate or evade the immune response. Indeed, the inability of most immune responses to eliminate parasites completely is due in large measure to the ability of the parasite to evade or influence the host immune responses directed against it. The multiplicity of immune effector mechanisms at the disposal of a mammalian host in particular dictates that a parasite must invest in a multiplicity of immune evasion strategies to thwart and subvert them (Mitchell, 1982).

When a parasite invades a host, there are several possible outcomes (Cox, 1982):

1. It may fail to become established due to innate host immunity.
2. It may establish and kill the host. This will tend to occur when there is a massive invasion, or the parasite can multiply within the host, or it is pathogenic and/or the host immune responses are inadequate or evaded. It is a viable strategy where host death or illness favours parasite transmission, provided the parasite reproduces or develops to the correct stage in time.
3. It may establish, but the host is able to overcome the infection, destroy the parasite and become immune to reinfection. This consequence is commoner amongst microparasites (bacteria, protozoans) but rare amongst helminths. It is again a viable strategy if the parasite is able to reproduce and disperse before being eliminated. This timing is crucial.
4. It may establish and the host respond but be unable to eliminate the infection, only reduce or control it. This situation is common in helminth infections, but is frequently unstable and may change to 2 or 3 or, in some cases, the parasite may establish and the host start to overcome the infection but instead of destroying the parasite becomes damaged itself, the response being misdirected to the host tissues and leading to immunopathological disorders (Cox, 1982).

The responses of invertebrates to parasites tend to be of an all or nothing nature: the parasite is recognised as self and accepted or as non-self and rejected. In general the parasites appear to elicit an initial response in all hosts, but in the preferred host the response is subsequently evaded whereas in the wrong host the parasite is encapsulated and killed (Lackie, 1980). Survival is possible in some relationships because the parasite surface, or the capsule surrounding it, is antigenically similar to the host, which thus does not recognise the invader as non-self. In other cases the parasite is recognised as non-self, but the effector response of the host is interrupted (Lackie, 1980). In most cases, however, it is not known how the host response is evaded, and there appears to be no effective resistance to infection in genetically susceptible host individuals.

Table 2.2: Parasite strategies for evading host immune responses

1. Reduced antigenic disparity
Molecular mimicry
2. Selection or change of site
Inhabiting an inaccessible/intracellular site
Inhabiting an immune-privileged site
Inhabiting lymphocytes
Moving away from the site of an immune response
3. Reduced or altered parasite antigenicity/immunogenicity
Population antigenic heterogeneity
Inert surface resistant to immune attack
Antigenic variation in surface coat
High surface turnover and shedding of surface antigens
Disguise with host antigens
4. Modification of induction or expression of host immune responses
Inhibiting cell or antibody binding
Inactivating host lymphocytes
Immunodepression
5. Modification of intra-macrophage environment

Source: after Cox, 1982 and Mitchell, 1982

Parasites of vertebrates, however, and especially of mammals have evolved a variety of strategies for evading host immune responses (Table 2.2), including reducing antigenic disparity between parasite and host, selecting an immunologically inaccessible site, altering their antigenicity or immunogenicity, modifying the induction or expression of the host immune response and even modifying the intra-macrophage environment.

The principle of reducing antigenic disparity is that the parasite achieves recognition as self and avoids stimulating host responses by presenting a surface to the host that appears to it as self, i.e. by molecular mimicry and antigen-sharing with the host. There is strong evidence that antigen-sharing does occur in some relationships, but it is not clear whether the shared antigens are true host antigens taken up by the parasite or imitation antigens synthesised by the parasite (Capron *et al*., 1968). It is easy to envisage strong selection pressure in favour of antigens that resemble those of the host (Sprent, 1969) since it will lead to diminished host responsiveness. Evolution along these or similar lines could proceed in the direction of peaceful coexistence and disarmament if the mimicry leads to ever closer antigenic convergence between parasite and host, but if the host is able to recognise the mimic and there is selection pressure on it to respond to the adaptations of the parasite then mimicry will merely provide a period of grace in the arms race.

The remaining strategies of evasion resemble a true arms race. The parasites provoke host responses, and respond to them in turn by counter-measures. Any one species of parasite may employ one or more methods of evasion. Small protozoans may avoid responses by living in intracellular sites (*Plasmodium*), by living in immune-privileged sites such as the eye and brain or by living within lymphocytes themselves (*Trypanosoma cruzi*). Larval nematodes (*Ascaris*) may change site and antigens, so that the response is directed to antigens no longer present and to a site where the parasite no longer lives. Species of *Leishmania* and *Trypanosoma* that live in macrophages may modify the macrophage function to prevent their being digested. (For details of these and subsequent examples see Cox (1982) and Mitchell (1982).)

Each life cycle stage of *Plasmodium* spp is antigenically distinct, each species is a complex of immunologically different strains, and each population exhibits antigenic heterogeneity, with each variant inducing specific immunity. *Trypanosoma brucei*, however, changes its antigenicity in the course of infection: it is capable of antigenic variation in its surface glycoprotein coat, and each variant provokes a specific antibody response. As one variant is destroyed, it is succeeded by another (Figure 2.4). *Schistosoma mansoni*, however, provokes immunological responses that render the host resistant to reinfection, yet itself survives in the face of the immune response it provokes. This situation of concomitant immunity is due to the accumulation of host antigens on the surface of the parasite (Smithers and Terry, 1969). The parasite thus disguises itself as host, its antigens are masked, host antibodies blocked and it is treated as self. The acquisition of the antibodies is a phenotypic character and takes time, so that new invaders are damaged by host antibodies before they have had time to pick up protective antigens. There is also increasing evidence that renewal of the schistosome surface may be a mechanism of immune evasion. The renewal involves recycling or shedding of surface membranes, membrane synthesis and turnover, and this latter is so rapid in schistosomes that it may be critical in protecting the cells from immune damage (Podesta, 1982). It may indeed be too rapid for parasite antigen masking by host

Figure 2.4: Antigenic variation in *Trypanosoma brucei*. Diagram illustrating the course of infection of *T. brucei* in a mammalian host. A, B, etc., refer to different antigen variants, and anti-A, anti-B, etc., to different and specific antibodies

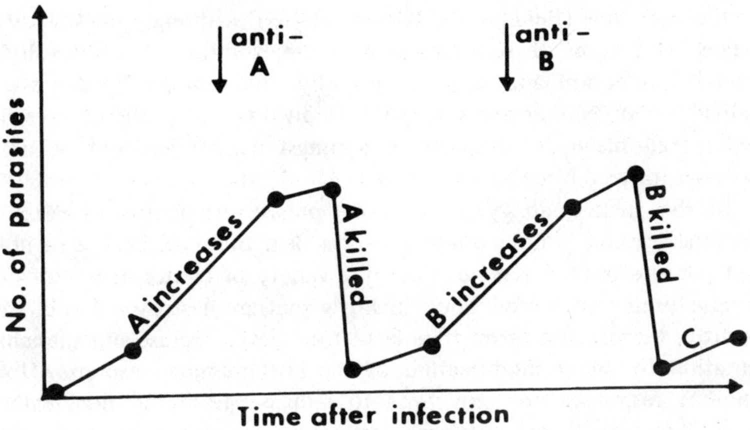

molecules to be an important means of immune evasion.

Many, perhaps even most, parasites are also capable of modification or regulation of the induction or expression of the host immune responses by a variety of means. Parasite-derived molecules have been shown to inhibit cell or antibody binding and to cause polyclonal stimulation of lymphocytes. Schistosome-derived molecules can also act as inhibitors of lymphocyte functions and induce significant and prolonged immunodepression in their hosts (Capron and Dessaint, 1982).

Parasites may also survive in hosts in which they provoke strong immune responses if their infection levels are below the threshold necessary to elicit an effective reaction. Thus heavy infections of *Trichuris muris* in mice normally provoke a response that leads to elimination of the parasite before it even becomes patent, but at low-level infections, below a threshold of 10-20 parasites per host, the parasites are not eliminated but are able to survive, mature and become patent. They provoke an immune response, however, and as long as they persist the host is resistant to new infections (Wakelin, 1973; Behnke and Wakelin, 1973). It appears that in small infections the antigenic stimulus is inadequate to elicit a response capable of expelling the parasite population but instead primes the immune system so that any additional stimulus results in an effective response. Parasite populations survive in the wild because transmission rates and so infection levels in wild mice are generally low (below the threshold), or because some host individuals are immunologically depressed by e.g. pregnancy and lactation and harbour heavy parasite burdens that are temporarily above the threshold level and provide enough eggs for the continuation of the life cycle (Behnke and Wakelin, 1973). Although not strictly a means of evasion of an immune response, survival at sub-threshold levels of infection and/or in temporarily immunologically depressed individuals or genetically susceptible individuals is probably a fairly widespread means of coexistence amongst nematodes, and possibly cestodes, of wild birds and mammals.

In the great majority of cases the precise mechanism of parasite immune evasion is not known, but it is clear that coexistence of host and parasite is rendered possible by a variety of tactics, ranging from precise timing of reproduction in hosts that are destroyed by, or can destroy, the parasite through molecular mimicry, disguise and antigenic variation to direct modification of the host immune response. Host immune responses not only tend to reduce parasite burdens rather than eliminate them, but can also serve as mechanisms regulating parasite population sizes and so stabilising the host-parasite relationship (Anderson, 1978; Anderson and May, 1978). Persistence of even pathogenic parasites in a host population is ensured by low transmission rates, the presence of resistant host genotypes and overdispersion. It is therefore possible for species of parasite to adopt a variety of strategies for coexistence, and it is hardly surprising that coevolution of host-parasite associations has proceeded along so many different pathways.

Resource Partitioning between Parasites

Intraspecific resource partitioning

All parasites represent a drain upon their host's resources, and these, even when renewable, are limited so that if parasite numbers rise there is a distinct possibility of the resources being insufficient. This possibility may be reduced by the immune and other defence responses of the host controlling the parasite burden, but where these responses are evaded or transmission rates increase there must be the potential for intraspecific competition.

In invertebrate hosts, the presence of large numbers of metazoan larvae generally results in smaller, stunted individuals and sometimes eventually in death of the host due to direct damage or depletion of

Figure 2.5: The crowding effect in adult cestodes. Changes in size and fecundity of the cestode *Hymenolepis diminuta* at different infection levels in rats

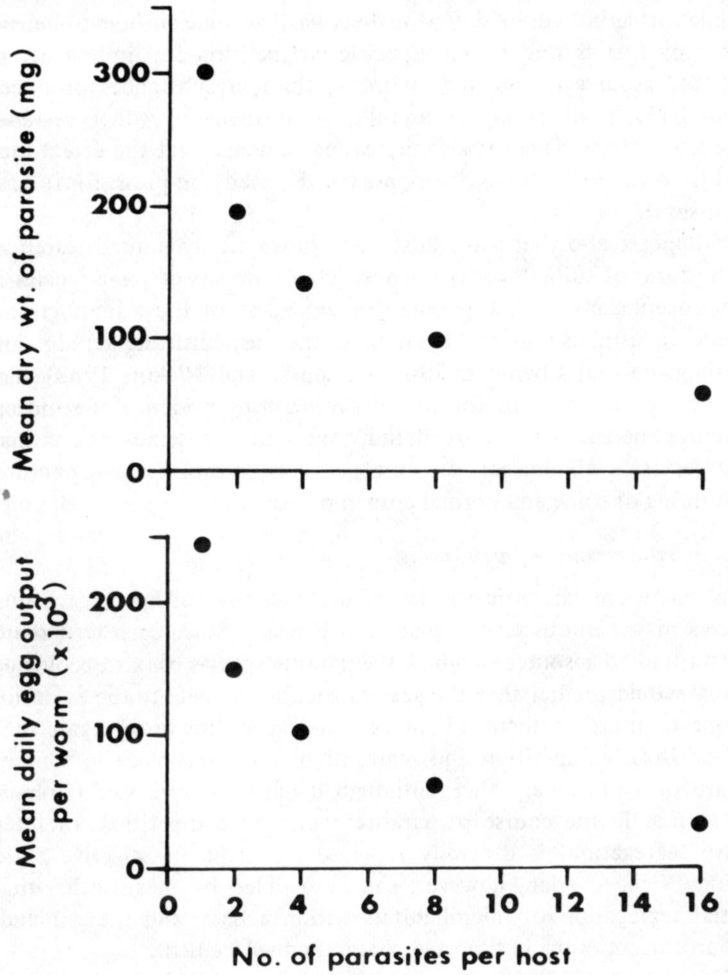

its reserves. In vertebrate hosts, heavy infections may provoke effective immune responses, or, particularly with cestodes, be manifested as a crowding effect, in which there is an inverse relationship between parasite density and average parasite size and fecundity (Figure 2.5). Overcrowding always results in an average decrease in individual parasite fecundity, but in some situations all parasites may be stunted whereas in others a few parasites may grow and reproduce normally whereas the remainder are stunted and retarded. The similarity in condition between crowded cestodes and those in hosts maintained on sub-optimal or carbohydrate-deficient diets has led some authors to believe that crowding is due to intraspecific competition for limited carbohydrate resources, mediated, perhaps, through substances produced by individuals inhibiting the uptake of nutrients by other cestodes (Roberts, 1980). There is also increasing evidence that the effect may be due to the host immune response (see Kennedy, in press, for further discussion).

It appears also that some hosts may provide only limited resources in the form of suitable space for restricted numbers of some species of acanthocephalans, and any parasites in excess of these numbers are unable to establish in the host, or if they establish are rapidly lost (Burlingame and Chandler, 1941; Uznanski and Nickol, 1982). The survivors grow and reproduce normally, with no evidence of stunting. Resource partitioning between individuals of the same species may thus result in elimination of some parasites, stunting of all parasites or stunting of some and normal growth of others.

Interspecific resource partitioning

Since there are far more species of parasites than of hosts, each host species may harbour several species of parasite. Since each host represents a limited resource in which the parasite species coexist, ecological theory would predict that the species should compete to the exclusion of one or more of them, or interact and specialise or segregate their niches. Both competition and segregation may take place in contemporary or evolutionary time, although it has been suggested (Holmes, 1973) that in the course of parasite evolution competition, or interactive segregation, is generally replaced by niche or selective segregation. Niche overlap, however, can be avoided by means other than spatial segregation of microhabitats within a host, and these include resource partitioning in time and character displacement.

Resource partitioning in time between species of parasites is not apparently very common, although it has been reported on occasion (Miller, 1952). The use of different age classes of host by closely related species of parasite in concurrent infections can also be regarded as a form of partial resource partitioning in time.

Simultaneous coexistence of species in the same host may also be possible if the species show morphological adaptations and/or character displacement so that they can occupy different sites or utilise different resources. Where different species of allocreadiid flukes coexist in the same host, it is generally found that they show differences in size of their oral sucker and often in addition occupy slightly different sites (Cannon, 1972). The several different species of tetraphyllidean cestodes that inhabit the spiral valve of rays all show different and specialised attachment mechanisms that enable them to occupy different tiers (Williams, 1968).

There is abundant evidence, however, that the commonest and most widespread method of interspecific coexistence is by resource partitioning in space (Holmes, 1973; Crompton, 1973). Parasites are not distributed at random throughout the body or any individual organ system of their host but each species occupies a characteristic site. Even along one or more resource gradients, such as the intestine or gills, species show niche segregation as a result of active site recognition and discrimination (Crompton, 1973), and each species occupies a characteristic site which may or may not overlap with that of another species (Figure 2.6). It is thus possible for species flocks to occur within hosts. Schad (1963) found that 8-10 species of nematode could coexist within the intestines of tortoises because their niches were separable on two resources, radial and longitudinal distribution. Hair and Holmes (1975) similarly found that up to 43 species of cestodes (77 per cent of which belonged to the single genus *Hymenolepis*) could occur in a population of ducks, and that each of the common species occupied a characteristic site along the intestine. In the great majority of cases species appear to exhibit selective site segregation (Holmes, 1973; Hair and Holmes, 1975). Examples of interactive site segregation, in which one species is displaced by the presence of another and its realised niche in the presence of the other species is not the same as its fundamental niche, are known. In some mixed infections of cestodes and acanthocephalans the specialist acanthocephalan displaces the more generalist cestode (Holmes, 1973; Kennedy, in press). There is general agreement, however, that most species exhibit resource partitioning in space by means of selective site segregation.

There is rather less agreement on the reasons and advantages of this, and on its evolutionary implications. Holmes (1973) and Hair and Holmes (1975) have argued that the selective segregation evident amongst contemporary parasites is a result of interactive segregation in the past. They believe that interspecific interactions led in the course of evolution to niche diversification and segregation, and thus interspecific competition for limited resources was minimised. They consider that contemporary parasite communities are mature and well-

Figure 2.6: Resource partitioning in space. Distribution of parasites in the intestine of flounders *Platichthys flesus*

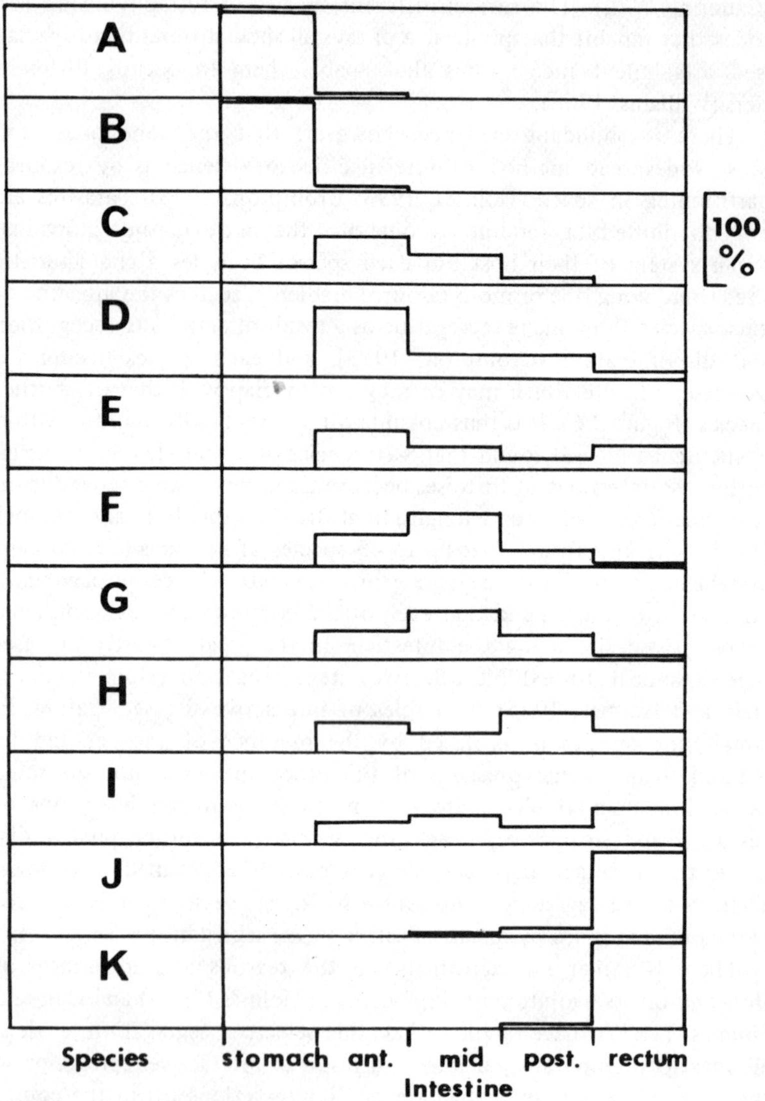

Source: from Mackenzie and Gibson (1970), by permission of the authors and Blackwell Scientific Publications Ltd.

established, the product of coevolution and selection, but can be modified by interactive segregation on occasions. A rather similar view, though of free-living communities, has been expressed by Lawlor and Maynard Smith (1976). They argue that communities are not arbitrary collections of species but assemblages that have existed over sufficiently long periods for each species to adapt to the other members. Such coevolution of potential competitors has led to species divergence and resource partitioning, and so to a reduction of competition and increased community stability. The demonstration by Hobbs (1980), that old communities of nematodes in pikas, *Ochotona princeps* and *O. collaris*, exhibit selective site segregation whereas recent members compete, tends to support Holmes' views. Price (1980), however, considers that there is no evidence that interspecific interactions have ever been involved in determining community structure and so they have played no evolutionary role. He believes that most parasite communities are in the early, non-interactive stages of development in which habitats are unsaturated and resources underutilised (McArthur and Wilson, 1967), and that species have not yet evolved to use them. He argues that niche expansion at high levels of infection (a widespread feature amongst parasites (Crompton, 1973)) suggests that there are few or no constraints on the parasites and that hosts are indeed unsaturated. Niche segregation is therefore due to factors other than competition. A similar view is held by Rohde (1979), who has provided evidence that many hosts are unsaturated with parasites, that resources are often underexploited and that many vacant niches for parasites exist in some hosts (Figure 2.7). He also provided evidence of narrow niche segregation in the complete absence of any other species utilising the same organ system of the same host species. He believes that the advantage of, and hence selection pressure for the evolution of, niche segregation is that it concentrates parasites in a restricted area and so increases the probability of mating in dioecious species and of cross-fertilisation between hermaphrodites, features that are necessary or desirable to maximise parasite reproductive potential. Thus, although there is disagreement on the role of competition in the evolution of parasite communities and on the selective pressures for the evolution of niche segregation, there can be no doubt that the resources of the host are partitioned out between parasite species and that if and when they are limited, selective site segregation is the commonest means of permitting them to be shared between parasite species.

Figure 2.7: Underexploitation of host resources by parasites. Relative species diversity of monogenean gill parasites of teleost fish in the Pacific and Atlantic oceans. The greater diversity at warmer latitudes and in the Pacific indicate that many niches on the gills are unfilled at colder latitudes and in the Atlantic

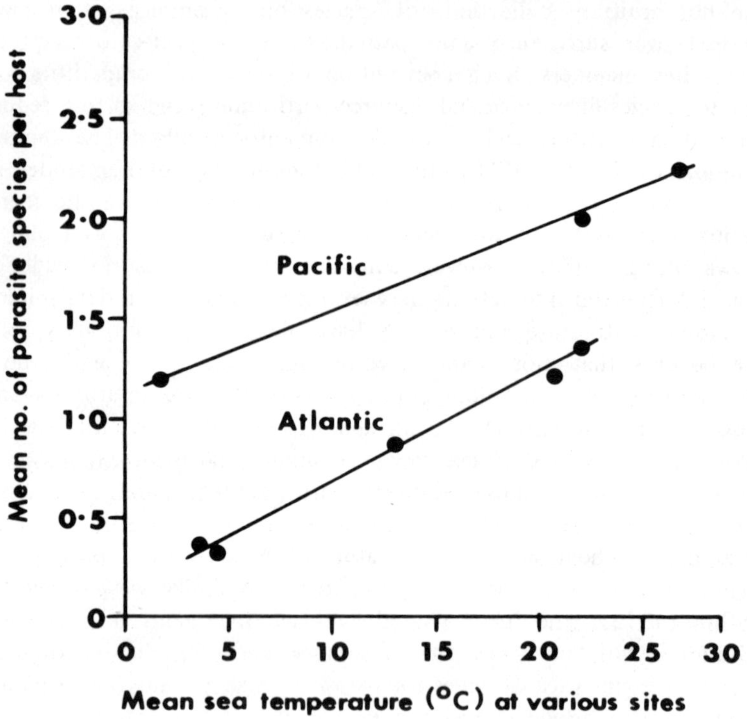

Source: from Rohde (1982), by permission of the author and the University of Queensland Press

Conclusions

Parasitism as a way of life presents many problems. In the course of its lifespan a parasite has to survive in at least two and often more different habitats, including at least one host species. Probabilities of infecting a new host individual are low, and after infection it has to share the host with other members of its own species and other species of parasite. At the same time, by living so intimately with the host it continually provokes and has to overcome the host's defences. Finally, for the parasite and its association to persist, it must also coevolve with at least one and often more species of free-living animal.

All these problems, however, have been overcome: by incorporation of resting stages into the life cycle, by adopting a strategy of *r* selection, by resource partitioning and by the employment of ingenious and sophisticated mechanisms at the molecular level to overcome host responses. The impact of parasitism on an individual host may in some cases be severe, but that on the host population is minimised by overdispersion. Parasites will always provoke host defences, but coevolution may proceed along many pathways. In the course of such coevolution the parasite can act as an important selection agent upon the host and vice versa. Coevolution cannot stop, however: the arms race must continue since survival of the parasite is only possible if neither partner gains a decisive advantage over the other, and so parasite evolutionary rates must perforce remain very high.

The benefits of parasitism to the parasite, however, are considerable. It has ready access to a cheap and renewable source of nutriment and energy, and in homoiotherms the advantage of a homeostatic environment without the cost of maintaining it. In addition, parasites can use the host for transport, locomotion and dispersal. The extent of this dependence on the host must make them the best scroungers in the business, and the success of this type of exploitation of one species by another can be judged by the widespread adoption of the parasitic habit by so many and such varied species. It is a salutary thought that if indeed parasitism is more common than all other feeding strategies combined (Price, 1980), then there are more species of scrounger on earth than of any other kind of animal and scrounging is the most widespread and successful way of life.

References

Anderson, R.M. (1978). The regulation of host population growth by parasitic species. *Parasitology 76*: 119-57

— (1982). Host parasite population biology. In D.F. Mettrick and S.S. Desser, (eds.), *Parasites – Their World and Ours*: 303-12. Amsterdam, Elsevier Biomedical Press

— and Gordon, D.M. (1982). Processes influencing the distribution of parasite numbers within host populations with special emphasis on parasite-induced host mortalities. *Parasitology 85*: 373-98

— and May, R.M. (1978). Regulation and stability of host-parasite population interactions. I. Regulatory processes. *J. Anim. Ecol. 47*: 219-47

— and May, R.M. (1982). Coevolution of hosts and parasites. *Parasitology 85*: 411-26

Bailey, G.N.A. (1975). Energetics of a host-parasite system: a preliminary report.

Int. J. Parasit. 5: 609-13

Behnke, J.M. and Wakelin, D. (1973). The survival of *Trichuris muris* in wild populations of its natural hosts. *Parasitology 67*: 157-64

Boray, J.A. (1969). Experimental fascioliasis in Australia. *Adv. Parasit. 7*: 96-210

Bradley, D.J. (1974). Genetic control of natural resistance to *Leishmania donovani* in mice. *Nature, Lond. 250*: 353

Burlingame, P.L. and Chandler, A.C. (1941). Host-parasite relations of *Moniliformis dubius* (Acanthocephala) in albino rats, and the environmental nature of resistance to single and superimposed infections with this parasite. *Am. J. Hyg. 33*: 1-21

Cannon, L.R.G. (1972). Studies on the ecology of the papillose allocreadiid trematodes of the yellow perch in Alqonquin Park, Ontario. *Can. J. Zool. 50*: 1231-9

Capron, A. and Dessaint, J.P. (1982). Mechanisms of immunogenesis. In D.F. Mettrick and S.S. Desser (eds.), *Parasites – Their World and Ours*, Vol II: 15-24. Amsterdam, Elsevier Biomedical Press

—, Biguet, J., Vernes, A. and Afchain, D. (1968). Structure antigénique des helminthes. Aspects immunologiques des relations hôte-parasite. *Path. Biol., Paris 16*: 121-38

Cox, F.E.G. (1982). Immunology. In F.E.G. Cox (ed.), *Modern Parasitology*: 173-203. Oxford, Blackwell Scientific Publications

Crofton, H.D. (1971a). A quantitative approach to parasitism. *Parasitology 62*: 179-93

— (1971b). A model of host-parasite relationships. *Parasitology 63*: 343-64

Crompton, D.W.T. (1973). The sites occupied by some parasitic helminths in the alimentary tract of vertebrates. *Biol. Rev. 48*: 27-83

Dawkins, R. and Krebs, J.R. (1979). Arms races between and within species. *Proc. Roy. Soc. (B) 205*: 489-511

Esch, G.W., Hazen, T.C. and Aho, J.M. (1977). Parasitism and r- and K-selection. In G.W. Esch (ed.), *Regulation of Parasite Populations*: 9-62. New York, Academic Press

Hair, J.D. and Holmes, J.C. (1975). The usefulness of measures of diversity, niche width and niche overlap in the analysis of helminth communities in waterfowl. *Acta parasit. Polon. 23*: 253-69

Hobbs, R.P. (1980). Interspecific interactions among gastrointestinal helminths in pikas of North America. *Am. Mid. Nat. 103*: 15-25

Holmes, J.C. (1973). Site selection by parasitic helminths: interspecific interactions, site segregation, and their importance to the development of helminth communities. *Can. J. Zool. 51*: 333-47

Kennedy, C.R. (1975). *Ecological Animal Parasitology*. Oxford, Blackwell Scientific Publications

— (1976). Reproduction and dispersal. In C.R. Kennedy, (ed.), *Ecological Aspects of Parasitology*: 143-60. Amsterdam, North-Holland Publishing Company

— (1982). Biotic factors. In D.F. Mettrick and S.S. Desser, (eds.), *Parasites – Their World and Ours*: 293-302. Amsterdam, Elsevier Biomedical Press

— (in press). General ecology. In C. Arme and P.W. Pappas (eds.), *The Biology of Cestodes*. London, Academic Press

Keymer, A.E. and Anderson, R.M. (1979). The dynamics of infection of *Tribolium confusum* by *Hymenolepis diminuta*: the influence of infective stage density and spatial distribution. *Parasitology 79*: 195-207

Lackie, A.M. (1980). Invertebrate immunity. *Parasitology 80*: 393-412

Lanciani, C.A. and May, P.G. (1982). Parasite-mediated reductions in the growth of nymphal backswimmers. *Parasitology 85*: 1-7

Lawlor, L.R. and Maynard Smith, J. (1976). The co-evolution and stability of competing species. *Am. Nat. 110*: 79-99

MacArthur, R.H. and Wilson, E.O. (1967). *The Theory of Island Biogeography*. Princeton, Princeton University Press

Mackenzie, K. and Gibson, D. (1970). Ecological studies of some parasites of plaice, *Pleuronectes platessa* (L.), and flounder, *Platichthys flesus* (L.). *Symp. Brit. Soc. Parasit. 8*: 1-42

Miller, R. (1952). A review of the *Triaenophorus* problem in Canadian lakes. *Bull. Fish. Res. Bd. Can. 95*: 1-42

Mitchell, G.F. (1982). Effector mechanisms of host-parasite immunity to parasites and evasion by parasites. In D.F. Mettrick and S.S. Desser (eds.), *Parasites – Their World and Ours*: 24-33. Amsterdam, Elsevier Biomedical Press

Pimentel, D. (1961). Animal population regulation by the genetic feedback mechanism. *Am. Nat. 95*: 65-79

— (1968). Population regulation and genetic feedback. *Science 159*: 1432-7

— and Belliotti, A.C. (1976). Parasite-host population systems and genetic stability. *Am. Nat. 110*: 877-88

—, Nagel, W.P. and Madden, J.L. (1963). Space-time structure of the environment and survival of parasite-host systems. *Am. Nat. 97*: 141-67

Podesta, R.B. (1982). Adaptive features of the surface epithelial syncitium favouring survival in an immunologically hostile environment. In D.F. Mettrick and S.S. Desser (eds.), *Parasites – Their World and Ours*: 149-55. Amsterdam, Elsevier Biomedical Press

Price, P.W. (1980). *Evolutionary Biology of Parasites*. Princeton, New Jersey, Princeton University Press

Roberts, L.S. (1980). Development of *Hymenolepis diminuta* in its definitive host. In H.P. Arai (ed.), *Biology of the Tapeworm* Hymenolepis diminuta: 357-423. London and New York, Academic Press

Rohde, K. (1979). A critical evaluation of intrinsic and extrinsic factors responsible for niche restriction in parasites. *Am. Nat. 114*: 648-71

— (1982). *Ecology of Marine Parasites*. St. Lucia, Queensland, University of Queensland Press

Schad, G.A. (1963). Niche diversification in a parasite species flock. *Nature, Lond. 198*: 404-6

Schom, C., Novak, M. and Evans, W.S. (1981). Evolutionary implications of *Tribolium confusum–Hymenolepis citelli* interactions. *Parasitology 83*: 77-90

Smithers, S.R. and Terry, R.J. (1969). The immunology of schistosomiasis. *Adv. Parasit. 7*: 41-93

Sprent, J.F.A. (1969). Evolutionary aspects of immunity in zooparasitic infections. In G.J. Jackson, R. Herman and I. Singer (eds.), *Immunity to Parasitic Animals*, Vol. I: 4-62. New York, Appleton Century Crofts

Uznanski, R.L. and Nickol, B.B. (1980). Parasite population regulation: lethal and sublethal effects of *Leptorhynchoides thecatus* (Acanthocephala: Rhadinorhynchidae) on *Hyalella azteca* (Aphipoda). *J. Parasit. 66*: 121-6

— and Nickol, B.B. (1982). Site selection, growth, and survival of *Leptorhynchoides thecatus* during the prepatent period in *Lepomis cyanellus*. *J. Parasit. 68*: 686-90

Wakelin, D. (1973). The stimulation of immunity to *Trichuris muris* in mice exposed to low level infections. *Parasitology 66*: 181-9

— (1978). Genetic control of susceptibility and resistance to parasitic infection. *Adv. Parasit. 16*: 219-308

Walkey, M. and Meakins, R.H. (1970). An attempt to balance the energy budget of a host-parasite system. *J. Fish. Biol. 2*: 361-72

Williams, H.H. (1968). The taxonomy, ecology and host specificity of some Phyllobothriidae (Cestoda: Tetraphyllidea), a critical revision of *Phyllobothrium* Beneden, 1849 and comments on some allied genera. *Phil. Trans. R. Soc. B 253*: 231-307

3 KLEPTOBIOTIC INTERACTIONS IN INVERTEBRATES

Fritz Vollrath

Interference or direct competition, unlike indirect competition, culminates in a contest over that resource which limits the contestants' growth, well-being or reproduction. The aim of the contest is to exclude a competitor from the resource; if he already holds it, to take the resource away from him. Kleptoparasitism is one form of direct competition (Curio, 1976).

Imagine a resource controlled by one animal, its owner. Another animal, the disputant, challenges the ownership. The takeover can be visualised as a point on a continuous scale which represents the interaction between owner and disputant (Figure 3.1). Piracy (acquisition by superior force) lies at one end, contest (acquisition by ritualised fight) in the middle, and pilfering (acquisition by stealth) and peculation (acquisition by disguise) at the other end. Which tactic is employed by the disputant will vary according to the circumstances.

Figure 3.1: Kleptobiotic interactions are a form of competition. The preconditions of equal or unequal competitive ability favour certain tactics employed by the animal which disputes the resource. If the disputant is much stronger than the owner it can take by force; if it is roughly equal in strength it has to contest; and if it is much weaker it will have to take by stealth or by misleading the owner

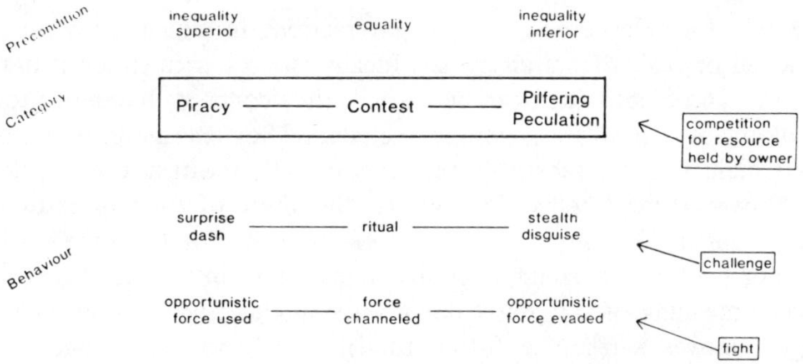

Piracy, peculation and pilfering all fall under the heading of kleptobiosis. So too does kleptoparasitism, which is a common term in the literature. However, its application in biology often appears to be determined only by anthropomorphisms, as is the use of its synonyms thief, robber, pirate, etc. The concept of kleptobiosis, nevertheless, has a viable biological basis (Figure 3.1) which may be used to distinguish 'klepto' interactions from other forms of competition. This distinction might be helpful if we want to group and compare the different ways animals might contest a resource.

Rationale

First I shall explain how I use the word/prefix 'klepto'. No strict borderline can be drawn to distinguish a kleptobiotic interaction from a contest between equals. But certain rules can be used to classify certain behavioural activities. These rules, however, should not be applied too strictly, as we shall see.

The prefix 'klepto' ('clepto') was first used in biology at the beginning of this century. Forel (1901) called cleptobiosis the theft by ants of food from larger ants. In the hymenopteran literature this spelling stuck (Wilson, 1971; Iwata, 1976). Wheeler in 1910 coined the word cleptoparasitism to describe the brood parasitism of some aculeate hymenopterans. Written with a 'k', kleptoparasitism was introduced to biology by Rothschild and Clay (1952) to describe the stealing of food from gulls (*Laridae*) by the arctic skua (*Stercorarius parasiticus*). The two different actions: (i) appropriation of stored brood provisions and (ii) the snatching of morsels of food, are still being linked to the different ways of spelling the word.

I will use 'kleptobiosis' for all interactions in which a resource is removed or stolen from an owner. Ideally the resource stolen is in the form of an object which can be carried, thus concepts like a territory, a colony or a nest are generally excluded. These can be usurped, but not stolen. I would substitute the term brood parasitism for Wheeler's cleptoparasitism. Forel's cleptobiosis, the theft of food or refuse in ants, is included in my definition of kleptobiosis (klepto is the correct form of the Greek spelling, from the Greek 'kleptein': to steal).

The meaning of the prefix 'klepto' seems clear: one organism physically removes a resource (often food) which another organism had gathered for its survival or reproduction. In most cases the klepto interaction is interspecific although it may also be intraspecific (see

Barnard, Ch. 4 this volume). To call the thief a (klepto) parasite implies not only that it decreases the fitness of the robbed but also that a co-evolved association exists with special adaptations of parasite and host toward one another. In defining parasite I follow, like Price (1977), *Webster's Dictionary* but slightly broaden their definition. Thus: 'A parasite is an organism living in, on, or habitually visiting another living organism, gaining from it part or all of a necessary resource; it is adapted to its host whom it causes some degree of real damage.' Experiments rarely measure harmful effects (Walkey and Meakins, 1970), therefore in practice a parasitic association is generally inferred from morphological adaptations of the 'parasite' or from specialised behaviour activities of 'parasite' and 'host' (see Kennedy this volume).

A kleptobiotic relationship is characterised by the interaction, direct or indirect, of a disputant with an owner. The owner acquires (collects, kills) a resource which incurs energy, time, risk or other costs. The thief removing a resource from the owner's control invests presumably less during the theft. Robbery may add to the total cost of the resource for the owner because (i) a pirating klepto may injure the owner; (ii) if a pilfering klepto is about, the owner has to be vigilant; (iii) the approach of a peculating klepto will only be detected by increased discrimination.

I distinguish these three main forms of kleptobiotic interactions: (i) Pirates are aggressive thieves and forcibly remove the resource from the owner. (ii) Pilferers are not aggressive but take the resource by stealth, often without direct interaction with the owner. (iii) Peculators always interact with the owner whom they deceive, as social parasites substituting for colony members or as inquilines offering drugs in exchange for food. The literature on ant societies provides many examples of peculation (Wilson, 1971).

Examples of Kleptobiosis

Many other resources besides food can be the object of interference competition: shells in hermit crabs (Hazlett, 1981b), parental investment in hymenoptera or scarabeid beetles (Iwata, 1976; Klemper, 1980), workforce in ants (Wheeler, 1910; Wilson, 1971), mating opportunity in most animals (Darwin, 1874). I shall draw briefly on some of these examples in the discussion, but the bulk of my examples of kleptobiotic competition will be for food and is selected from one animal group, the spiders. I know them best. They are particularly vulnerable to kleptobionts because a spider in its web cannot run away or hide its food.

Before I present some examples of kleptobiotic feeding interactions, I want to emphasise two basic points: (i) kleptobiosis is theft from an owner, not from a primary producer; and (ii) it is a subject of competitive interactions. One example will demonstrate these points. Bumblebees (*Bombus terrestris*) may rob nectar from flowers. The nectar is produced by the flower to attract pollinators, and the morphology of both plant and pollinator ensures the transfer of pollen. Insects which are not legitimate pollinators (to which the flower has adapted) may visit to take nectar without pollinating. Short-tongued bumblebees (*Bombus lucorum*) cut holes in the base of the florescents of comfrey (*Symphytum officinale*); nectar is sucked from the outside, no pollen is transferred (Brian, 1957). The purpose of the nectar is foiled. The resource, in this example nectar, is produced in order to attract the resource 'visitors', i.e. it is destined to be exchanged for labour. It is taken without the required exchange taking place, it is misappropriated. The flower's safeguard, its morphology, is circumvented. This example is not a case of kleptoparasitism because: (i) *Bombus* steals from a primary producer. *Bombus* does not compete with the flower for the pollinator *Apis*, instead (ii) it competes indirectly rather than directly with *Apis* for the resource nectar.

Food kleptobiosis appears to be a common form of direct competition and often the relationship is more than accidental, the thief showing distinct morphological and behavioural adaptations. Jeanne (1972) describes how the polistine wasp *Mischocyttarus drewseni* pounces on a prey insect entangled in a spider's web and chews it to a ball, ready for transportation to its nest. The wasp hangs from the spider's web by her hindlegs while handling the prey and seems to have no difficulty in moving from the web onto the surrounding vegetation. Occasionally a foraging wasp lands on the lower part of an orb web. Using the sticky spiral threads as a ladder it climbs up towards the centre, where the spider sits. The spider shakes the web violently, an anti-predator reaction as well as a sign of intraspecific aggression (Hoffmaster, 1982). It jumps off when the wasp approaches even closer. Some individuals of *Mischocyttarus drewseni* appear to specialise in searching for prey trapped in spider webs (Jeanne, 1972). The wasp may even be morphologically adapted for walking in sticky spider webs (Jeanne, 1972). Its mid- and hindlegs show an asymmetry of the tarsal lobes which may enable it to move over fine silk strands. The wasp not only steals prey items from the spider's web, if it can catch the owner of the web, it will prey on it.

In a similar way to the wasps, scorpionflies (*Panorpa* spp) invade spider webs from the vegetation or land directly on trapped prey (Thornhill, 1975; Nyffeler and Benz, 1980). Occasionally they actually remove the prey packet from the spider's web but generally the prey is consumed then and there. If challenged by the owner of the web, the *Panorpa* may either flee or lash its abdomen at the spider, defending its life and booty. If itself in danger of falling prey to the spider, a fluid regurgitated from *Panorpa's* midgut deters the attacker. The same fluid contains enzymes which dissolve the spider's silk, allowing the escape of an entangled *Panorpa* (Thornhill, 1975). Nevertheless, feeding on spiders' prey appears to be a risky way of life; more than half the mortality of American *Panorpa* observed by Thornhill was in the webs of spiders. Nyffeler and Benz (1980) observed much lower predation in European scorpionflies.

The mode of prey acquisition is similar in wasp and *Panorpa*, both animals seem adapted to cope with the spider and its silk. Both approach the resource directly and the defence of the spider is met with aggression. Yet a major difference exists between the two. The wasp is the more dangerous to the spider. It is generally larger than the spider, its sting can be deadly. It is nearly always victorious in stealing prey and driving the spider away. The wasp is an established pirate. The scorpionfly is often smaller than the spider and may flee when challenged. Its defence seems highly unpleasant for the spider but it is certainly not deadly. The scorpionfly has a 50-50 chance of perishing during a robbery attempt (an average for several species, some have a better chance of survivorship, Thornhill 1975). *Panorpa* spp are initially pilferers: if discovered they may escape, become prey, or fend the spider off and thus become pirates. What they do depends apparently on the motivation and the size of the spider. *Panorpa's* success is conditional on its adversary; the wasp's is not.

Wasps and scorpionflies are not the only animals which aggressively take prey hanging in a spider's web. Neotropical dragonflies and damsel flies (Odonata: *Mecistogaster* sp, *Megalosprespus* sp) and hummingbirds (Trochilidae: *Phaethornis longirostris*) search the vegetation for spider webs and examine all objects hanging in the silk. The edibles, spider or prey, are carefully removed (Vollrath, 1977). Many species of wasps, damselflies and hummingbirds are known to be regular predators of spiders (Bristowe, 1941). It would not be farfetched to suppose that these predators often take prey items left behind by the fleeing spider. A transition from predator to thief can easily be envisaged. To be able to hover in flight is a preadaptation towards taking objects from spider webs. The ability of many wasps to hunt on foot may be a similar preadaptation to invade a web. Yet the animal holding the resource, the spider, is not without defence. However, if defence of the resource is more costly than the acquisition of a new resource, the spider ought to give up speedily. Rather than sustain injury the spider yields the resource if challenged. But if pursued it will defend its life dearly, with costs to the attacker. Thus if piratism yields a higher reward or incurs a lower cost for the attacker than predation, selection might favour specialisation towards piratism in predators.

Scorpionflies, on the other hand, are scavengers which feed on dead and decaying animals. Competition for natural food seems to be intense (Thornhill, 1975) and spiders' prey may be one of few alternative food items available in the natural habitat. The spider's partly digested prey insect must be highly attractive to a foraging *Panorpa*, which takes the risk of being caught and eaten. This has presumably favoured the evolution of the defence mechanisms which deal efficiently with the silk and the spider.

The intrusion into spider webs of the scorpionfly may be compared to the foraging behaviour of some hemipteran bugs (Nabidae, *Arachnocoris*). Like *Panorpa* they feed on the spiders' prey or on small insects trapped in the webs and are apparently adapted to cope with the spiders' silk. Sometimes they are caught by the spiders (Vollrath, unpublished). The rocking movement during locomotion may mask their activity and location on the web (Robinson, personal communication). Members of another, smaller hemipteran genus (Miridae, *Phylinae*) inhabit the webs of the social spider *Anelosimus eximius* (Vollrath, in preparation). They are minute compared with their hosts and move freely about in the tangled strands of the web. Prey, overpowered by the spiders, is approached carefully. The bug searches for a spot not occupied by feeding spiders, inserts its proboscis and sucks. Dead spiders are also consumed. If a spider is alerted to its presence, the bug often 'freezes'; vibrations which might reveal its presence are avoided. Alternatively it may flee swiftly. All developmental stages of this hemipteran live in the webs of *Anelosimus* and seem highly specialised: I have never found them in the webs of other spiders. These animals clearly pilfer. Mirids as well as nabids are pilfering kleptobionts. Their behaviour, stealthy movements and retreat if attacked by the host, suggest an inability openly to challenge the ownership of the disputed prey. All four species discussed so far deprive their hosts of food when they steal, all show adaptations to their way of life; the relationship they have with their hosts is therefore kleptoparasitic.

Often the complex and vast webs of *Anelosimus eximius* also harbour larvae and imaginos of a noctuid moth (*Neopalthis madates*, Robinson, 1977; Vollrath, in preparation). The adult moth seems to leave the web soon after emergence but returns for oviposition. I have never seen the adult feed in the web. It seems well adapted to cope with the dense network of threads in the *Anelosimus* web, negotiating the strands, walking or flying without attracting the spiders. The adult moths are often found resting, hanging from a spider thread. The larvae move about the centre of the communal web (Robinson, 1977), constructing paths or tunnels of silk as they go along. They feed on the carcasses of the prey insects, which have apparently been abandoned by the spiders. With their strong mandibles they bite into the insect cuticle, crushing parts which the spiders were unable to feed on. Unlike the spiders they also seem to be able to ingest and digest chitinous material. The larvae move about the web carefully, avoiding give-away vibrations. When attacked by the spiders, which seems a rare occurrence in natural conditions, they attempt a rapid retreat into a safe part of their tunnel. If this is not possible, they wipe their body with their mouthparts, depositing a brown regurgitate distasteful to the spider. This behaviour is reminiscent of the defensive 'tobacco juice' of *Panorpa* and some grasshoppers.

The noctuid larvae might be regarded as pilfering. However, they mostly, if not exclusively, feed on prey abandoned by the spiders. Thus they might be called commensals (neither benefiting nor harming their hosts). The social spiders keep their web meticulously clean. They defecate over the rim of the web and dead spiders, exuviae and spent prey are thrown out (Vollrath, in preparation). This can be a risky endeavour since spiders near the rim often fall prey to foraging hummingbirds, dragonflies or wasps (Vollrath, in preparation). Thus by eating the carcass the larva eliminates the necessity of a potentially dangerous trip for the spider. The relationship might, in this light, be called mutualistic (beneficial for both sides). On the other hand, the spider's behaviour of removing carcasses may have evolved in order to remove the food of the lepidopteran, the larvae or adults may, unknown to us, affect the colony negatively. Clearly, before the relationship can be classified, a more intensive study is necessary.

Aggregations of social animals seem to provide many opportunities for symboints as well as kleptobionts. The best-studied examples are ant societies (Wilson, 1971). Very common in an ant colony are various invertebrates, called symphiles (Wilson, 1971). These animals either mingle with the foraging ants or gain access to the colony proper. Generally they beg and receive food from the ants. Frequently they offer an appeasement substance, produced in special glands and eagerly taken up by the ants. Most of these symphiles are highly specialised in their behaviour as well as in their morphology and are found only in association with certain ants. Adult symphiles are mostly peculating; the ants are deceived and accept the intruder as a genuine member of the colony. Most symphiles might be called kleptobionts, because the food solicited was collected by the ant. Although the individual ant is not harmed, the colony is deprived of food and the relationship might be called kleptoparasitic. The larvae of some symphiles feed on the brood of the host. Thus this stage is purely predatory or (if the colony is taken as a superorganism) parasitic. Overall, the relationship is a mixture of kleptoparasitic and predatory/parasitic traits.

Not only social spiders and insects have their symphiles; solitary animals are often accompanied by small attendants which generally pilfer food. The lack of social food exchange (trophallaxis) between host individuals precludes peculation and the size difference between host and attendant precludes piracy.

Robinson and Robinson (1977a) observed small milichiid flies (*Phyllomyza* sp) landing and resting on the cephalothorax of the large orb spider *Nephila clavipes*. When the spider caught an insect and enveloped it with digestive fluid, the flies moved onto the prey and drank the juices. The 'dipso parasites' or 'bibio commensals' (as Robinson calls them) clearly pilfer. Stealth is not necessary, because the spider is quite defenceless against their presence on its unprotected back. Frequent turning of the prey in its mouthparts makes feeding difficult for the flies but does not drive them off. The same or similar flies parasitise other spiders (McMillan, 1965; Sivinski and Stowe, 1980) and assassin bugs (*Rediviidae*; Robinson and Robinson, 1977a). They may even follow robber flies (*Asilidae*) around, feeding on the prey the robber fly carries (Vollrath, unpublished).

A highly mobile host like a fly can only be kleptoparasitised by either an equally mobile klepto or conversely by an 'ecto'-kleptoparasite, i.e. a parasite which, similar to a tick or mite, clings to the outside of the host. This apparently is rare. Much more common are kleptobiotic relationships between stationary hosts like a web spider or an ant colony and a wide variety of 'guests'. These may be highly mobile and visit many hosts. They may be as sedentary as the host itself, often with special adaptations and living their entire life on one or a few host individuals.

Hermit crabs are not sedentary. Some species in fact travel considerable distances (Hazlett, 1981a). They all, however, carry like a second skin their shell, a controlled microhabitat and a firm substrate in an often inhospitable environment. The shell may not only house the crab, its owner, but may provide a base for bryozoans or cnidarians, living in a mutualistic relationship with the crab (Jensen and Bender, 1973; Hazlett, 1981b). Shells housing *Pagurus bernhardus* are often also inhabited by the annelid *Nereis fucata* which partakes at the meals of its host (Gilpin-Brown, 1969). The shells of a hermit crab may also have amphipods (Taylor, 1979) or pocellid crabs (Telford and Daxboeck, 1978). Both are pilfering kleptoparasites which snatch small morsels from their host's meal.

A comparable relationship exists between the tiny symphytognatid spider *Curimagua bayano* and the large mygalomorph spider *Diplura* sp (Vollrath, 1978). *Curimagua* lives in the web of its host. The guests also ride on the cephalothorax of *Diplura*, climbing down over the huge chelicerae (mouthparts) of their host onto its prey. The host seems unaware of the guests, even when they crawl over its eyes. *Curimagua* appears to be highly specialised; I have never found it anywhere but in association with one *Diplura* species in one area of Panama and another in the Amazon region. It is likely that the characteristic mouthparts of *Curimagua bayano* are adapted to its pilfering lifestyle, only feeding on liquid food, predigested by the host.

Other spiders (Mysmenidae; Platnick and Shadab, 1978) also frequent the webs of *Diplura*. They search for tiny prey trapped but neglected by the host but are also attracted to the host when it feeds (Kirkendall, personal communication). The smallest of the three Panamanian diplurid species (*Ischnothele* sp) is host to a small mysmenid (*Mysmenopsis ischnamigo*). The medium-sized *Diplura* sp is host to the small as well as a slightly larger mysmenid (*M. dipluroamigo*). In addition to these two, the webs of the largest *Diplura* harbour a third, even larger, mysmenid (*M. gamboa*). This shows that the associates can remain only if they are small enough to escape either the attention or the pursuit of the host (see also Barnard, Ch. 4 this volume). It is likely that the largest associate present is still below the threshold of the host's predatory behaviour.

The mysmenids roam the extent of the diplurid web, feeding on small insects which are either not perceived or else neglected by the host. This activity constitutes a commensal relationship. However, they also approach the feeding host and may mount and share its prey.

This behaviour elicits a reaction in the host who attempts to shake the mysmenids off by frequently turning the prey packet in its mouthparts. Clearly, those of the kleptobionts which succeed in sharing the host's prey are pilfering. Intraspecific interactions ensure that only individuals of the largest klepto species present share. Thus in the webs of the different sized diplurids different sized mysmenids become klepto-parasites, the smaller species are 'merely' kleptobionts.

This problem of the changing aspect and complexity of kleptobiotic interactions formed one part of my doctoral thesis (Vollrath, 1977). I studied the relationship between a host spider *Nephila clavipes* and two kleptobiotic spiders: *Argyrodes elevatus* and *A. caudatus*. I shall use this example to demonstrate some interesting aspects and general problems of interactions between kleptobionts (or kleptoparasites) and their hosts.

A Detailed Example

Some species of *Argyrodes* (Theridiidae) can regularly be found in the webs of other spiders of a wide range of genera and families (Exline and Levi, 1962; Brignoli, 1966; Vollrath, 1977). Other *Argyrodes* are free-living, some feed on spiders (Eberhardt, 1979; Smith Trail, 1980; Wise, in press; Vollrath, 1977 and in press). The kleptobiotic species are, as we shall see, ecologically and behaviourally most interesting (Robinson and Olazarri, 1971; Robinson and Robinson, 1973). They can be loosely classified into two groups: the specialists and the generalists. Of the 50-odd neotropical *Argyrodes* two are presently known to be specialists, about ten to be generalists (Vollrath, unpublished). The specialists are found in the webs of large araneid spiders such as *Araneus* spp, *Cyrtophora* spp, *Argiope* spp and *Nephila clavipes* (for list see Vollrath, 1977). The generalists frequent a multitude of additional hosts from several families.

In my study in the Panamanian rainforest the specialist thief was *Argyrodes elevatus*, the generalist was *A. caudatus*. Both are often found in large numbers in the webs of the *Nephila clavipes* (Araneidae). Here they not only prey upon small insects entangled and neglected in the web, but they also steal prey which have already been caught and often predigested by the host spider (Figure 3.2). Females of *A. elevatus* pillaging the prey stores of their host *N. clavipes* are astonishingly adept at stealing and the success rate of some individuals can be consistently high. The prey stolen are as a rule much bigger than the thief. *Nephila* hangs prey insects from the hub at the centre of the web. It is apparent to the observer that the klepto is informed about the availability of prey and the activities of the host (Wiehle, 1928; Robinson and Olazarri, 1971). Fine threads constructed by the klepto connect its resting place, in the outskirts of the complex host web, with the hub and several radii. These serve as signal lines.

The Raiding Behaviour of the Klepto

In a series of experiments the *modus operandi* of *A. elevatus* was studied (Vollrath, 1977, 1979a, b, in preparation a). The host was enticed to catch prey held to the web on the tip of a vibrator. In 86 per cent of the trials (n=201) the kleptos, caught in the wild and kept for about 10 trials each, started a raid. Comparative trials simulating a prey escape (n=50), in which the prey was withdrawn the moment the host tried to bite or wrap it, elicited only 9 per cent 'klepto reaction'. Control experiments in the dark had comparable results; thus optical perception was excluded and it was assumed that the pattern of vibrations generated during the host's run for prey informed the kleptos about the activities and the success of their hosts. When analysing the behaviour of the kleptos during a raid and relating into the prey-catching sequence of the host it was found that the host's movements of wrapping the prey were of crucial significance. Those movements triggered a raid. Dead pieces of prey presented directly to the mouthparts of the host may or may not be wrapped. If they were wrapped (n=10), the kleptos always went for a raid. If they were not wrapped (n=10), the kleptos always stayed put. Measurements of the vibrations generated by the host during prey capture revealed a difference between the wrapping sequence and other activities (Figure 3.3). They are a reliable signal. A prey insect which has not been wrapped is held in the chelicerae. *Nephila* carries it around and it cannot be stolen. The wrapping vibrations signal the presence of prey available to the klepto. Normally only large prey (insects larger than a housefly) are wrapped and hung from the hub.

The spiral thread is sticky in the capture area of the web, not so on the hub. In my experiments a large proportion of the host's prey was hung from the hub so that the middle of the prey packet was nearly level with the borderline between the sticky and non-sticky spiral. This is an incidental result of the host's behaviour during the suspension of the prey. Highly successful kleptos nearly always moved along this borderline, waving their long legs and thus covering a path 1-2 cm wide,

Figure 3.2: (a) *Argyrodes gibbosus*, the European kleptoparasite, carefully approaches its araneid host *Cyrtophora citricola* feeding on a cricket. No web separates host and klepto. (b) Wrapping a grasshopper with the hindlegs, a marked *Nephila clavipes* hangs from the hub by its frontlegs. On the right two kleptoparasitic *Argyrodes elevatus* can be seen. One had employed the direct approach and attempts to cut loose the first prey packet. The other klepto approaches using the delayed tactic

(a)

(b)

Figure 3.3: Vibrations in the web of *Nephila* generated by the movements of the host and recorded on a photodetector sensing the positions of a light beam reflected off the web (Vollrath, 1979). The left diagram shows an entire sequence of prey capture: the spider resting at the hub, the rush to the prey, the bite and struggle to pull it from the web, the return to the hub, the wrapping sequence, the suspension of the prey involving a 180-degree turn of the host, the securing with a thread and the resumption of the resting position. The diagrams on the right are enlarged versions of the activities rest and wrapping, showing the bursts of vibrations characteristic of the wrapping behaviour. An enlarged section of these shows more clearly the 'twin peaked wave' with a cycle of circa 0.5 Hz. The wrapping behaviour of *Argiope argentata* (an alternative host) generates a similar wave (Vollrath, 1979b)

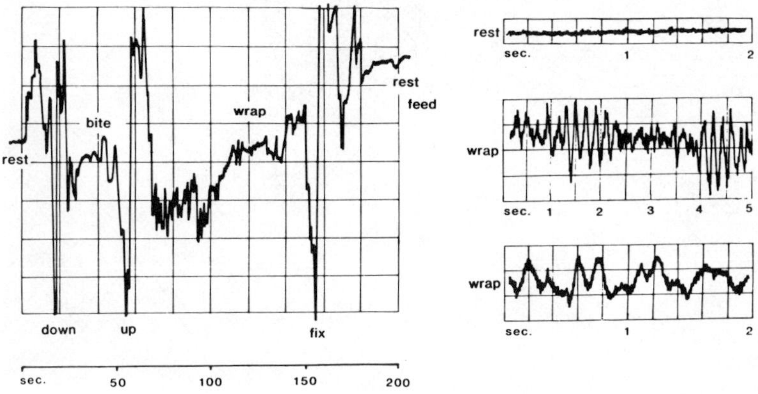

where 80 per cent of the prey hung (Vollrath, 1977).

The Host's Anti-klepto Behaviour

The prey is suspended from a thread, hanging down from the slightly sloping plane of the web (Figures 3.2, 3.4a). *Nephila* moves about on the same (lower) side of the orb. Stolen prey transported on this side is easily discovered by *Nephila* who often accidentally disrupts the thief when she returns from catching a prey. Prey moved up through the plane of the web is hardly ever recovered once it has passed through the hole in the orb bitten by the klepto. The fine mesh size of the *Nephila* web allows *Argyrodes* to pass easily, *Nephila* is too big and has to run around the outside of the large web, which can easily be 50 cm in diameter. Undisturbed *A. elevatus* females have a centre of operation connected by the signal lines to the host's web. There the klepto hangs when resting and there she returns with her stolen prey. Often this centre is above the upper plane of the web, near the hub and opposite the host's resting place. Thus the klepto's access to and retreat from the hub is short, direct and not obstructed by the host.

Figure 3.4: (a) Side and front view of the web of *Nephila*. (b) Diagram demonstrating the prey-catching behaviour of the host spider (P=prey)

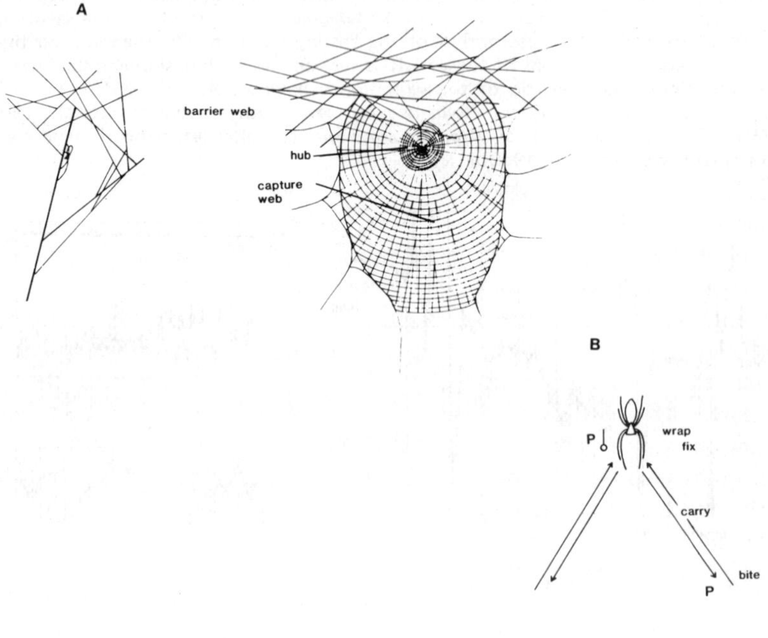

The presence of a klepto moving on the orb or the hub may be sensed by the host who reacts aggressively. The klepto retreats when the web is shaken vigorously. If shaking has no effect *Nephila* may rush towards the disturbance, thus forcing the klepto to flee. Often the klepto simply drops, connected to the web only by its dragline which, if discovered, is cut by the host. The correct timing of a raid reduces the host's ability to sense the klepto's presence. When the host is active during prey capture, transport and wrapping, it is oblivious to subtle movements in the web. A klepto moving during this time does not alert the host. Direct aggression is one way of deterring kleptos from stealing and may even drive them off the web. *Nephila* has another way of ridding itself from the kleptos, it goes elsewhere.

Kleptos can at times occur in large numbers; I observed up to 45 *A. elevatus* in one *Nephila* web. They may be quite a drain on the prey captured by their host (Vollrath, 1979, 1981; Rypstra, 1981). The host may move if the number of kleptos active in its web is large (Rypstra, 1981). The construction of a new web some distance away gives the host several days' respite before it is colonised again (Robinson and Robinson, 1973; Vollrath, 1977, 1981). The move is probably triggered by a decline, due to theft, in the number of prey *Nephila* consumes. It is possible, however, that the decision to move is also influenced by the disturbing presence of many kleptos active on the hub. If they reduce their activity to quick raids, only when food is available and the host insensitive, this disturbance is kept low.

The Klepto's Two Approach Tactics

Argyrodes elevatus employs two clearly distinguishable stealing tactics (Vollrath, 1979a, b). In the *direct approach* the klepto reaches the hub in less than five minutes after the first prey hit the web. Very few kleptos require longer than this. If, on the other hand, the *delayed approach* is taken, the klepto does not proceed onto the hub before the second prey has been attacked by the host, which in my experiments was 10 to 15 minutes later (Figure 3.4b, 3.5). In this approach the klepto either stirs briefly during the host's capture of the first prey and then resumes its resting position or it moves forward onto the nearest web radius, where it waits. Then, as soon as the host runs towards the second prey, the klepto rushes towards the hub, reaches it quickly (often within seconds) and immediately commences its frantic search. In my experiments the success rate was significantly different between the two tactics, the probability of carrying the prey away was P=0.47 with 'direct approach' and P=0.28 with 'delayed approach' (P = 0.05, X^2=5.24). The direct approach was used more often (64 per cent) than the delayed approach (36 per cent, n=799 trials).

Figure 3.5: Some prey capture sequences of *Nephila* and the corresponding raiding sequences of *A. elevatus*

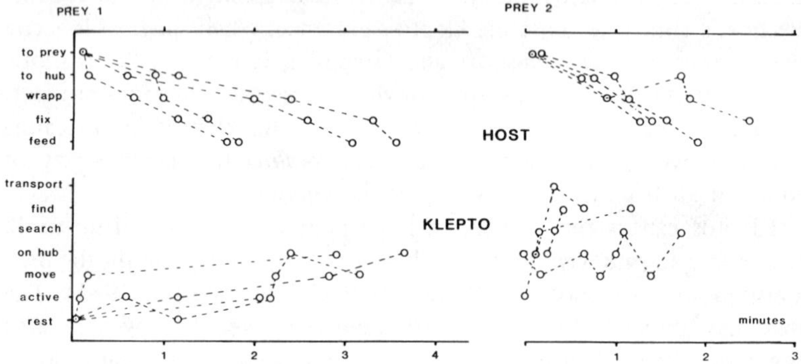

We might assume that after a long period without food the klepto is motivated to attempt an early and direct approach towards stored prey. After feeding, the threshold for the direct approach may be increased to the level of the delayed approach and either approach may be used. An experiment in which the sequential employment of both tactics was analysed gave no conclusive results (Table 3.1). The trend suggests, however, that after a successful raid the next approach is chosen randomly. After an unsuccessful raid the direct approach is more likely to be employed than the delayed approach. More experiments are needed to confirm this interpretation.

Nephila webs in the wild may have several to many adult *A. elevatus* (Vollrath, 1981) and stored prey nearly always attracts one or more thieves. Alternation between both approach tactics may reduce the probability of a klepto meeting another klepto attempting to steal the same prey. Interactions between kleptos may become quite violent and often attract the host. This may result in a loss of stealing opportunity to both contestants. It also makes the host aware of the presence of kleptos. Costs and benefits accrue from both tactics: (i) The delayed approach has a lower rate of success which clearly is a disadvantage. (ii) If, during the direct approach, no second prey is caught by the host, the klepto has ventured out in vain. Abandoning the resting position and moving about may attract predators. It may also alert the

Table 3.1: Does the tactic used on one raid influence the tactic used on the next? The numbers indicate the approach used by the klepto *Argyrodes elevatus* during the second raid

first trial	second trial: direct	second trial: delayed
success	17	17
failure	29	17

in *Nephila clavipes*

first trial	second trial: direct	second trial: delayed
success	25	17
failure	47	18

in *Argiope argentata*

host. Avoiding this is clearly an advantage of the delayed approach.

The Raids Improve with Experience

It is possible that individual preferences predetermine the probability with which a certain klepto uses either of the two approach tactics. Most kleptos used both approaches equally but a few kleptos employed one of the two tactics more often. It is possible that experience during its first raids predisposes a klepto towards one particular approach tactic. In order to study the development of the raiding techniques, female *A. elevatus* were reared from the egg in small vials, fed on dead fly pupae and deprived of any experience with the host or its web. These females, when adult, were released in *Nephila* webs. Their behaviour during the first prey capture sequences of the host spider displayed both eagerness and ignorance. In response to the wrapping stimuli the klepto eagerly moved around the host's web, frantically waving its legs in the typical searching behaviour. Yet it appeared to be ignorant of its orientation in the web. The structure of the orb which guides the experienced klepto to the hub and prey (Vollrath, 1977) did not seem to provide any clues. Over a number of trials most individuals found their orientation and acquired the ability of timing the raid correctly; most became quite proficient thieves (Figure 3.6). They did not seem to prefer or dislike the approach tactic they employed on their first raid.

The unexpected result of this experiment was that the kleptoparasite learns how to behave correctly in the web of its host. However, not only the orientation and the timing of the approach behaviour is important but also the speed of the actual theft. The time it takes to remove the prey from the hub is a crucial factor determining success. The klepto has only a limited amount of time before the host deals with the prey and resumes its alert position (Figure 3.5). After this, stealing is a risky as well as a generally fruitless undertaking. We would

Figure 3.6: Each point represents, in successive trials, the final activity of one *A. elevatus* at the end of each trial. Prior to the first trial, the kleptos had had no experience with the host's prey capture. They had been set into the web 24 hours before. On a few occasions some kleptos did not respond, these points are left out. The activities are ranked and inclusive. Those above the dotted line represent kleptos active on the hub, only there can the klepto be successful. A X^2 analysis of the six quadrants gives an X^2 of 9.6. This is significant at the 0.01 level

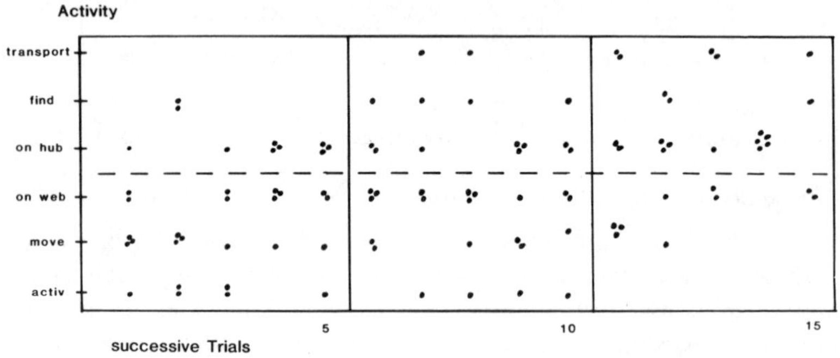

expect that naive kleptos learning to steal would show an improvement in the speed with which prey is located and carried away. Such an improvement was found (Figure 3.7). The trend in the two graphs is suggestive, although the small and highly uneven sample sizes preclude powerful statistical tests. More trials are needed to confirm this suspicion.

Invertebrates are often thought of as small automata, too specialised in their habits and too shortlived to indulge in the luxury of learning, of experience modelling or altering a behaviour pattern. The classic study of Tinbergen (1951) on the orientation toward her nest of *Philanthus* or that by Baerends (1941) on the brood provisioning of *Ammophila* show that some invertebrates do have an amazing capability for learning certain things. If the environment is sufficiently variable, the response of the animal has to be adaptable. The prey-catching sequence of *Nephila* and its storage of prey in the web are stereotyped behaviour patterns (Robinson and Mirick, 1971). In Panama, however, the klepto *Argyrodes elevatus* may also be found (if much less frequently) in the webs of another orb web-building spider, *Argiope argentata* (Robinson and Olazarri, 1971). *Argiope* occupies a different type of web from *Nephila*'s and its prey catching and storage behaviours are also different (Robinson, 1975; Vollrath, 1979a, b). It is probably advantageous for a klepto dealing with different hosts to be

Figure 3.7: Scattergrams representing two activities of *A. elevatus* in the web of *Nephila*. The upper graph shows the mean time it took kleptos to find prey, starting at the moment the second prey was presented. The different symbols represent the two tactics of approach. The diameter of a symbol denotes the sample size, the bar the standard error (max n=9, min n=2). The lower graph depicts the mean length of time kleptos required to remove prey from the hub after they had located it

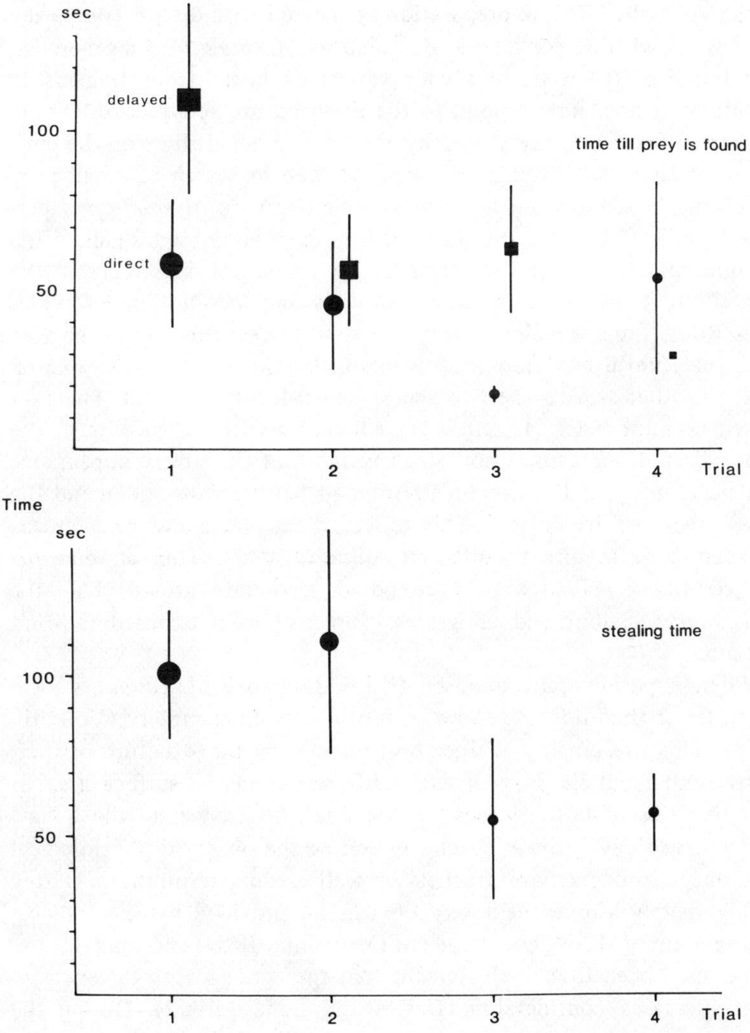

able to learn its orientation and the timing of a raid, adapting both to the idiosyncrasies of different hosts. Experiments to test this hypothesis are under way.

Ecological Adaptations

Argyrodes elevatus is not only behaviourally well adapted to its life as a kleptobiont but also shows morphological and physiological adaptations (Vollrath, 1977, in preparation a). These I shall discuss comparing *A. elevatus* with its congeneric *A. caudatus*. *A. caudatus* is a generalist, it is found in the webs of a wide variety of hosts, from araneids to theridiids. It does not respond to the prey-capture activities of its host and rarely feeds on prey stored by the host spider. It lives on the perimeter of the host's web and roams the web in search of small prey insects neglected by the host, in a similar way to *Mysmenopsis* described earlier. *A. caudatus* also feeds as a predator on small, often juvenile spiders and can accordingly be found moving freely in the vegetation. *A. caudatus* in its morphology and locomotion resembles *A. elevatus*. Long, slender forelegs are used to feel the way in the host web, the careful and slow movements generating few give-away vibrations. In other aspects the two species are radically different. The dark brown or matt black *A. caudatus* is highly cryptic when at rest, the shiny silver *A. elevatus* is not. *A. elevatus* owes the silvery appearance of its abdomen to lacunae of air trapped between the inner and the outer layers of its cuticle. This reflects light and allows *A. elevatus* to keep cool in direct sunlight (Vollrath, 1977). Thus *A. elevatus* can colonise *Nephila* webs in exposed secondary growth habitats, *A. caudatus* cannot and emigrates from host webs exposed to wind and sun.

When experimentally exposed to low humidity, *A. caudatus* loses water faster than does *A. elevatus*, it also dies more rapidly (Vollrath, 1977). This presumably is due to differences in the structure of their opisthosomal cuticle as well as to differences in the surface area to volume ratios of both species. In the field, differences in the behaviour and ecology of both species underline the observed physiological differences. Spiders have tracheal as well as lung respiration. During resting metabolism enough oxygen can be provided by the trachea. During activity this is not sufficient (Miyashita, 1969) and lung respiration is used even though this means that the water loss to the environment increases considerably (Davies and Edney, 1952). During the rainy season, when the environmental humidity is high, *A. elevatus* is day active, *A. caudatus* night active. During the dry season in exposed habitats *A. elevatus* is active at night (when the humidity is high) but *A. caudatus* is absent (Vollrath, 1977, in preparation a).

A. elevatus (the specialist) is *r* selected (Stearns, 1976): many offspring are produced in a short time. It may produce one egg case every five days, each containing about 37 eggs. Its egg sacs are hung in the host's web. *A. caudatus*, on the other hand, as a generalist has *K* selected reproduction. It guards its camouflaged egg case which is hung in the vegetation away from the host web. Guarding means that *A. caudatus* must wait at least the 21 days it takes for the young to develop and hatch, before it can accrue resources for the next batch. With circa 27 eggs, its egg case contains significantly fewer eggs than the egg case of *A. elevatus* ($P \leqslant 0.001$, Moore's (1957) two sample t-test based on range, Vollrath, 1977).

Competition Between the Kleptobionts

The two species of *Argyrodes*, together with at least four other species, are found in the webs of *Nephila clavipes* (Vollrath, 1977). In my study site in Panama the *Nephila* population is cyclical (Vollrath, 1980 and in preparation b). A similar cycle is followed by the *A. elevatus* population. *A. caudatus* is much less cyclical than *A. elevatus*. It is not so dependent on one host, and as a generalist may visit many other host spiders. Its population in the webs of *Nephila* builds up slightly, following the build-up of the *Nephila* population. The population of *A. elevatus*, on the other hand, follows the *Nephila* population from zero, but with a short delay. After the numbers of *A. elevatus* in the *Nephila* webs have overtaken the numbers of *A. caudatus*, the population of *A. caudatus* experiences a depression whereas the population of *A. elevatus* keeps on growing, only to break down with the host population later in the season. This cycle is found twice annually in Panama (Vollrath, 1980 and in preparation b).

The population dynamics of *A. elevatus* and *A. caudatus* cohabiting with *Nephila* suggests that individuals of *A. elevatus* drive individuals of *A. caudatus* out: *A. elevatus* is dominant over *A. caudatus*. This hypothesis is supported by observations of the orientation of both species in the *Nephila* web. Individuals of *A. elevatus* are on the average found closer to the host than individuals of *A. caudatus* (Vollrath, 1977 and in preparation a). *A. elevatus* for much of its foraging depends on signals generated by the host, *A. caudatus* does not. Proximity and correct placing in the host web is more important (even crucial) for *A. elevatus* than it is for *A. caudatus*. The outer perimeter of the host web, on the other hand, could be a dangerous place, predation seems higher here than in the centre of the web (unpublished observations of foraging *Trypoxylon* wasps). If both species co-occur, *A. elevatus* moves closer to the centre of the web (and the host), but *A. caudatus* moves further out (Figure 3.8). This displacement appears to be the result of aggressive interactions in which *A. elevatus* maintains the upper hand (Vollrath, in preparation a).

Figure 3.8: Graph showing the spatial displacement of both *Argyrodes* species when inhabiting the same web (Vollrath, in preparation c). The circles and triangles represent the mean distance to the host *Nephila clavipes*; the filled symbols representing animals at rest, the open symbols active animals. The standard deviation of the samples is shown by the length of the bars

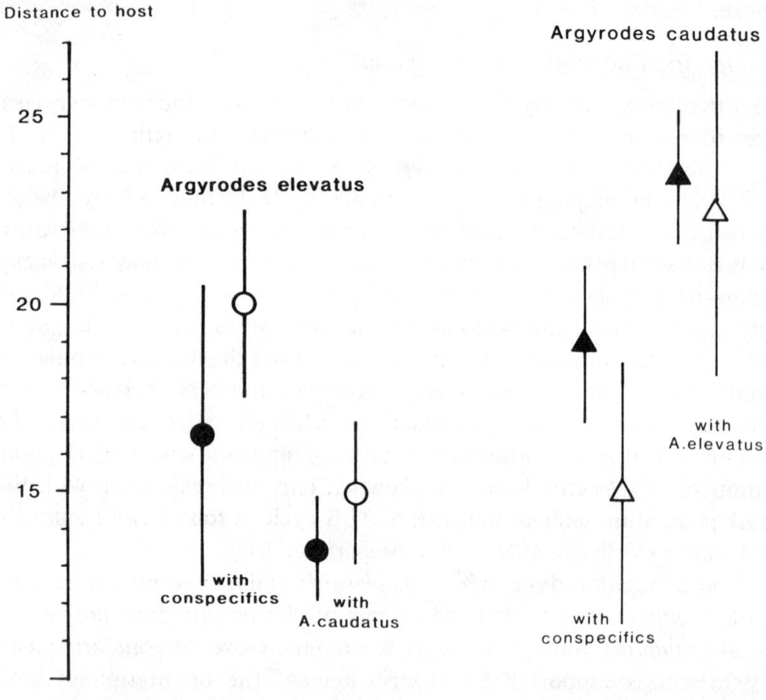

A. elevatus is clearly a parasite kleptobiont. It removes prey insects, in which *Nephila* has invested energy and time (web silk, web construction, capture, digestive enzymes). *A. caudatus*, on the other hand, takes prey neglected by the host. In orb web-building spiders like *Nephila* these small insects might have been ingested by the host when it eats its old web before building a new one. Thus some small amount of energy is sometimes lost to the host under certain conditions perhaps representing the energetic equivalent of a few eggs. The relationship between *A. caudatus* and orb spiders is therefore kleptoparasitic. Most other host spiders do not eat their webs, in which case the small, neglected insects eaten by *A. caudatus* would have escaped or rotted away and the host would not have experienced a loss. Then the relationship is kleptobiotic.

Both species of kleptos cut host threads when moving around in the web, often simply to create free spaces and thus to facilitate movement. Alternatively parts of the web are cut down and ingested; the amino acids are presumably 'recycled' to assist the klepto's own silk production. This habit might, under certain conditions, create considerable holes in the capture area of the web and may thus bring about a decline in the capture rate. Silk theft may, since ultimately the host's prey consumption is curtailed, be called a parasitic behaviour (Kullman, 1959). Both species of *Argyrodes* will also attack and eat the young of their host (Vinson, 1863). They have been found to open the egg case of *Argiope argentata* and to eat the eggs (Vollrath, 1977).

The Host's Response

After this relatively detailed description of the klepto's behaviour, physiology and ecology, the host's reaction is discussed in five brief points: (i) I have mentioned that *Nephila* may move from its web site if large numbers of kleptos steal its food. Another possible adaptation of *Nephila* which restricts the harmful effects of kleptos is (ii) *Nephila*'s habit of executing the second prey capture sequence faster than the first (Figure 3.5 and Vollrath, 1977). This leaves the klepto less time for a raid. (iii) If a prey packet has been stolen, *Nephila* often climbs up into the barrier web and searches there for the lost prey, shaking the threads vigorously. Sometimes this behaviour is successful and the prey recovered. It is, however, also employed if no kleptos are present in the web. It may have evolved as a response to kleptoparasite activity and became fixed as a stereotypic behaviour which is executed under certain conditions. (iv) Robinson and Robinson (1973) suggested that *Nephila* stores its prey near the hub as an adaptation to kleptoparasitism. Thus presumably control over the captured prey is increased and loss reduced. This hypothesis is unlikely to be correct; in my experiments *Nephila* lost marginally more prey to the kleptos than did *Argiope* who stores the prey in the capture area of its web (Robinson and Olazarri, 1971; Vollrath, 1979a). (v) The deposit of *Nephila* egg cases far away from the web may be another adaptation against klepto action. I have seen *A. caudatus* and *A. elevatus* take eggs and spiderlings from the egg sacs of *Argiope argentata* which are kept in the web.

The possibility that the relationship has some side-effects, which may be beneficial for the host, shall not be excluded. *A. elevatus* consumes the stolen prey in the barrier web. There *Nephila* males wait for the demise of the dominant male, which guards the female (see Arak, Ch. 6 this volume, for detailed discussion of satellite strategies). These satellite males have no access to prey, in contrast to the guarding male, which shares the prey of the females. They can feed on silk, which they take from the barrier web (Vollrath, 1977). *Nephila* males are, however, bigger and stronger than the kleptos. On several occasions I have seen them drive a klepto away from its booty and consume the stolen prey themselves. We may speculate that the klepto's activity supports in this way some of the attending *Nephila* males. *Nephila* females are receptive up to the point of egg laying and they are generally inseminated by many males (Christenson and Goist, 1979; Vollrath, 1980). The female's offspring will presumably have a wider range of genotypes, if more males mate with her. This might be beneficial considering the cyclic environment Panamanian *Nephila* inhabit (Vollrath, 1980, in press).

Stages in the Evolution of Kleptoparasitism

A comparison of the examples demonstrating prey theft by spiders might give a solid base for speculations about the evolution of *Argyrodes* from an independent web builder or hunter to a kleptoparasite dependent on other spiders. Many web spiders aggregate in areas of high prey density (Turnbull, 1964; Blanke, 1972) and areas with abundant structural support for their webs (Schaefer, 1978). Aggregation may result in dense clusters of webs where the threads of the different webs interconnect. In these circumstances a spider hunting in its own web might accidentally invade the snare of another individual and treat the resident as an invader. A fight will ensue which either might win and depending on the outcome the web may change ownership. In the dense network of threads connecting clustered webs, vibrations of prey struggling in one web are transmitted to a neighbour which is attracted and may trespass, searching for the prey but attempting to avoid the resident. If the resident is occupied in other matters (mating, for example), the intruder might be able to retreat unchallenged with the stolen prey. Intruder as well as prey might be below the threshold for attack, if the resident is much larger. On the other hand, the resident may respond to the thief; or resident and intruder attack the same prey and fight when they meet. The different modes of trespassing or invasion of a neighbour's snare are frequently observed in dense aggregations of web spiders (Lubin, 1974; Buskirk, 1975a).

One can easily imagine that spiders which are larger, stronger or in any other respect superior in fighting ability, stay after an invasion. They usurp the webs of inferior conspecifics and of those heterospecifics whose webs they can inhabit. Usurpation of webs which are successful in trapping many prey seems to be a common trait in gregarious spiders: *Cyrtophora citricola* has a hierarchy of web occupancy correlated with the capture rate of the web (Rypstra, 1979); *Metabus gravidus* has a hierarchy of occupancy linked to web location (within the aggregation) which determines the trapping success (Buskirk, 1975b).

To construct a web costs the spider time and energy, it costs additional energy to synthesise the construction material, the silk (Witt, Reed and Peakall, 1968). A good web site is only found by trial and error (Turnbull, 1964) and the quality of a new site is not easily evaluated (Vollrath, in preparation c). Usurpation of a successful web ought to be cheaper than building, considering the investment of energy and time in synthesising silk and waiting for prey. Yet a usurped web may not catch high numbers of prey every day. Moreover, it is a delicate structure which deteriorates with time. A decision has to be made by the usurper either to rebuild the web or move on to another. Selection might favour an intruder who, rather than usurp and operate a web, only feeds briefly on the usurped prey and then moves on. However, spiders moving between webs may have a high risk of being eaten (Vollrath, in preparation c). The spider is safest in a web where many potential predators cannot operate, and others are detected early and can be avoided. Web-builders change their web site only if the food supply is very low or drops dramatically (Janetos, 1982; Vollrath, in preparation c); moving web seems to be a relatively rare occurrence. Usurpers would have to move often, whenever the state of the ephemeral host web requires it. The move could be a lengthy one if another suitable host web is not found rapidly. Stealing undetected from the webs of much larger spiders reduces the frequency with which the invader has to search anew for a host web since the web is maintained by the resident host and may be 'active' over a very long timespan.

Webs with a large capture area catch presumably more prey. The intruder in a larger web should benefit by gaining more prey. Web diameter is generally correlated with the size of its builder and occupant (Witt, Reed and Peakall, 1968). It would clearly be profitable for a small spider to invade the web of a large spider. The host, however, in this case is too large to be attacked, and the hunter turns its attention toward small prey insects struggling in the host web. Small size, combined with stealthy movements, aids the intruder, especially if it is close to, or below, the sensory or profitability threshold of the host. In this case the intruder may move about carefully and steal prey insects ignored by the resident. This I call pilfering. We can imagine a further specialisation, from pilfering small prey in the webs of larger spiders (like *Argyrodes caudatus*) to pilfering large prey (like *A. elevatus*). Large prey can only be stolen after its capture by the resident since the prey is more powerful than the klepto.

Kleptoparasites might have evolved from primarily facultative, then obligatory usurpers, who turned pilferers. An alternative evolutionary pathway can be drawn: all but a few spiders readily attack and prey upon conspecifics. Some species, like *Theridion* spp, *Achaearanea* spp and *Pholcus* spp, appear to be quite proficient at invading webs and killing residents (Bristowe, 1958). Some species, like *Ero fulcata* and *Portia fimbriata*, search for the webs of their victims, which they kill very efficiently (Bristowe, 1958; Jackson and Blest, 1982). The stealth which is necessary for a successful catcher of other spiders to prevent detection and flight of the quarry, also allows the hunter to move about freely and undetected in the webs of larger spiders.

This is the evolutionary path I presume *Argyrodes* to have taken. *Argyrodes* almost certainly evolved from web-building theridiid stock: most theridiids are web-builders and many *Argyrodes* 'still' surround their egg sac with a space web and wrap their prey. Yet some *Argyrodes* (which now include the former genera *Rhomphaea* and *Ariamnes,* Exline and Levi, 1967) are more or less specialised spider hunters. *A. attentuatus* displays a distinctive hunting behaviour; suspended on a single strand it waits for wandering spiders to use this line as a bridge. Then it spreads its legs apart, basket-like, recoils and jumps forward to grab the victim (Vollrath, 1977; Eberhardt, 1979). *A. trigonum* invades the webs of other spiders and preys upon the residents (Wise, in press); this behaviour has been described of other *Argyrodes* as well (Smith Trail, 1981). *A. projiciens* occupies the webs of the social theridiid *Anelosimus eximius* and preys upon the juvenile colony members (Vollrath, in press).

No *Argyrodes* is known to construct and operate a feeding web.

If *Argyrodes* is indeed a monophyletic group, we may assume that recent ancestors of the kleptoparasitic *A. elevatus* were predators, the more distant ancestors pirates or usurpers. Some habits of the kleptobiotic species show evidence of predatory behaviour, pirating and usurpation: *A. caudatus* invades the webs of small spiders. Generally, however, the resident flees and the intruder is left with the prey and the web, both of which it consumes (Vollrath, 1977). *A. caudatus* as well as *A. elevatus* join and follow the swarms of newly-hatched spiderlings, which stay together for some time. Both feed voraciously on the spiderlings. I also observed both species feeding on mature *Nephila clavipes*, *Argiope argentata* and *Eriophora fuliginea*. These orb spiders are all much larger than the kleptos (50 times their weight). It seems that the kleptos approach the host during its moult, when it is soft and helpless, bite it and commence feeding.

All stages of specialisation, from an independent facultative thief to a completely dependent, obligatory symbiont, can be found among spiders. Juveniles of the araneid *Nuctenea* spp can catch prey in their own webs but sometimes invade the webs of larger orb spiders where they remove small prey (Schmidt, 1980; Nyffeler and Benz, 1980; Nyffeler, 1981). *A. caudatus* removes neglected prey insects from the webs of many host species. *A. elevatus* steals prey packets from the host's mouthparts. *Mysmena* spp of different sizes are found in the webs of different sized hosts. And finally, *Curimagua* is so highly specialised that it appears to be quite incapable of existing without its one host.

Other Examples: Kleptobiosis and Parental Investment

Food can be stolen for immediate consumption, as it was in the case of the kleptobiotic wasps, scorpionflies and spiders. But food can also be used to feed one's young or to attract a mate. In this form food is not the proper term for the resource; we better call it parental investment and male attractant.

The male of the hanging fly *Bittacus apicalis* offers an insect to the female (Thornhill, 1979). The size of the prey determines the success of sperm transfer. Instead of foraging for prey, the males may try to rob other males or copulating pairs of their prey which they, in turn, use to attract a female. This interaction might be called kleptobiotic. *Bittacus* males specialising in stealing from other males may be called kleptoparasites. It does not matter that the prey is not consumed but used to attract females. Stealing prey is, of course, only a feasible alternative to catching if enough males do catch flies. Since these males provide the only influx of attractant for the population, it is clear that the ratio of hunters to thieves determines, among other things (size for example), what a particular *Bittacus* male should do. This is obvious and is true also for my other examples in which the animal has the option of hunting or stealing. When the animal is so specialised that it must take a resource collected by others, an arms race may ensue (Dawkins and Krebs, 1979), with adaptations by the klepto and counter-adaptations by the owner (see Barnard, Ch. 4 this volume).

Parental investment encompasses, in addition to physical resources, a wide range of parental behaviour patterns. A behaviour cannot be stolen or usurped (there is one remarkable exception, as we shall see later) but its product can. Products of parental investment, destined for the protection and provision of brood, are the burrows dug by some wasps (Specidae) and beetles (Scarabeidae), as well as the food stashed away there. The burrows can be usurped and the provisions removed. Let us consider the usurpation of a nesting hole first. The wasp *Sphex ichneumoneus* excavates a tube which is to be used to store katydid grasshoppers as larval provisions (Brockmann and Dawkins, 1980). It may happen that another *Sphex*, after a fight, takes possession of this tube. If the tube had already been stocked with provisions, the aggressor appropriates these along with the tube. The intruder does not feed on the pirated goods but instead fills the tube with more provisions and finally deposits her own egg. Brockmann and Dawkins (1980) analyse the interaction between the excavating and provisioning females and conclude (i) that the usurpers are not females specialised in seeking out the tubes of other females, but (ii) a female encountering a tube may assume it to be vacant and begin to provision, even if the tube already contains prey. Sometimes both females meet during provisioning. They contest each other's presence and fight over possession. The fight between the two owners must be called a contest between equals. It appears to be decided by the staying power of one fighter, generally the female which contributed more prey. This is not necessarily the female who did the digging. Females appear not to be specialised usurpers or diggers. A tube, if suitable, is treated as a feature of the habitat. Empty (old or abandoned) nests are not rare and the wasp appears to have no behaviour pattern, other than fighting, to deal with the contingency of a shared and usurped tube. In this case, the usurpation of another animal's parental investment appears to be a coincidence, due to limitations of the wasp's sensory system and behavioural responses.

Such interactions between conspecifics might evolve, given the appropriate selection, into a relationship similar to the following example of two heterospecific wasps: *Stizoides unicinctus* and *Prionyx atratus* (Iwata, 1976). *Stizoides* seeks out a *Prionyx* excavating her nest with a paralysed spider to provision it nearby. *Stizoides* watches closely and, after *Prionyx* has finished her work and left, digs up the nest, enters, eats the host's egg and oviposits her own. This is definitively a parasitic behaviour. Instead of competing directly with the owner, as is the case with *Sphex*, *Stizoides* appropriates the completed and abandoned nest of *Prionyx* and substitutes her egg. Unlike *Sphex*, females of *Stizoides* and *Prionyx* are not competing for the resource 'tube and provisions'. The resource sought out and acquired by the disputant is 'tube *cum* provisions'. The aggressor depends on the completed and provisioned nest.

These examples can be compared to the theft of broodballs and the usurpation of brood provisions by scarabeid beetles larvae. Klemper (1980) studied the activities of larvae of the scarabeid beetle *Aphodius rufipes* which seek out the broodballs of another species *Geotrupes spiniger*. *Geotrupes* larvae feed in deep tubes dug by the mother underneath the dung pad and filled with dung. *Aphodius* larvae dig their own tubes; the older the instars, the deeper it digs. A third instar larva of *Aphodius* might thus chance upon a *Geotrupes* tube and feed on *Geotrupes'* stored dung. *Geotrupes'* eggs are destroyed by chance. So are the larvae, they are attacked and killed like any other larva which responds aggressively towards burrowing *Aphodius* larvae. None of the behaviour patterns of the *Aphodius* larva shows specific adaptations toward brood parasitism (Klemper, 1980). The brood provisions are not moved, and only partly consumed before the intruder moves on. *Aphodius* larvae compete with *Geotrupes* larvae for food provided by the *Geotrupes* mother, both larvae compete directly for food to be consumed. Instead of a parent disputing parental investment, the resource food is disputed by the offspring. However, the food has been provided for the resident offspring, so the offspring can be called the owner.

Compare these examples with the brood parasitism in some bees. *Nomeda* females invade brood cells of *Andrena* and oviposit there (Iwata, 1976). *Nomeda* larvae kill the host larvae and grow on the host's brood provisions. *Nomeda* larvae have been placed in their position by the mother. The mother usurped the brood cells, she is dependent on heterospecific parental investment. Her behaviour is adapted to deal with the host, who might defend the nest. The larvae also have an adaptation to parasitic life: they kill the host larvae. The *Nomeda/Andrena* relationship is similar to the interaction which mother and offspring of the cuckoo (*Cucculus*) and the widow bird (*Vidua*) have with their host species. Both are also dependent on another's nest and provisioning. But in the case of the birds the interactions between parent and offspring usurpers (parasites), parent resident (host) and offspring resident (competitor) are more intricate. Particularly in the widow bird, where the parasites have special adaptations to fool the host parents into feeding them (Nicolai, 1974). All these examples can be used to show how parasitism of the resource 'parental investment' (here burrowing, nest building and provisioning behaviour) can be analysed as kleptobiotic behaviour, separating the different aspects: usurpation, theft and rivalry.

A kleptobiont of parental investment misappropriates maternal or paternal behaviour patterns. In the case of the cuckoo the parent is forced, with manipulation of its sensory system, to work for the parasite (Dawkins and Krebs, 1979). Such manipulation is parasitisation of behaviour. Some ants have taken parasitisation of behaviour further, they abduct heterospecifics and use these 'slaves' in their own colony. A raid to take slaves proceeds like this (for references see Wilson, 1971): members of the aggressor species assemble inside their home colony, march to the colony singled out to be raided, engage the defending workers and soldiers in a fight, enter the foreign colony and steal larvae and pupae which are carried home. These, when mature, accept the alien species and work in its colony. They themselves may even participate in slaving raids. Force is used in a slaving raid which is thus a form of piracy. Defending ants are often killed. The resource stolen is potential labour.

In another form of slave taking, a reproductive of the active species gains entrance to the colony of the host species, seeks out the queen and kills her. Consequently the workers of the host species accept the alien reproductive as their queen and tend to her eggs, larvae and pupae. In this relationship force is used only against the host queen and success for the intruder depends on a mixture of piratism and peculation. The host colony generally is doomed except in those cases where the workers begin laying eggs. At first sight it appears that only the resource labour is usurped. However, the physical structure of the host colony is also occupied. It may, after one generation, be the sole remains of the host. Both, labour and living space had been produced by the host colony.

In this chapter I have reviewed kleptobiotic interactions between invertebrates. The examples illustrate how interactions which are at first opportunistic and sometimes ineffective may evolve into finely-tuned relationships. Any animal controlling a resource is vulnerable to others contesting its ownership. Especially vulnerable are holders of resources which impair their mobility. They are open to exploitation both from pirates which force access to the resource and pilferers which gain access deceitfully. Pilferers may evolve to become peculators, no longer relying on secrecy, but instead on the manipulation of host behaviour to provide the resource.

Acknowledgements

I am indebted to M.H. Robinson who pointed *Argyrodes* out to me. I am grateful to O. V. Helversen, D. Kaziti, H. Levi, G. Osche, M.H. and B. Robinson, P. Weygoldt and D. Windsor who all provided much encouragement and help during the lengthy course of the study. M. Ridley kindly read the manuscript. The study was financed by DAAD, DTG and GRAFÖG stipends from the West German government.

References

Baerends, G.P. (1941). Fortpflanzungsverhalten und Orientierung der Grabwespe, *Ammophila campestris. Tijd. Entomol. 84*: 68-275

Blanke, R. (1972). Untersuchungen zur Ökophysiologie and Ökethologie von *Cyrtophora citricola* in Andalusien. *forma functio 5*: 125-206

Brian, A.D. (1957). Differences in the flowers visited by four species of bumblebees and their causes. *J. Anim. Ecol. 26*: 71-98

Brignoli, P.M. (1966). Le società eterotipiche degli Araneidi (I). *Rend. Acad. Naz. 16*: 219-46

Bristowe, W.S. (1941). The comity of spiders (II). *Ray Society 126*: Reprint Johnson, London (1968)

— (1958). *The World of Spiders*. London, Collins

Brockmann, H.J. and Dawkins, R. (1980). Joint nesting in a digger wasp as an evolutionary stable preadaptation to social life. *Behav. 71*: 203-45

Buskirk, R.E. (1975a). Aggressive display and orb defence in a colonial spider, *Metabus gravidus. Anim. Behav. 23*: 560-7

— (1975b). Coloniality, activity patterns and feeding in a tropical orb-weaving spider. *Ecology 56*: 1314-28

Christenson, T.E. and Goist, K.C. (1979). Costs and benefits of male-male competition in the orb-weaving spider *Nephila clavipes. Behav. Ecol. Sociobiol. 5*: 87-92

Curio, E. (1976). *The Ethology of Predation*. Berlin, Springer

Darwin, C. (1874). *The Descent of Man and Selection in Relation to Sex*. 2nd ed. London, Murray

Davies, M.E. and Edney, E.B. (1952). The evaporation of water from spiders. *J. Exp. Biol. 29*: 571-82

Dawkins, R. and Krebs, J.R. (1979). Arms races between and within species. *Proc. Roy. Soc. Lond. B 205*: 489-511

Eberhardt, W.G. (1979). *Argyrodes attenuatus*: a web that is not a snare. *Psyche 86*: 407-13

Edmunds, J. (1982). The number of instars occurring during the development of *Argiope flavipalpis* and *Argiope trifasciata* in Ghana, West Africa. *Bull. Brit. Arachnd. Soc. 5*: 294-303

Exline, H. and Levi, H. (1962). American spiders of the genus *Argyrodes*. *Bull. Mus. Comp. Zool. 127*: 75-202

Forel, A. (1901). Fourmis termitophages, lestobiose, *Atta tardigrada*, sous-genres d'*Euponera*. *Ann. Soc. Ent. Belg. 45*: 389-98, 3 figs

Gilpin-Brown, J.B. (1969). Host-adoption in the commensal polychaete *Nereis fucata*. *J. mar. Biol. Ass. UK, 49*: 121-7

Hazlett, B.A. (1981a). Daily movements of the hermit crab *Clibanarius vittatus*. *Bull. Mar. Sci. 31*: 177-83

— (1981b). The behavioural ecology of hermit crabs. *Ann. Rev. Ecol. Syst. 12*: 1-22

Hoffmaster, D.K. (1982). Responses of the spider *Argiope aurantia* to low frequency phasic and continuous vibrations. *Anim. Behav. 30*: 123-7

Iwata, K. (1976). *The Evolution of Instinct: Comparative Ethology of Hymenoptera*. New Dehli, Amerind

Jackson, R.R. and Blest, A.D. (1982). The biology of *Portia fimbriata*, a web-building jumping spider from Queensland: utilization of webs and predatory versatility. *J. Zool. Lond. 196*: 255-93

Janetos, A.C. (1982). Active vs sit-and-wait predators: a simple model. *J. theor. Biol. 95*: 381-6

Jeanne, R.L. (1972). Social biology of the neotropical wasp *Mischocyttarus drewseni*. *Bull. Mus. Comp. Zool. 144*: 63-150

Jensen, K. and Bender, K. (1973). Invertebrates associated with snail shells inhabited by *Pagurus bernhardus*. *Ophelia 10*: 185-92

Klemper, H.G. (1980). Kleptoparasitic behaviour of *Aphodius rufipes* larvae in nests of *Geotrupes spiniger* Marsh. *Ecol. Entomol. 5*: 143-51

Kullmann, E. (1959a). Beobachtungen und Betrachtungen zum Verhalten der Theridiide *Conopista argyrodes* Walck. *Mitt. Zool. Mus. Berlin 35*: 275-92

— (1959b) Über parasitares Verhalten der Spinne *Theridium tepidariorum* Koch. *Verh. Deut. Zool. Ges. Munster*: 332-42

— (1960). Beobachtungen an *Theridium tepidariorum* Koch als Mitbewohner von *Cyrthophora* Netzen. *Deut. Ent. Z. 7*: 146-64

Lubin, Y.D. (1974). Adaptive advantages and the evolution of colony formation in *Cyrtophora*. *Zool J. Linn. Soc. 54*: 321-39

McMillan, R.P. (1965). Observations on flies of the family Milichiidae cleaning *Araneus* and *Nephila* spiders. *West. Austral. Nat. 13*: 96

Miyashita, K. (1969). Effects of locomotory activity, temperature and hunger on the respiratory rate of *Lycosa T-insignata*. *Appl. Ent. Zool. 4*: 105-14

Moore, P.G. (1957). The two sample t-test based on range. *Biometrica 44*: 482-9

Nicolai, J. (1974). Mimicry in parasitic birds. *Sci. Amer. 231*: 92-8

Nyffeler, M. (1980). Juvenile Kreuzspinne als Nahrungsdieb im Netz einer adulten Kreuzspinne *Araneus quadratus*. *Mitt. Ent. Ges. Basel 31*: 90

— and Benz, G. (1980). Kleptoparasitismus von juvenilen Kreuzspinnen und Skorpionsfliegen in den Netzen adulter Spinnen. *Rev. Suisse Zool. 87*: 907-18

Platnick, N.I. and Shadab, M.U. (1978). A review of the spider genus *Mysmenopsis*. *Am. Mus. Nov. 2661*: 1-22

Price, P. (1977). General concepts on the evolutionary biology of parasites.

Evolution 31: 405-20

Robinson, M.H. (1975). The evolution of predatory behaviour in araneid spiders. In G. Baerends, C. Beer and A. Manning (eds.), *Function and Evolution in Behaviour*: 293-312. Oxford, Clarendon Press

— (1977). Symbioses between insects and spiders: an association between lepidopteran larvae and the social spider *Anelosimus eximius. Psyche 83*: 225-32

— (1980). The ecology and behaviour of tropical spiders. In K. Gruber (ed.), *Proceed. 8. Internatl. Arach. Kongress*: 13-32. Wien, Egermann

— and Mirick, H. (1971). The predatory behaviour of the golden-web spider *Nephila clavipes. Psyche 78*: 123-39

— and Olazarri, J. (1971). Units of behaviour and complex sequences in the predatory behaviour of *Argiope argentata. Smiths. Cont. Zool. 65*: 1-36

— and Robinson, B. (1973). The stabilimenta of *Nephila clavipes* and the origins of stabili mentum-building in araneids. *Psyche, 80*: 277-88.

— and Robinson, B. (1977a). Associations between flies and spiders: bibio commensalism and dipso parasitism. *Psyche 84*: 150-7

— and Robinson, B. (1977b). The ecology and behaviour of *Nephila maculata*, a supplement. *Smiths. Cont. Zool. 218*: 1-22

Rothschild, M. and Clay, T. (1952). *Fleas, Flukes and Cuckoos*. London, Collins

Rypstra, A.L. (1979). Foraging flocks of spiders. *Behav. Ecol. Sociobiol. 5*: 291-300

— (1981). The effect of kleptoparasitism on prey consumption and web relocation in a Peruvian population of the spider *Nephila clavipes. Oikos 37*: 179-82

Schaefer, M. (1978). Some experiments on the regulation of population density in the spider *Floronia bucculenta. Symp. Zool. Soc. Lond. 42*: 203-10

Schmidt, G. (1980). Spinnen, alles wissenswerte über Lebensweise, Sammeln, Haltung und Zucht. *Lehrmeister-Bücherei 108*. Albrecht Philler Verlag

Sivinski, J. and Stowe, M. (1980). A kleptoparasitic cecidomyiid and other flies associated with spiders. *Psyche 87*: 337-48

Smith Trail, D.S. (1980). Predation by *Argyrodes* on solitary and communal spiders. *Psyche 87*: 349-55

Stearns, S.C. (1976). Life-history tactics: a review of the ideas. *Quart. Rev. Biol. 51*: 3-47

Taylor, P.R. (1979). An association between an amphipod, *Liljeborga* sp and the hermit crab, *Pagurus hemphilii. Mar. Behav. Physiol. 6*: 185-8

Telford, M. and Daxboeck, C. (1978). *Porcellana sayana* symbiotic with *Strombus gigas* and three species of hermit crabs in Barbados. *Bull. Mar. Sci. 28*: 202-5

Thornhill, R. (1975). Scorpionflies as kleptoparasites of web-building spiders. *Nature 258*: 709-11

— (1979). Male and female sexual selection and the evolution of mating strategies in insects. In. M.S. Blum and N.A. Blum (eds.), *Sexual Selection and Reproductive Competition in Insects*: 81-122. New York, Academic Press

Tinbergen, N. (1951). *The Study of Instinct.* Oxford, Oxford University Press

Turnbull, A.L. (1964). The search for prey by a web-building spider, *Achaearanea tepidariorum. Can. Entomol. 96*: 568-79

Vinson, A. (1863). *Aranéides des iles de la Réunion, Maurice et Madagascar*. Paris, Rovet

Vollrath, F. (1976). Konkurrenzvermeidung bei tropischen kleptoparasitischen Haubennetzspinnen der Gattung *Argyrodes. Ent. German. 3*: 104-8

— (1977). Zur Ökologie und Biologie von kleptoparasitischen *Argyrodes elevatus* und synöken *Argyrodes*-Arten. Dissertation, Universität Freiburg, FRG

— (1978). A close relationship between two spiders: *Curimagua bayano* synecious on a *Diplura* species. *Psyche 85*: 347-53

— (1979a). Behaviour of the kleptoparasitic spider *Argyrodes elevatus. Anim. Behav. 27*: 515-21

Vollrath, F. (1979b). Vibrations, their signal function for a spider kleptoparasite. *Science 205*: 1149-51

— (1980). Male body size and fitness in the web-building spider *Nephila clavipes. Z. Tierpsych. 53*: 61-78

— (1981). Energetic considerations of a spider parasite–spider host system. *Revue Arachnol. 3*: 37-44

— (1982). Colony foundation by the social spider *Anelosimus eximius. Z. Tierpsych. 60*: 313-24

— (in press). Relative and absolute growth in *Nephila clavipes*. In O. Kraus (ed.), *Taxonomy, Biology & Ecology of the Araneae*

— (in preparation a). Ecology of the kleptoparasitic spiders *Argyrodes elevatus* and *Argyrodes caudatus*

— (in preparation b). Seasonality of the orb spider *Nephila clavipes* in Panama

— (in preparation c). Decisions to change web site in an orb web-building spider

Walkey, M. and Meakins, R.W. (1970). An attempt to balance the energy budget of a host-parasite system. *J. Fish. Biol. 2*: 361-72

Wheeler, W.M. (1910). *Ants, Their Structure, Development and Behaviour*. New York, Columbia University Press

Wiehle, H. (1928). Beiträge zur Biologie der Araneen, insbesondere zur Kenntniss des Radnetzbaues. *Z. Morph. Okol. Tiere 11*: 115-51

Wilson, E.O. (1971). *The Insect Societies*. Belknap Press of Harvard University Press

Wise, D.H. (in press). Predation by a commensal spider, *Argyrodes trigonum*, upon its host: an experimental study. *J. Arachnol*

Witt, P.N., Reed, C.H.F. and Peakall, D.B. (1968). *A Spider's Web*. New York, Springer

4 THE EVOLUTION OF FOOD-SCROUNGING STRATEGIES WITHIN AND BETWEEN SPECIES

C.J. Barnard

Next to true parasitism, food-scrounging, particularly piracy or kleptoparasitism (Rothschild and Clay, 1952), but also various forms of dietary and foraging strategy imitation (Krebs *et al.*, 1972; Balph and Balph, 1979; Barnard and Sibly, 1981), is probably the most familiar form of exploitation between organisms. Although this is no doubt partly because of its prevalence among bird species (Rand, 1954; Grant, 1971; Brockmann and Barnard, 1979; Rohwer and Ewald, 1981; Barnard and Sibly, 1981) and scavenger/carnivore mammals (van Lawick and van Lawick-Goodall, 1970; Schaller, 1972; Kruuk, 1972), scrounging relationships have been recorded across a wide range of taxonomic groups including mammals, birds, reptiles, fish, insects and arachnids. The aim of this chapter, however, is to examine the evolutionary origins and consequences of food-scrounging rather than detail its occurrence and various forms. Most of the examples I shall use come from vertebrates, partly because the most detailed accounts of scrounger relationships come mainly from vertebrates, particularly birds, and partly because Vollrath deals specifically with invertebrate examples in the preceding chapter.

The chapter falls into three sections. The first discusses the selective pressures favouring scrounger strategies, the second looks at the evolutionary consequences of scrounging within populations and communities, particularly the counter-adaptive 'arms races' (Dawkins and Krebs, 1979) caused by the impact of scroungers on their 'hosts', and the third speculates about the long-term outcome of arms races for host and scrounger lineages. Many of the examples which will be used in the discussion of counter-adaptation involve interspecific kleptoparasitic relationships (as in Vollrath's (Ch. 3 this volume) definition) again because they have been most thoroughly studied to date.

Selective Pressures Favouring Scroungers

Finding food along with avoiding becoming food is the highest priority maintenance behaviour of most animals. With the exception of some parasites and sessile groups, however, animals cannot simply reach out and grab food the minute they feel hungry. Finding and dealing with food items is a time- and energy-consuming business. In many cases, the most serious problem facing an animal is locating food in the first place. The time and energy costs of searching for prey which employ a variety of effective anti-predator strategies like crypsis, mimicry and crepuscular habits (Edmunds, 1974; Harvey and Greenwood, 1978, for detailed reviews) may be very substantial. Selection should therefore favour any foraging strategy which reduces searching costs relative to food intake (see Curio (1976) for an excellent review). A convenient way of doing this, at least in environments where food items are patchily distributed and cryptic or concealed, is to use the presence of other foragers (con- or heterospecific) as indicators of the best places to start feeding. Predators tend to spend more time where food supplies are richer (Fiscus and Baines, 1966; Hassell and May, 1974; Smith and Sweatman, 1974a, b) and therefore provide a reliable guide to good feeding areas.

Where prey are difficult to find, however, using other foragers merely to locate rich food supplies may leave substantial searching costs still to be overcome. An obvious way to reduce them is to allow other individuals to find food items and then steal them. This is what is meant generally by kleptoparasitism. As Brockmann and Barnard (1979) pointed out, however, it is difficult to make a sharp distinction in terms of the cost to recipients between kleptoparasitism and other forms of feeding interaction like area-copying (Barnard and Sibly, 1981). Usurping food items directly (by stealing them from the procurer) or indirectly (e.g. by taking them from the same area in which the procurer is feeding) can both reduce another individual's chances of survival. The train of counter-adaptive events set in motion by kleptoparasitism might equally well be set in motion by apparently less costly interactions. The term kleptoparasitism is also usually applied to *inter*specific interactions. As Brockmann and Barnard (1979) showed, however, and as we shall now see, very similar relationships occur *within* species.

Costs and Benefits of Intraspecific Food-scrounging

While several studies have pointed to the importance of feeding interactions within single species aggregations (Krebs *et al*., 1972; Murton, 1971; Feare *et al*., 1974; Barnard, 1978; Pitcher *et al*., 1982), it has become clear only recently that certain individuals within these aggregations may actually *specialise* in obtaining food by interaction. Two species in which scrounger strategies appear to be clear cut are the New World Harris sparrow (*Zonotrichia querula*) and the Old World house sparrow (*Passer domesticus*). Studies of these species by Rohwer and Ewald (1981) and Barnard and Sibly (1981) have revealed important effects of scrounger individuals on the evolution of a status signalling system in one case, and on the build-up and stability of feeding flocks in the other. Furthermore, the studies illustrate nicely the environmental conditions under which scrounging is likely to be profitable.

Rank-dependent Scrounging in Harris Sparrows. The scrounger strategy among Harris sparrows emerged from Rohwer's (1977) and Rohwer and Rohwer's (1978) studies of the black bib status cue birds to determine priority of access to limited food supplies (high-ranking birds had larger bibs). When Rohwer tried to manipulate bib size by enlarging the bibs of subordinate birds with black dye, he found that birds with artificially high apparent status suffered increased aggression. If increased apparent status was backed up by increased testosterone levels (which make birds more aggressive), however, the deception worked (Rohwer and Rohwer, 1978). Since production of the small amounts of extra black pigment and testosterone needed to boost a bird's social status is physiologically very inexpensive, the question immediately arises as to why cheats do not occur in the sparrow populations. Rohwer and Ewald (1981) suggested a possible answer. Their hypothesis depended on a subtle producer/scrounger relationship between dominant and subordinate birds which made it *advantageous* for the latter to display their low status.

Observation of Harris sparrow flocks showed that subordinate birds did not feed less efficiently than dominants, as might have been expected if they were always displaced from good feeding areas. The reason appeared to be that dominants sometimes allowed a number of subordinates to feed with them on rich food supplies to indicate the whereabouts of food. A small number of subordinates were then defended against use by other dominants. The evidence for this 'shepherding' effect was basically fourfold:

(i) Dominant birds on patchy and concealed ('cached') seed supplies (where area-copying would be profitable) tended not to search for themselves but instead displaced subordinates (Figure 4.1a).

(ii) Dominants displaced subordinates which were feeding more frequently than those which were not (Figure 4.1b). Furthermore the degree of rank difference between attacker and recipient mattered only for non-feeding subordinates (Figure 4.1b). The probability of displacement was much higher between closely ranked birds (i.e. dominants, since dominants were the attackers).

Figure 4.1: (a) High-ranking (dominant) Harris sparrows displaced other birds more often on patchily distributed (open circles, dashed line) than on diffusely distributed (solid circles and line) seed supplies. Furthermore, most aggression was directed towards other dominants (dominance increases with rank number). (b) Dominant Harris sparrows were more likely to displace feeding birds (open circles, dashed line) than non-feeding birds (solid circles and line). Displacement was more or less independent of rank difference between birds when the displaced bird was feeding, but only similarly ranked non-feeding birds were displaced. (c) Shows that, although displacements were likely between similarly ranked birds, they were most likely to occur between dominants. (d) Subordinates found more seeds dispersed from 'cached' supplies than dominants, perhaps as a result of spending less time aggressively defending supplies

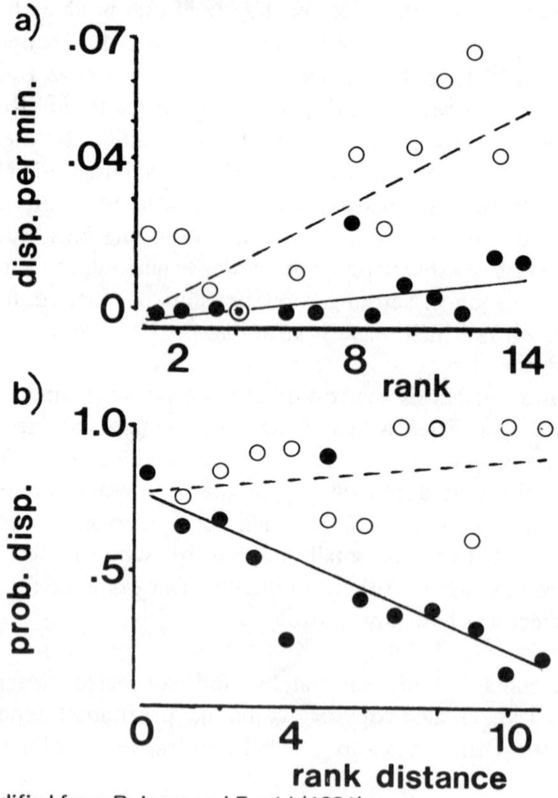

Source: modified from Rohwer and Ewald (1981)

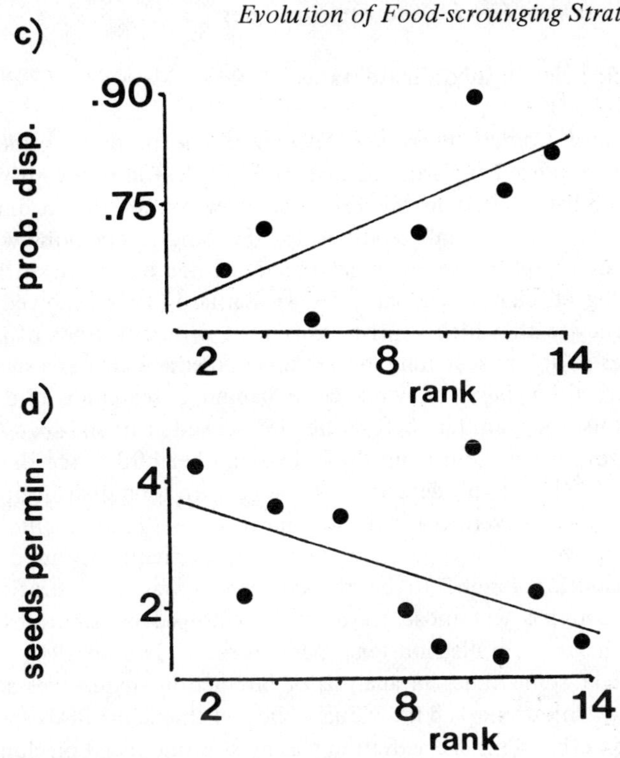

Source: modified after Rohwer and Ewald (1981)

(iii) Displacement rates were higher between dominants than between subordinates (Figure 4.1c) as expected if dominants were defending food supplies and feeding subordinates against intrusion by other dominants.
(iv) Subordinates did marginally better than dominants on 'cached' seed supplies (Figure 4.1c) as might be expected from the fact that dominants spend more time defending food supplies.

Subordinate birds thus appeared to gain an advantage as long as they were providing useful information about food to dominants. This depended on the distribution and visibility of food supplies and on the benefit of the information to the dominant relative to the cost of defending feeding subordinates against other dominants. The producer/scrounger relationship within flocks thus resulted in a frequency-dependent advantage to being subordinate and hence an advantage to advertising clearly subordinate status.

Searchers and Copiers in House Sparrows. The producer/scrounger relationship reported by Barnard and Sibly (1981) in house sparrows was very similar to that in the Harris sparrow except that it did not appear to depend on rank relationships. Feeding interactions within flocks are an important way in which individual house sparrows increase their feeding efficiency (Barnard, 1978). Barnard (1978) showed that the frequency with which captive birds used different kinds of interaction (*area-copying*: searching in the same immediate area as a successful forager; *following*: following close behind a searching bird and sometimes usurping an item before the bird pecked at it; and *snatching*: forcibly usurping an item from the bill of another bird – see Barnard and Sibly (1981) for full descriptions) varied with food distribution on a grid where birds were searching for mealworms (*Tenebrio molitor*) concealed in wells. The type of interaction most frequently used on a given distribution tended to be the one which was most profitable. Thus area-copying was most prevalent on clumped distributions and snatching on uniform distributions. While most birds used all types of interaction at some time, an analysis of interaction frequencies across a number of colour-marked individuals, showed that some birds specialised in interacting. This specialisation was most pronounced on clumped food distributions and is represented in Figure 4.2. Interactors (open circles) tended to move about rapidly between other foragers, sitting still on the grid between bursts of movement, to interact at a high rate and take a large proportion of their prey by interaction. Because area-copying was the commonest form of interaction, Barnard and Sibly (1981) called the specialist interactors '*copiers*' and the birds which predominantly searched for their own food '*searchers*'.

Since copiers appeared to depend on searchers for a large part of their food intake, Barnard and Sibly predicted frequency-dependent benefits to searching and copying determined by the relative frequency of the two strategies within a flock. Copiers should do better when there were more searchers and vice versa. An important first step towards testing this prediction was to show that the two strategies did not change as a function of flock composition. Analysis of variance showed no significant variation in the proportion of prey taken by interaction by copiers or searchers with changing searcher:copier (S:C) ratios in the flock. However, while frequency-dependent benefits did appear to accrue (Figure 4.3a), their direction was opposite to that predicted. Copiers did better when there were *fewer* searchers

Figure 4.2: The distinction between 'searchers' (solid circles) and 'copiers' (open circles) in feeding flocks of captive house sparrows

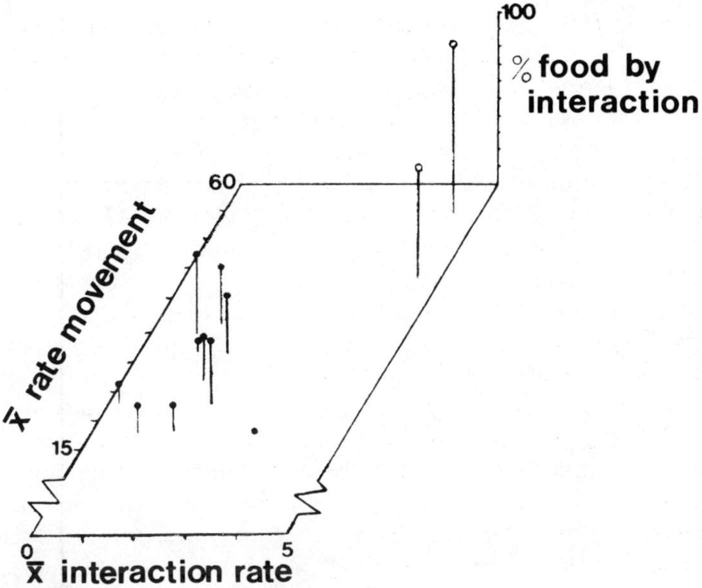

Source: Barnard and Sibley (1981)

(although they did better when there was at least one searcher present than when there were only copiers in the flock). The reason appeared to be that searchers (as might be expected) were more efficient at exploiting the patchily distributed prey because they showed more area-restricted searching (Tinbergen, 1960; Smith, 1977a, b; Barnard, 1978) after a find. When a lot of searchers were present, the limited food supply tended to be depleted in the time taken for copiers to obtain items by interaction.

Figure 4.3a shows a point of equal payoff for each type of individual at a ratio of about 0.7S:1C. Barnard and Sibly therefore predicted that this would be a point of unstable equilibrium in flock composition. The probable departure rate of copiers should increase with the S:C ratio beyond 0.7S:1C and that of searchers should increase with decreasing S:C ratio. However, they also predicted a point of stability in flocks of searchers only, since searchers were not dependent on copiers for finding food. Figure 4.3b shows that these predictions were borne out. Flocks were most persistent when the S:C ratio was in the region of 0.7S:1C and when flocks consisted of searchers only.

The searcher/copier relationships within house sparrow flocks

Figure 4.3: (a) Mean prey (mealworm) capture rate by 'searchers' (solid bars) and 'copiers' (dashed bars) in flocks with different 'searcher': 'copier' (S:C) ratios. (b) Flock stability (mean ± s.e. persistence time) in relation to the S:C ratio. 'C' indicates the apparent cross-over point for relative 'searcher' and 'copier' capture rates in a

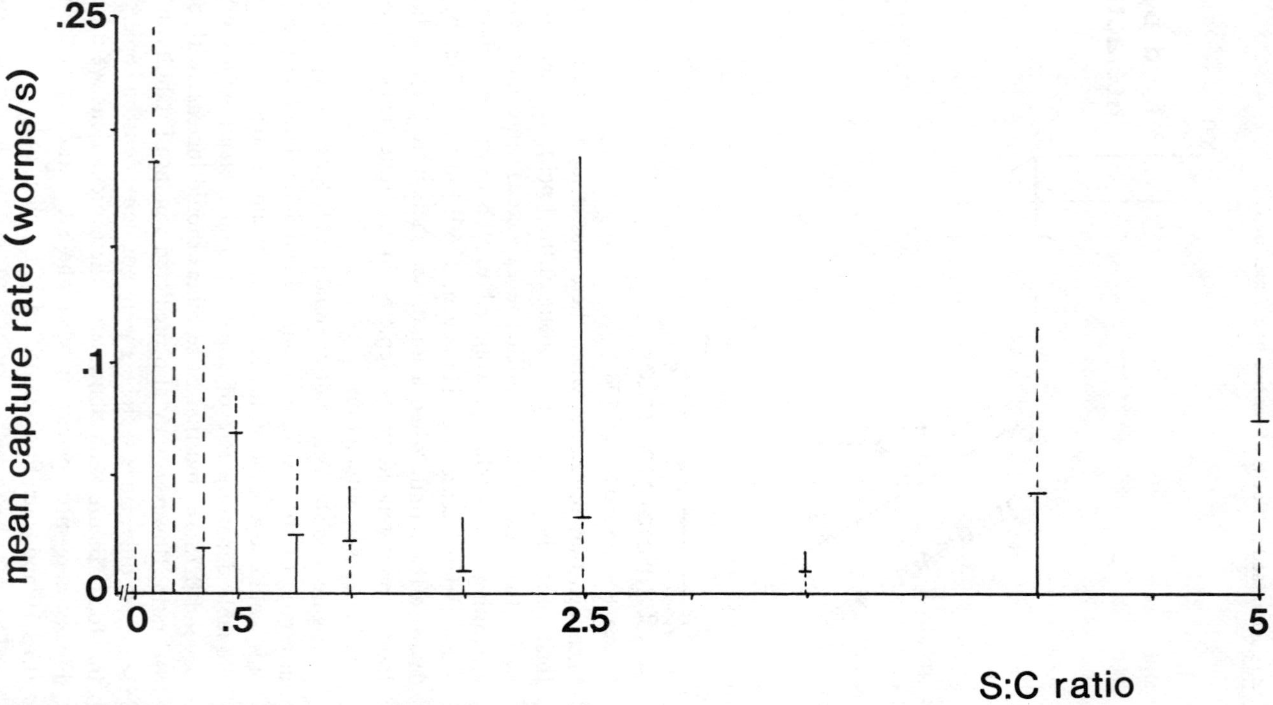

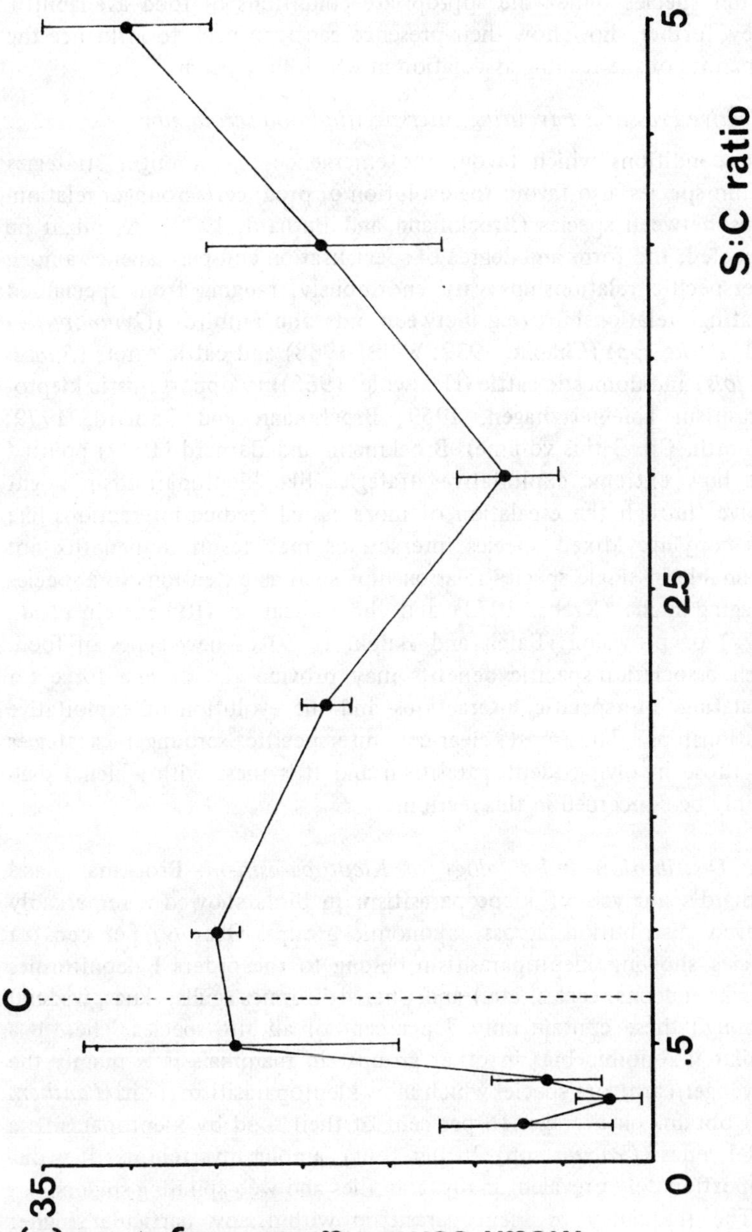

Source: Barnard and Sibly (1981)

illustrate again how scrounger strategies can evolve and be maintained within species under the appropriate conditions of food availability. They further show how their presence can feed back to influence the dynamics of the feeding association in which they occur.

Selective Pressures Favouring Interspecific Food-scrounging

The conditions which favour the emergence of scrounger strategies within species also favour the evolution of producer/scrounger relationships between species (Brockmann and Barnard, 1979). As might be expected, the form and degree of specialisation and permanence among interspecific relationships vary enormously, ranging from specialised 'beating' relationships (e.g. between ants and antbirds (*Gymnopithys* and *Alethe* spp) (Chapin, 1939; Willis, 1968) and cattle egrets (*Bubulcus ibis*) and domestic cattle (Heatwole, 1965)) to opportunistic kleptoparasitism (Meinertzhagen, 1959; Brockmann and Barnard, 1979; Vollrath, Ch. 3 this volume). Brockmann and Barnard (1979) pointed out how extreme exploitative strategies like kleptoparasitism might evolve through the escalation of more casual feeding interactions like area-copying. Mixed species interactions may result in benefits not attainable in single species relationships such as extensions of a species foraging niche (Krebs, 1973) and the indication (Rubenstein *et al*., 1977) or provision (Balph and Balph, 1979) of new types of food. Such association-specific benefits may provide the driving force for escalating interspecific interactions and the evolution of exploitative relationships. The most clear-cut interspecific scrounger strategies are those involving kleptoparasitism and it is these with which I shall mainly be concerned in this section.

The Distribution and Ecology of Kleptoparasitism. Brockmann and Barnard's analysis of kleptoparasitism in birds showed a remarkably uneven distribution across taxonomic groups. Over 60 per cent of species showing kleptoparasitism belong to the orders Falconiformes (hawks, falcons, eagles, etc.) and Charadriiformes (gulls, skuas, waders) although these contain only 7 per cent of all bird species. There is a similar taxonomic bias in other groups: in mammals it is mainly the scavenger/carnivore species which show kleptoparasitism (lions (*Panthera leo*) obtain, on average, 16 per cent of their food by kleptoparasitism and hyenas (*Crocuta* spp) 26 per cent), among invertebrates it is disproportionately prevalent in dipteran flies and web-spinning spiders.

The frequency of kleptoparasitism within any particular higher taxonomic group does not appear to be a function simply of the number of species within that group. Rather there is a number of well-defined ecological factors associated with groups which make them more or less likely to contain kleptoparasites. Following Brockmann and Barnard (1979), these can be grouped into three categories: (a) the tendency to associate closely (including for predation) with con- and/or heterospecifics; (b) the characteristics of potential host populations and food supplies; and (c) the life-style of the potential kleptoparasite. Since these have been discussed at length by Brockmann and Barnard, the main factors are summarised in Table 4.1. The points that emerge from the table are that kleptoparasitism is most likely where some kind of close association between individuals has been favoured for some reason, where food items are relatively large and profitable and where they are being procured at a high rate by potential hosts. It is also more likely to occur as one of several alternative feeding strategies in species like gulls, egrets and hyenas which are dietary opportunists. The importance of food supply characteristics and behavioural flexibility underlines the homology between inter- and intraspecific food-scrounging strategies. However, even within established associations, there may be more fine-grained factors which determine the emergence, form and timing of kleptoparasitism. One of these is age. Burger and Gochfeld (1981), for instance, found age-dependent success rates and frequencies of different kleptoparasitic strategies in gulls (*Larus* spp) stealing from each other. Adults tended to be kleptoparasites more often and victims less often than young birds. However, age effects depended to some extent on the food supply. Age differences were not apparent when gulls were feeding on a rubbish dump, but they were when they were feeding on offal or kleptoparasitising starlings (ring-billed gulls (*L. delawarensis*)). A study of young herring gulls (*L. argentatus*) suggested that the age effect was due partly to learning and developing skills (particularly holding food items in flight).

On the face of it, kleptoparasitism and other scrounging strategies are cheap and easy ways of obtaining food. As long as feeding conditions and potential host species are appropriate, they may be good ways of reducing foraging effort. However, as with most purely exploitative relationships, their very convenience and profitability create the selective pressures which may militate against their spread through the community (see Sibly, Ch. 9 this volume). In the next section, I shall discuss (a) the impact of scroungers on their hosts and the countermeasures which hosts may take to reduce it, and (b) the effects of competition between scroungers exploiting a given host population.

Table 4.1: Ecological and behavioural factors favouring kleptoparasitism (from discussion in Brockmann and Barnard, 1979)

Food and host availability		Behavioural factors
1) Host concentration. Kleptoparasitism is common where potential hosts are aggregated as in breeding colonies (e.g. gulls, auks, digger wasps) and feeding groups (e.g. lapwing flocks, eider rafts)	4) Food predictability. Kleptoparasitism is more likely where the habits of the host make food availability temporally and spatially predictable (e.g. puffins repeatedly return to the same nest burrow where skuas and gulls lie in wait)	7) Behaviour of parasite. Many klepto-parasites are dietary opportunists and piracy occurs as one of a flexible range of feeding strategies when appropriate food items are made available. The ability to move quickly and with agility is also a common feature of klepto-parasites
2) Food availability. In some species, kleptoparasitism occurs only at times when hosts are making large quantities of food available, e.g. systematic piracy of auks by skuas, gulls and terns occurs only in the nesting season when hosts are feeding chicks	5) Food detectability. Kleptoparasitism is more likely where food items are easily detectable thus reducing searching time costs. Long prey-handling times make predators particularly vulnerable to piracy	8) Behaviour of host. Kleptoparasitism is more likely where there is little defence or retaliation by hosts and where there is little likelihood of costly chases or host escape. See also 1)
3) Food quality. Many kleptoparasites appear to attack selectively host individuals which are carrying large or high-quality food items (e.g. gulls selectively attack swans bringing mussels rather than weed to the water surface)	6) Food shortage. Kleptoparasitism may be a 'best of a bad job' strategy adopted during times of food shortage. In birds, for instance, it is often subordinate individuals which show kleptoparasitism	

The section deals mainly with kleptoparasitic relationships.

Arms Races Between Producers and Scroungers

Clearly, a feeding strategy like kleptoparasitism is likely to have a serious impact on host individuals. Hosts are spending time and energy obtaining food and are losing sometimes a substantial proportion of it to kleptoparasites. Cheetahs (*Acinonyx jubata*), for instance, lose about 12 per cent of their kills to other carnivores, and overwintering lapwings (*Vanellus vanellus*) feeding on agricultural pasture in Britain lose 67 per cent of earthworms over 80 mm in length to black-headed gulls (*Larus ridibundus*) (Barnard and Stephens, 1981). This impact is likely to set up a counter-adaptive arms race between host and kleptoparasite, selection acting on the host lineage to minimise the impact of the kleptoparasite and on the kleptoparasite lineage to counter the host and maximise its attack efficiency. I shall discuss the possible long-term outcomes of such arms races later. First I shall consider the types of counter-adaptive strategies employed by hosts and kleptoparasites.

Strategies Employed by Hosts and Kleptoparasites

From studies published so far, there appear to be three broad types of strategy employed by hosts to reduce the effects of kleptoparasitism. These are (a) *evasion*, where the host attempts to avoid attack altogether; (b) *retaliation*, where the host attempts to deter the kleptoparasite and/or reclaim stolen items; and (c) *toleration/compensation*, where the host simply tolerates losses to kleptoparasites or compensates for them other than by retrieval.

Evasion Strategies. Under certain conditions, it may be possible for host individuals to avoid attack, although the time and energy costs of avoidance may be extremely high. A nice series of counter-adaptive strategies based on host evasion, however, has been revealed in mixed associations of lapwings, golden plovers (*Pluvialis apricaria*) and black-headed gulls. Lapwings and golden plovers attempting to prevent loss of earthworm prey to the kleptoparasitic black-headed gulls feeding with them may be forced into long and expensive aerial chases (Källander, 1977, 1979; Barnard *et al.*, 1982). Estimates of energy expenditure by birds during flight suggest an increase of over 15X the basal metabolic rate (King, 1974). Källander (1977) calculated that 31.4 per cent of a kleptoparasitic gull's daily energy expenditure may be spent on aerial chases of lapwings and golden plovers. Figure 4.4 shows that, in golden plovers, chases may last up to 25 seconds even though the gulls' success rate drops off at about 12 seconds. Furthermore, even if birds manage to evade one gull in a chase, they often lose their prey to another when they return to the flock (Källander, 1977).

Figure 4.4: The probability of successful kleptoparasitism (solid lines) of golden plovers by black-headed gulls as a function of chase duration (flight time). The broken line shows the number of attacks of each duration recorded. Flight times are lumped in 2-second intervals

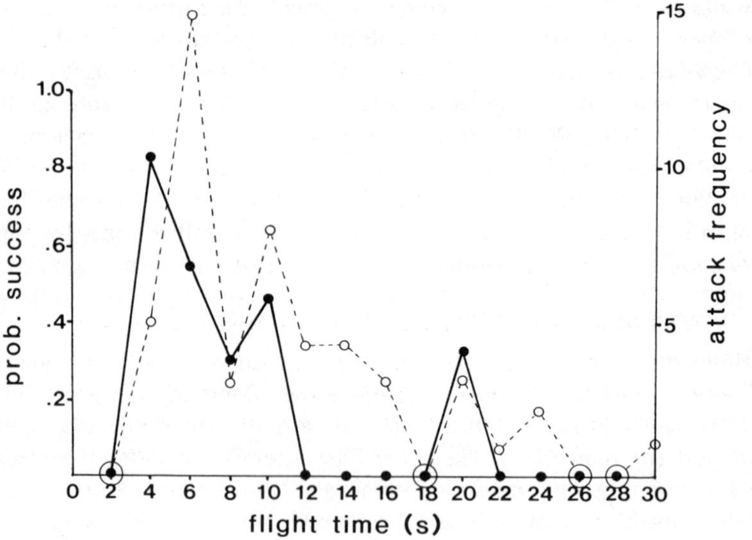

Source: from Thompson, D.B.A. (1983). Foraging strategies of golden plovers (*Pluvialis apricaria*) in mixed species flocks, Ph. D. thesis University of Nottingham

Since chases are also expensive for gulls, we might expect them to counter-adapt by attacking earlier and reducing the risk of the target taking off. As the time taken to extract a worm from the ground is by far the greater part (65-70 per cent) of handling time, however, attacking too early would risk scaring a target off before the worm was made available. Gulls appear to solve the problem by cueing into and preferentially attacking lapwings or plovers which adopt a tell-tale crouching posture immediately before pecking (Barnard and Stephens, 1981). Crouching provides an extra few seconds of orientation time before attack and means the gull has a better chance of reaching the target bird before the worm is fully extracted and can be removed. Attacks against lapwings which crouch are four times more successful than those against birds which do not (Barnard and Stephens, 1981). Sometimes lapwings are attacked when they first show the jerking movements associated with extracting worms. Occasionally they are even attacked when they peck unsuccessfully at the ground (Källander, 1977; personal observation).

Although crouching enhances feeding efficiency in both lapwings and golden plovers (Barnard and Stephens, 1981; Barnard *et al.*, 1982), birds of both species reduce the frequency of crouching before pecking when gulls arrive in the flock. As a result, the feeding efficiency of lapwings but, interestingly, not golden plovers drops significantly. The reduction in crouching appears to be a means of reducing vulnerability to attack rather than the result of other, mutually incompatible, activities increasing in frequency (Barnard and Stephens, 1981). It also results, incidentally, in smaller and less vulnerable worms being taken because crouching is associated with taking deeper worms which tend to be larger (Thompson and Barnard, in press). A particularly astonishing observation is that lapwings and plovers may actually *discard* the large worms which are the gulls' principal target after having extracted them at great expense from the ground (Thompson and Barnard, in press). Similar rejection of prey has been reported by Kushlan (1978) in New World herons feeding with kleptoparasitic egrets. In these instances hosts apparently attempt to avoid attack at the point where they procure vulnerable prey. If lapwings and golden plovers continue to handle prey, handling times per mm of prey are significantly shorter in the presence of gulls.

Another good example of counter-adaptive strategies based on host evasion comes from Grant (1971), Nettleship (1972) and Arnason and Grant (1978), who studied arctic skuas (*Stercorarius parasiticus*) and gulls (*Larus* spp) kleptoparasitising puffins (*Fratercula arctica*). Skuas attacked puffins returning to their nesting colony with fish held transversely in their bill, while gulls (mainly herring and great blackbacked (*L. marinus*) gulls) attacked them at the colony and also stole already kleptoparasitised prey from skuas. The studies detailed a range of strategies adopted by kleptoparasites and apparently counter-adaptive evasion strategies employed by the puffins. These are summarised in Table 4.2. Avoidance responses were most in evidence during the fledgling period when adult puffins were feeding large chicks as well as themselves and losses to kleptoparasites were likely to have the most serious consequences. At other times of the year, when pressure

Table 4.2: Attack and evasion strategies by gulls, skuas (kleptoparasites) and puffins (hosts). Four pairs of apparently counter-adaptive strategies can be distinguished. From discussion in Grant (1971), Nettleship (1972) and Arnason and Grant (1978)

Kleptoparasite attack strategy	Puffin evasion strategy
1. (Skuas) – Puffins were mostly chased far (>25 m) from the colony where they were more likely to drop food items	Puffins turned back out to sea when they spotted a skua and headed in elsewhere
2. (Skuas) – Low-flying puffins were attacked preferentially (75% of attacks were against birds flying below 100 m and 47% against those below 50 m). The reason appeared to be that puffins were more likely to drop fish when they were low	Puffins tended to fly in at heights above 100 m (54% of birds) and sometimes to circle high before coming in to land
3. (Skuas) – attacks were more effective against solitary puffins	Breeding puffins (those bringing in food) tended to fly back with groups of non-breeders (not carrying food)
4. (Gulls) – gulls waited on the slopes outside puffin burrows and attacked birds as they landed or slowed down to land	Puffins preferred to nest in steep faces where launching attacks was more difficult

was presumably lower, avoidance was less apparent.

Similar direct evasion strategies have been recorded in other groups. Leopards (*Panthera pardus*), for instance, drag their prey up into trees to avoid loss to hyenas. Interestingly, the tendency to climb trees with prey is absent in leopard populations which are not sympatric with hyenas as, for example, in Sri Lanka (Kruuk, 1972). Evasion strategies are also evident intraspecifically. The house sparrows studied by Barnard and Sibly (1981) handled given-sized mealworms more quickly in larger flocks (Figure 4.5) where there were more likely to be copiers present.

Retaliation Strategies. A more direct way of reducing the impact of kleptoparasitism is to reclaim stolen items from kleptoparasites. Retaliation against kleptoparasites may have been an important factor in the evolution of web structure in certain neotropical spiders (Vollrath, 1979; Ch. 3 this volume).

Figure 4.5: Reduction in handling time per item (standard-sized mealworm pieces) with increasing flock size in two groups of captive house sparrows

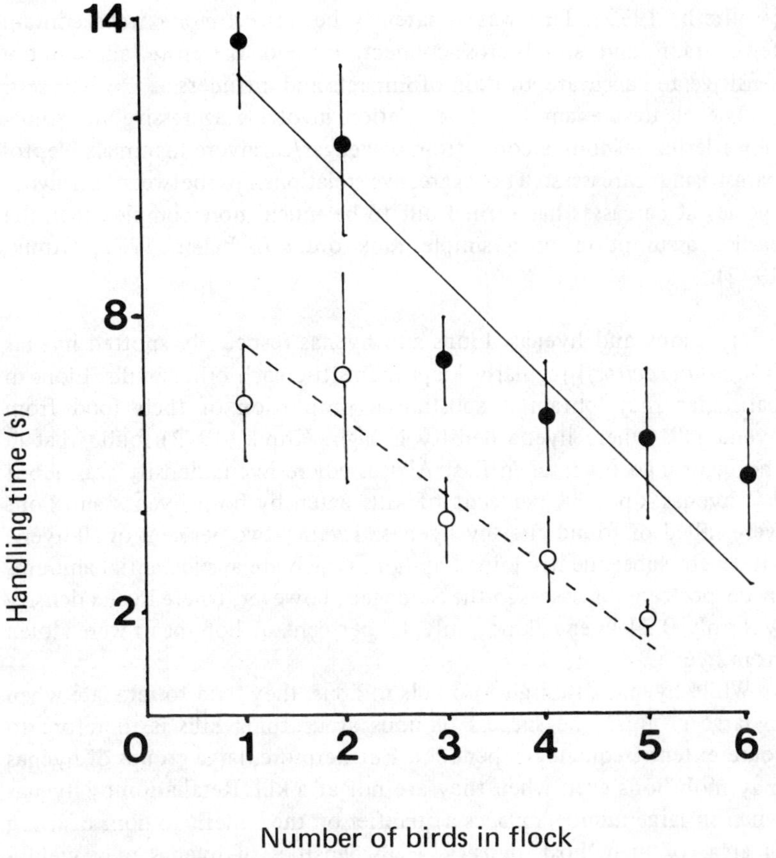

Source: Barnard (1978)

Two genera of orb-web spiders, *Argiope* and *Nephila*, are klepto-parasitised by spiders of another genus, *Argyrodes*. *Argyrodes* spins thin strands between the radii and frame threads of the host web and detect prey hitting the web by vibration transmission. When vibrations are detected, the kleptoparasite enters the host web and attempts to remove the item. Normally, there are 2-5 *Argyrodes* per host web, but as many as 40-5 have been recorded.

The impact of prey hitting the web is, of course, also detected by the host who can move in to intercept the kleptoparasite. The host may reach the prey first or it may retrieve it from a successful kleptoparasite. An analysis of retrieval success in *Argiope* and *Nephila*, however, showed that *Argiope* was almost 5X better at retrieving than *Nephila* (Vollrath, 1979). This was apparently because *Argiope* webs contain fewer radii and spiral cross-connections and therefore allow more sensitive and accurate location of impacts and intruders.

The clearest examples of retaliation involving aggressive and sometimes lethal responses come from scavenger/carnivore mammals kleptoparasitising carcasses. The aggressive relationships between carnivore species at carcasses has turned out to be much more complex than the earlier assumption of a simple rank order (Schaller, 1972; Kruuk, 1972).

(a) Lions and hyenas. Lions and hyenas (especially spotted hyenas (*Crocuta crocuta*)) regularly kleptoparasitise each others' kills. Lions in particular may obtain a substantial proportion of their food from hyena kills where hyena density is high. Kruuk (1972) found that in the Ngorongoro Crater in East Africa, where hyena density was about 1.7 hyenas/km^2, 84 per cent of kills eaten by both hyenas and lions were killed or found first by hyenas. Twenty-two per cent of all hyena kills were subsequently joined by lions which ate a 'substantial amount' in 63 per cent of cases. In the Serengeti, however, where hyena density was only 0.24 hyenas/km^2, only 18 per cent of lion meals were stolen from hyenas.

While hyenas often give up kills to lions, they tend to retaliate when in large groups. The success of lions at usurping kills is therefore to some extent frequency-dependent. Furthermore, large groups of hyenas may mob lions even when they are not at a kill. Retaliation by hyenas when in large numbers places a qualifier on the benefit to lions of living in areas of high 'host' density. High densities of hyenas may yield a lot of convenient kills, but their presence at each kill in large numbers reduces the kill's accessibility to lions.

(b) Hyenas and hunting dogs. Hyenas capitalise heavily on hunting dog (*Lycaon pictus*) kills (Kruuk, 1972). They often follow dog packs, wait until the first two or three dogs have caught prey then move in and attack. The relationship between hyenas and hunting dogs is more complex than that between lions and hyenas. As well as being frequency-dependent (occurring mainly when large numbers of hunting dogs relative to hyenas are present at a kill), aggressive retaliation by dogs may depend on their hunger state, previous experience and the behaviour of attacking hyenas (Kruuk, 1972). Of particular interest with

Figure 4.6: Factors determining the success of kleptoparasitism and other interactions between lions, spotted hyenas and hunting dogs. Solid arrows indicate predominant direction of interaction

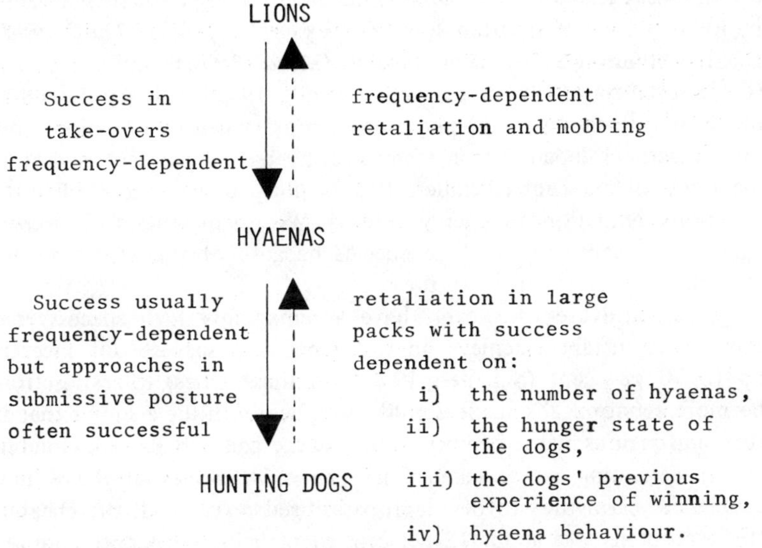

Source: from data and discussion in Kruuk (1972), Schaller (1972) and van Lawick and van Lawick-Goodall (1970)

regard to the last factor is the tendency for adult hyenas to adopt a submissive posture (crouching with ears laid flat back) as they approach dog kills. The posture is maintained until the hyena is within striking distance of the carcass whereupon it rushes in, seizes a mouthful of meat and retreats. Young hyenas, however, maintain the submissive posture while at the kill and feed unimpeded with the dog pack. The factors affecting host retaliation in lion/hyena and hyena/hunting dog relationships are summarised in Figure 4.6.

Another example of aggressive retaliation comes from the mustelids. Wolverines (*Gulo gulo*) use their well-developed 'stink' glands to dispel pumas (*Panthera concolor*) and brown bears (*Ursus actos*) which sometimes steal their kills.

Toleration/Compensation Strategies. In some cases it may be more economical for a host to tolerate the impact of a kleptoparasite or to compensate in some way rather than to take evasive or retaliatory action. It is clear, however, that the ability of a host to sustain klepto-parasites will be limited by the amount of extra work it is forced to do. In most carnivores the tolerable increase is likely to be small because hunting efficiency is often low. For example, wolves (*Canis lupus*) achieve about 7.8 per cent success (Mech, 1970), solitary jackals (*C. mesomelas*) about 16 per cent (Kruuk, 1972), solitary lions 8-19 per cent (increasing to 30 per cent or more when two or more lions hunt together) depending on technique, hyenas 32 per cent, and hunting dogs 49 per cent (Schaller, 1972). Pillaging may therefore cause a serious reduction in energy returns. We might, therefore, expect aggressive retaliation in these species because of the effort needed to replace lost items in the future.

Some carnivores, however, have a remarkably high success rate. Pumas, for instance, achieve up to 90 per cent success and cheetahs up to 70 per cent (Schaller, 1972). In these cases, toleration may be more economical since it is unlikely to be worth the energy expenditure and serious risk of injury (van Lawick and van Lawick-Goodall, 1970) defending an attacked kill. The only species of these 'high efficiency' carnivores to be kleptoparasitised regularly is the cheetah. While cheetahs lose about 12 per cent of their kills to kleptoparasites, particularly hyenas, they have never been observed to defend kills or to retaliate. Instead, kills are abandoned. One reason for the cheetah's relatively low loss rate may be that its range of prey includes many species which are smaller than those normally taken by e.g. hyenas.

In some cases a kleptoparasite may be tolerated because it takes prey which are relatively unprofitable to the host and which therefore do not represent a serious loss. This is possibly the case in the lapwing/golden plover/gull relationship studied by Barnard and Stephens (1981) and Barnard *et al.* (1982). Gulls take predominantly the larger worms ($>$ 80 mm in length) but these turn out to be the sizes which are least profitable and least frequently taken by lapwings and golden plovers (Figure 4.7). As we saw earlier, however, both species respond to the presence of gulls by taking smaller worms (because they reduce orientation time prior to pecking). These small worms ($\leqslant$16 mm) are also unprofitable and as their frequency in the diet increases, individual net rate of energy intake decreases. Birds appear to make up for this, however, by pecking more often and moving about more rapidly (Barnard and Stephens, 1981). The enforced shift to less profitable worms may therefore be compensated by an increased ingestion rate, although this ignores the increased

Figure 4.7: Range of prey (earthworm) sizes taken by lapwings and golden plovers (histograms and dots above the abscissa) and stolen from them by black-headed gulls (histograms below the abscissa). Above the abscissa, the histograms show the percentage of worms taken which fell into each of six 16 mm size classes when gulls were present, while the dots show the percentages taken in the absence of gulls (see text)

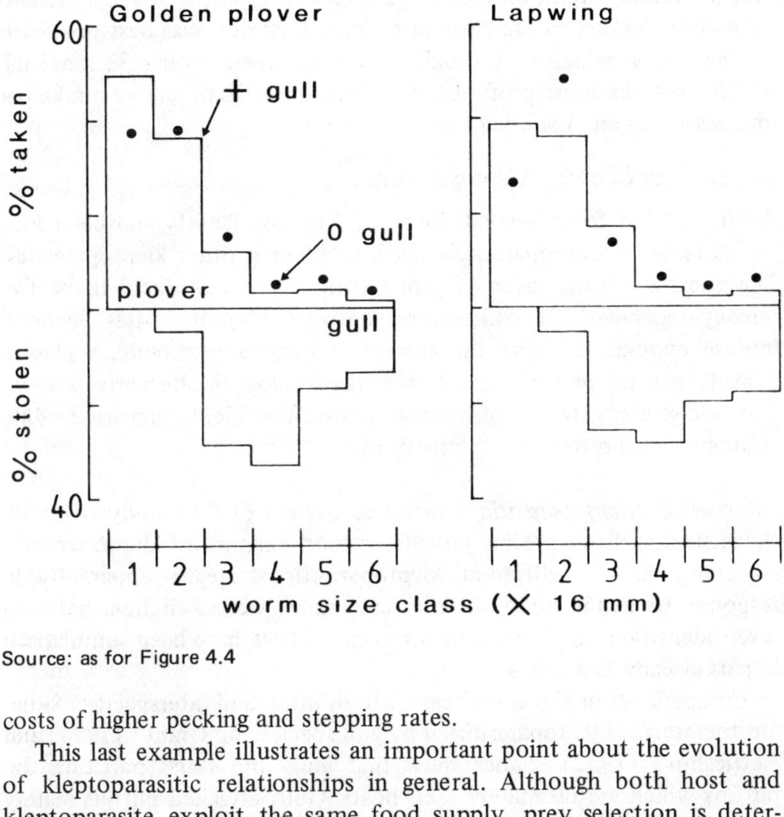

Source: as for Figure 4.4

costs of higher pecking and stepping rates.

This last example illustrates an important point about the evolution of kleptoparasitic relationships in general. Although both host and kleptoparasite exploit the same food supply, prey selection is determined ultimately by the host. The extent to which a kleptoparasite can exploit a given host is therefore determined by the magnitude of the discrepancy between host and kleptoparasite optima. If it is small, the overlap between the distribution of items taken by host and kleptoparasite great and the cost of kleptoparasitism relative to alternative feeding strategies low, then a compromise diet and host specialisation may be the optimal solution for the kleptoparasite. The opposite conditions are likely to select for generalist kleptoparasitism of several hosts or for the abandonment of kleptoparasitism for alternative feeding strategies. That kleptoparasites are sensitive to changes in the distribution of prey items taken by their hosts is shown clearly by the responses of black-headed gulls stealing earthworms from lapwings. Barnard *et al.* (1982) found that the number of gulls in a flock was best predicted by the rate at which lapwings were capturing worms over 30 mm in length, the size range over which gulls stole most successfully. The number of lapwings in a flock, however, was best predicted by the rate at which they caught worms between 16 and 32 mm long which were the most profitable for them in terms of energy intake per unit searching and handling time.

Competition Between Kleptoparasites

Assuming that feeding conditions and host availability and behaviour are suitable, a kleptoparasite's main problem is other kleptoparasites. The success of any given kleptoparasitic strategy is likely to be frequency-dependent. If competition between kleptoparasites becomes intense enough to make the current strategy unprofitable, a kleptoparasite can do one of three things: (a) employ an alternative kleptoparasitic strategy; (b) employ an alternative non-kleptoparasitic feeding strategy; or (c) restrict the competition.

Alternative Kleptoparasitic Strategies. Grant's (1971) study of arctic skuas stealing from puffins provides a good example of kleptoparasites switching between different kleptoparasitic strategies apparently in response to competition. In this case skuas were switching between seven identified kleptoparasitic strategies. These have been summarised in part already in Table 4.2.

Competition in the skuas' case is both intra- and interspecific. Skuas are themselves kleptoparasitisied by gull species and Grant's (1971) and Nettleship's (1972) studies show that gulls and skuas partition the puffins which are ultimately their hosts. Gulls attacked puffins mainly at the colony rather than away from it. Pairs established nesting sites near puffin colonies so that both adults could attack puffins without straying too far from their eggs. Skuas, on the other hand, tended to chase puffins away from the colony. Even so, the strategy adopted by skuas away from the colony was influenced strongly by the presence of gulls on the ground. While skuas appeared to obtain an advantage by attacking low-flying puffins (because they were more likely to drop food), food dropped by these low-fliers was likely to hit the ground before the skuas could catch it. Food hitting the ground was usually lost to gulls. Interestingly, therefore, Grant (1971) found that the tendency to attack low-flying puffins reflected the presence or absence of gulls. During the puffin breeding season when gulls were abundant, skuas often attacked high-flying puffins; at the end of the breeding season, when there were far fewer gulls, skuas tended to attack low-fliers. Arnason and Grant (1978) observed skuas apparently driving some high-flying puffins to low altitudes where food was more likely to be dropped.

When competition for available incoming puffins was high, the returns for solitary kleptoparasites decreased and skuas frequently switched to attacking in groups. Groups comprised 2-6 birds and were significantly more effective in eliciting a drop than single birds (Arnason and Grant, 1978). Interestingly, while success rate went up with group size, it went down per individual (see also Hatch (1975) for group chases by laughing gulls (*Larus atricilla*)). A more careful analysis, however, showed that birds which were in the best position within the group when food was dropped did better with increasing group size, although it was not clear what the 'best' position was. Hatch (1975) found that it tended to be 'joiner' gulls rather than the initial attacker which did better in laughing gulls. Grant (1971) calculated that a group size of two was optimal for the skuas in his study. An alternative to joining group chases in skuas appeared to be chasing puffins nearer to the colony ($<$ 50 m out). Exactly which strategy skuas chose out of the range open to them appeared, therefore, to depend on the amount of food involved, the chances of success, the presence of secondary kleptoparasites and the number of other skuas employing each strategy.

Alternative Non-kleptoparasitic Strategies. Clearly, another option open to a kleptoparasite under conditions of severe competition, or low food availability for other reasons, is to abandon kleptoparasitism altogether and adopt a different feeding strategy. Since kleptoparasitism is often just one of many feeding strategies employed by opportunistic predators like house sparrows and hyenas, switching between kleptoparasitic and non-kleptoparasitic strategies is commonplace. However, what I am more concerned with here are changes of this sort which are the result of decisions to abandon kleptoparasitism as a primary means of obtaining food because it is temporarily unprofitable.

Källander (1977) discusses alternative non-kleptoparasitic feeding strategies in black-headed gulls which otherwise obtained food by stealing from lapwings feeding on pasture in southern Sweden. Although gulls could obtain more than their estimated daily energy requirement by stealing worms from lapwings, they did so only under a narrow range of weather conditions (moderately wet). If it was too dry, worms were not available to lapwings or, therefore, to gulls; if it was very wet, worms were sufficiently available for gulls to take them independently of lapwings. However, even on days when kleptoparasitism was evident, gulls still employed other feeding strategies. Over the six observation days detailed by Källander (1977), 30.6 per cent of recorded gulls were kleptoparasitising lapwings, 36.7 per cent were following ploughs and taking prey turned up in the soil, 18.4 per cent were searching for worms on pasture themselves by walking along probing the ground, and 14.3 per cent were searching aerially for worms in pasture. The total number of gulls observed over the six days was 6,601. Although Källander was unable to provide conclusive data quantifying the relative profitability of each strategy, it is possible that changes from one to another reflected temporal changes in the payoffs accruing from each. Andersson (1976) observed great skuas (*Stercorarius skua*) changing between kleptoparasitising gannets (*Sula bassana*) and puffins and predating puffins and kittiwake (*Rissa tridactyla*) chicks. Once again, however, the factors influencing changes of strategy were not clear.

An important point concerning changes of strategy is that they may occur in most members of a population at some time or another, or particular strategies may be associated with particular individuals or particular stages of development. In skuas, terns (*Sterna* spp) and frigate birds (*Fregata* spp), there is evidence for the former (Grant, 1971; Arnason and Grant, 1978; Dunn, 1973; Diamond, 1975), while in hummingbirds (Trochilidae) and jackdaws (*Corvus monedula*) it has been recorded as characteristic of particular individuals (Young, 1971; Birkhead, 1974). Examples of both have been recorded in gull populations (Corkhill, 1973; Hatch, 1975). Kleptoparasitism as a characteristic of particular age groups has been recorded in gulls by Pettit (1952) and Burger and Gochfeld (1981). The distinctions made above have important implications for the evolution of kleptoparasitism and other scrounging behaviour as an alternative strategy within populations. Parker considers this again in Chapter 5. For an excellent review of alternative strategies in general see Krebs and Davies (1981).

Restricting Competition. In some cases, defending a host against other kleptoparasites may be more economical than switching to an alternative feeding strategy.

In many associations, hosts are defended in patches in an analogous way to the defence of other patchily distributed food supplies. We have already seen an intraspecific example in the 'shepherding' of subordinate Harris sparrows by dominants. Interspecifically, Ingolfsson (1969) reported glaucous gulls (*Larus hyperboreus*) stealing bivalves from eider duck (*Somateria mullissima*) and defending rafts of eider against attacks by other glaucous gulls. Ingolfsson regarded the defence as territoriality because bouts of defence were not necessarily followed by a reward but appeared to be geared to longer-term benefits (see Davies (1976) and Caraco (1979) for long-term benefits of territoriality).

Similar host defence occurs in black-headed gulls attacking lapwings and golden plovers. When only one gull is present in a flock, its attacks tend to range over the whole flock (Barnard, in preparation). As more gulls arrive, however, the flock is partitioned into defended gull territories so that each gull attacks only a proportion of the flock (see also Källander, 1977). The territories are mobile and apparently move with varying local host densities within the flock. Eventually an equilibrium number of gulls is reached, the size of which is likely to be determined by a complex interaction of factors relating to flock composition, worm availability, field size and weather (Barnard *et al.*, 1982; Thompson and Barnard, in press).

Where food supplies are unsuitable for territorial defence (e.g. if they are too large to defend), protracted fights may ensue. A good example is the conflict between lions and hyenas when they simultaneously usurp a carcass from hunting dogs. Depending on the relative numbers of lions and hyenas contesting the kill, such conflicts can be long drawn out, injurious and sometimes lethal (van Lawick and van Lawick-Goodall, 1970).

The Outcome of Arms Races

Having discussed the nature of counter-adaptive arms races in some examples of feeding producer/scrounger relationships, it is worth considering briefly how such arms races might end. Dawkins and Krebs (1979) have discussed a number of factors which might influence the outcome of a race in favour of one lineage or another and I shall use their classification as a basis for discussion. It is beyond the scope of this chapter to consider the nature of arms races *per se.* Readers wishing to know more are referred to Dawkins and Krebs' excellent review and to Dawkins (1982).

In Dawkins and Krebs' (1979) terminology, the producer/scrounger relationships discussed here are *asymmetric* intra- and interspecific races in that the coevolving lineages in each case use a different adaptive currency (producers evolve adaptations to reduce the impact of scroungers while scroungers evolve adaptations to counter them and improve their scrounging efficiency). There are at least three major factors which might influence the outcome of races.

Relative Evolutionary Rates

Lineages may differ in the rate at which they evolve (Stanley, 1979). Races between producers and scroungers of different species may be biased if one lineage has an intrinsically higher rate of evolution. This is particularly likely when lineages are phylogenetically widely separated as, for instance, in hummingbird/spider relationships, but may also be true of relationships between more closely related species which have very different ecologies (e.g. cattle egret/tern relationships).

Several factors may introduce an asymmetry in the relative rates of evolution of producer and scrounger lineages. Four important factors are population size, the amount of variation within populations, the tendency for populations to become fragmented, and generation time. In many cases, the asymmetries may ultimately favour producer lineages. The first two factors, population size and available variation, tend to correlate positively. There is likely to be more variation on which selection can act in large populations. This has led to the general assumption that rates of evolution will be higher in large populations. While this is likely to be true, because the probability of variation appropriate to impinging environmental pressures being present increases with the amount of available variation, it does not preclude natural selection from being a major evolutionary force in small populations The generalisation that evolutionary change in large populations is due mainly to natural selection, while that in small populations is due more to random factors like genetic drift, is fallacious. We can usefully consider naturally selected arms races between lineages existing in large and small populations, even though small populations are more liable to random sampling effects and, ultimately, extinction.

As we have seen, scroungers compete with one another for access to producers with the result that producer populations may become partitioned out among scrounger individuals. Such partitioning may be maintained by competitive spacing resulting in (a) small populations of scroungers and (b) scrounger life-history strategies which tend to the K extreme of the r/K continuum. If such scroungers exploit large producer populations which tend towards r-selected life-history strategies and therefore have relatively short generation times and the potential for higher rates of evolution, they may be out-evolved. It is important to point out, however, as do Dawkins and Krebs (1979), that arms races may themselves cause changes in the rate of evolution of competing lineages.

Selection Pressure and the Cost of Adaptation

Adaptations are costly. They use up metabolic resources. It is reasonable to suppose that resources used up in one adaptation are not available or are available only in reduced quantities for other adaptations. The extent to which a lineage can invest in any one adaptation is therefore likely to be compromised by the strength of selection for other adaptations. In producer/scrounger relationships, this may determine the degree to which scroungers can specialise on particular producers. A generalist scrounger involved in counter-adaptive arms races with several phylogenetically distantly related producers is unlikely to show fine-grained adaptation to any one of them because selection is likely to favour a range of competing adaptations. Exploitation of more closely related producers, however, may mean that similar adaptations are selected for in each case. Scroungers may therefore become more specialised with respect to each producer. It is interesting that the species which have specialised most as kleptoparasites are those which exploit a phylogenetically narrow range of hosts. Good examples are the frigatebirds which steal mainly from boobies (Sulidae). Specialisations for kleptoparasitism in frigatebirds include small, unwebbed feet, a vestigial uropygial gland which results in a greatly reduced amount of oil in the feathers, and a wingspan to bodyweight ratio which is greater than that in any other bird. These features greatly facilitate the aerobatic flight on which the birds depend for stealing food, but they also mean the birds can no longer enter the water from which most of their food originates (Nelson, 1967). While the range of producers exploited is likely to influence the degree to which scroungers can specialise, so also is the intensity of selection acting on the producer/scrounger relationships. The intensity of selection will determine not only how far the arms race in a particular case will proceed, but, because it is unlikely to be equal in the two lineages, it may also determine the winner of the race. The lineage with the most to lose or gain will be selected to proceed furthest with the arms race and channel more of its resources into counter-adaptation. Dawkins and Krebs (1979), following Slatkin, term this the 'Life-Dinner Principle' by analogy with a predator chasing prey. Selection is likely to act more intensely on adaptations by the prey to outrun the predator rather than vice versa because the prey has its life to lose (and therefore its total future reproductive potential), whereas the predator misses only a meal if it fails and is still likely to survive and reproduce.

In producer/scrounger races, the Life-Dinner Principle could favour either lineage depending on relative payoffs and the availability of alternative strategies. Scroungers are likely to be at an advantage if their impact on the producer population is not severe because selection for producer counter-adaptation would be weak. They are also likely to be at an advantage if the payoff from scrounging is high and/or the availability of alternative strategies low, since more resources would be channelled into the race. The kleptoparasitic relationship between lapwings and black-headed gulls is a good example where the payoff to the scrounger (gull) is high and the scrounger's impact on the producer (lapwing) is low. Gulls can obtain well in excess of their daily energy requirement solely by stealing earthworms from lapwings (Källander, 1977; Thompson, in preparation) and most of their foraging day is spent simply standing still. Lapwings, on the other hand, lose only relatively unprofitable (very large) worms to gulls, although attacks are apparently sufficiently costly to select for counter strategies (see above).

The Life-Dinner Principle may favour producers if scrounger impact is severe, resulting in a strong directional selection for counter-adaptation, if the payoff to scroungers is low and/or the availability of alternative producers or resource accrual strategies is high. The relationship between the brood parasitic European cuckoos and their small passerine hosts appears to be an example where severe impact on the host has resulted in the latter ultimately winning the race. One response of passerine hosts to cuckoo parasitism is to recognise and remove cuckoo eggs from their nests. This has resulted in selection for detailed host egg mimicry in the cuckoos. A survey of the range of passerine species which are parasitised by cuckoos revealed that these species tended to be poor egg discriminators, whereas those which were not parasitised are often good discriminators (Wyllie, 1981). While it is possible that cuckoos have simply avoided discriminating hosts, it is also possible that the non-parasitised discriminators are the victors of past arms races against cuckoos which have since moved on to naive new hosts.

Kleptoparasites in mixed seabird colonies provide examples of low scrounger payoffs and either generalist scroungers exploiting a wide range of hosts or opportunist individuals/species for which scrounging is one of a range of resource accrual strategies. Gulls and skuas steal food from a wide variety of species within colonies, including auks, other gulls and skuas, terns and procellariiforms. In many cases, individual success rates are low. Hatch (1975), for instance, found that the probability of successful theft in individual laughing gulls stealing from terns varied from 10 per cent to about 37 per cent, depending on gull group size. At certain times of the year (e.g. early summer), the average success rate was only 17 per cent. As well as being generalist kleptoparasites within colonies, gulls and skuas also adopt non-kleptoparasitic feeding strategies. Many gull species, for instance, catch fish for themselves or forage on nearby pasture (Sibly and McCleery, 1983). In some cases, gulls and skuas are also predators of their erstwhile hosts (Andersson, 1976).

Learning Ability

There is a close and often-stated analogy between learning and natural selection as processes leading to adaptation (Pringle, 1951). Natural selection is an adaptive process operating across generations, learning is an adaptive process operating within generations. Because of this difference in time-scale, however, better learners may experience longer periods of grace in an arms race because more slowly adapting lineages take longer to counter their adaptations (Dawkins and Krebs, 1979).

In many interspecific food-scrounging relationships, the evolutionary dice may be loaded unevenly in favour of scroungers because scroungers are often opportunists which tend to have good learning ability. Selection for adaptive flexibility in foraging strategy is likely to mean that opportunists can capitalise quickly on changes in the nature of food availability. One type of change which can rapidly be exploited, therefore, is behavioural or other change in a producer population. Many producers, on the other hand, are dietary specialists (Brockmann and Barnard, 1979) which may have highly stereotyped foraging behaviour (see also Vollrath, Ch. 3 this volume). Stereotyped behaviour patterns occur largely independently of short-term environmental change and are modified only by selection acting across generations. It may be, therefore, that scroungers abound because many producers are attempting to outwit them on the wrong time-scale.

The factors discussed above indicate some of the ways in which producer/scrounger arms races may favour one lineage or the other. However, as Dawkins and Krebs (1979) point out, there need not necessarily be an endpoint to an arms race in terms of a winner and loser. There may be cyclical endpoints which end as oscillations between two or more unstable equilibria, or coevolution may ultimately produce some kind of compromise and hence stability (Pimentel and

Stone, 1968). The latter might occur, for instance, where a scrounger maintains itself with little impact on its host and the host tolerates a small loss if it is protected by inter-scrounger conflict from invasion by a more costly, non-coadapted scrounger. Finally, producer/scrounger arms races may continue because populations of producers only are evolutionarily unstable. While producers may fare better in the absence of scroungers, they are always open to invasion by them. Scroungers are therefore Hobson's choice for producers and are likely to arise repeatedly in producer populations even though arms races may eliminate particular scrounger strategies.

Acknowledgements

I thank Fritz Vollrath, Richard Sibly, Dennis Lendrem and Des Thompson for helpful comments on the manuscript.

References

Andersson, M. (1976). Predation and kleptoparasitism by skuas in a Shetland seabird colony. *Ibis 118*: 208-17

Arnason, E. and Grant, P.R. (1978). The significance of kleptoparasitism during the breeding season in a colony of arctic skuas, *Stercorarius parasiticus. Ibis 120*: 38-54

Balph, D.F. and Balph, M.H. (1979). Behavioural flexibility in pine siskins in mixed species foraging groups. *Condor 81*: 211-12

Barnard, C.J. (1978). Aspects of winter flocking and food-fighting in the house sparrow (*Passer domesticus domesticus* L). Unpubl. D. Phil. thesis, University of Oxford

— (1982). Social mimicry and interspecific exploitation. *Am. Nat. 120*: 411-15

— and Sibly, R.M. (1981). Producers and scroungers: a general model and its application to feeding flocks of house sparrows. *Anim. Behav. 29*: 543-50

— and Stephens, H. (1981). Prey size selection by lapwings in lapwing/gull associations. *Behaviour 77*: 1-22

—, Thompson, D.B.A. and Stephens, H. (1982). Time budgets, feeding efficiency and flock dynamics in mixed species flocks of lapwings, golden plovers and gulls. *Behaviour 80*: 44-69

Birkhead, T.R. (1974). Utilization of guillemot *Uria aalge* colonies by jackdaws *Corvus monedula. Ornis Scand. 5*: 71-81

Brockmann, H.J. and Barnard, C.J. (1979). Kleptoparasitism in birds. *Anim. Behav. 27*: 487-514

Burger, J. and Gochfeld, M. (1981). Age-related differences in piracy behaviour of four species of gulls, *Larus. Behaviour 77*: 242-67

Caraco, T. (1979). Time budgeting and group size: a test of theory. *Ecology 60*: 618-27

Chapin, J.P. (1939). The birds of the Belgian Congo. Part II. *Bull. Amer. Mus. Nat. Hist. 75*: 352

Corkhill, P. (1973). Food and feeding ecology of puffins. *Bird Study 20*: 207-20

Curio, E. (1976). *The Ethology of Predation*. Berlin, Springer-Verlag

Davies, N.B. (1976). Food. Flocking and territorial behaviour of the pied wagtail (*Motacilla alba yarrellii*) in winter. *J. Anim. Ecol. 45*: 235-54

Dawkins, R. (1982). *The Extended Phenotype*. London, Freeman

— and Krebs, J.R. (1979). Arms races between and within species. *Proc. Roy. Soc. Lond. B. 205*: 489-511

Diamond, A.W. (1975). Biology and behaviour of frigatebirds *Fregata* spp. on Aldabra Atoll. *Ibis 117*: 302-23

Dunn, E.K. (1973). Robbing behaviour of roseate terns. *Auk 90*: 641-51

Edmunds, M. (1974). *Defence in Animals*. London, Longman

Feare, C.J., Dunnet, G.M. and Patterson, I.J. (1974). Ecological studies of the rook (*Corvus frugilegus* L.) in northeast Scotland: food intake and feeding behaviour. *J. Appl. Ecol. 11*: 867-96

Fiscus, C.H. and Baines, G.A. (1966). Food and feeding behaviour of Steller and Californian sealions. *J. Mammal. 47*: 195-200

Grant, P.R. (1971). Interactive behaviour of puffins (*Fratercula arctica* L.) and skuas (*Stercorarius parasiticus* L.). *Behaviour 40*: 263-81

Harvey, P.H. and Greenwood, P.J. (1978). Anti-predator defence strategies: some evolutionary problems. In J.R. Krebs and N.B. Davies (eds.), *Behavioural Ecology: an evolutionary approach*: 129-58. Oxford, Blackwell

Hassell, M.P. and May, R.M. (1974). Aggregation of predators and insect parasites and its effects on stability. *J. Anim. Ecol. 43*: 567-94

Hatch, J.J. (1975). Piracy by laughing gulls *Larus atricilla*: an example of the selfish group. *Ibis 117*: 357-65

Heatwole, H. (1965). Some aspects of the association of cattle egrets with cattle. *Anim. Behav. 13*: 79-83

Ingolffson, A. (1969). Behavior of gulls robbing eiders. *Bird Study 16*: 45-52

Källander, H. (1977). Piracy by black-headed gulls on lapwings. *Bird Study 24*: 186-94

— (1979). Skrattmasen (*Larus ridibundus*) som kleptoparasit pa tofsvipa (*Vanellus vanellus*). *Fauna och Flora 74*: 200-7

King, J.R. (1974). Seasonal allocation of time and energy resources in birds. In R.A. Paynter, Jr. (ed.), *Avian Energetics. Publ. of the Nuttall Ornith. Club 15*: 4-70

Krebs, J.R. (1973). Social learning and the significance of mixed-species flocks in chickadees (*Parus* spp). *Can. J. Zool. 51*: 1275-88

— and Davies, N.B. (1981). *An Introduction to Behavioural Ecology*. Oxford, Blackwell

—, MacRoberts, M.H. and Cullen, J.M. (1972). Flocking and feeding in the great tit (*Parus major*) – an experimental study. *Ibis 114*: 507-30

Kruuk, H. (1972). *The Spotted Hyena: a study of predation and social behavior.* Chicago, University of Chicago Press

Kushlan, J.A. (1978). Nonrigorous foraging by robbing egrets. *Ecology 59*: 649-53

Lawick, H. van and Lawick-Goodall, J. van (1970). *Innocent Killers*. London, Collins

Mech, L.D. (1970). *The Wolf: the ecology and behavior of an endangered species*. Garden City, New York, Natural History Press

Meinertzhagen, R. (1959). *Pirates and Predators*. London, Oliver and Boyd

Murton, R.K. (1971). Why do some birds feed in flocks? *Ibis 113*: 534-6

Nelson, J.B. (1967). Etho-ecological adaptations in the great frigatebird. *Nature, Lond. 214*: 318

Nettleship, D.N. (1972). Breeding success of the common puffin (*Fratercula arctica* L.) on different habitats at Great Island, Newfoundland. *Ecol. Monogr. 42*: 239-68

Pettit, R.G. (1952). Comparative aggressiveness of the first-year and adult black-headed gull. *Brit. Birds 45*: 333-4

Pimentel, D. and Stone, F.A. (1968). Evolution and population ecology of parasite-host systems. *Canad. Entomol. 100*: 655-62

Pitcher, T.J., Magurran, A.E. and Winfield, I.J. (1982). Fish in larger shoals find food faster. *Behav. Ecol. Sociobiol. 10*: 149-52

Pringle, J.W.S. (1951). On the parallel between learning and evolution. *Behaviour 3*: 174-215

Rand, A.L. (1954). Social feeding behaviour of birds. *Fieldiana: Zool. 36*: 1-71

Rohwer, S. (1977). Status signalling in Harris sparrows: some experiments in deception. *Behaviour 61*: 107-29

— and Rohwer, F.C. (1978). Status signalling in sparrows: experimental deceptions achieved. *Anim. Behav. 26*: 1012-22

— and Ewald, P.W. (1981). The cost of dominance and advantage of subordination in a badge signalling system. *Evolution 35*: 441-54

Rothschild, M. and Clay, T. (1952). *Fleas, Flukes and Cuckoos*. London, Collins

Rubenstein, D.I., Barnett, R.J., Ridgely, R.S. and Klopfer, P.H. (1977). Adaptive advantages of mixed-species feeding flocks among seed-eating finches in Costa Rica. *Ibis 119*: 10-29

Schaller, G.B. (1972). *The Serengeti Lion: a study of predator-prey relations*. Chicago, University of Chicago Press

Sibly, R.M. and McCleery, R.H. (1983). The distribution between finding sites of herring gulls breeding at Walney Island UK. *J. Anim. Ecol. 52*: 51-68

Smith, J.N.M. (1974a). The food searching behaviour of two European thrushes I. Description and analysis of search paths. *Behaviour 48*: 276-302

— (1974b). The food searching behaviour of two European thrushes II. The adaptiveness of the search patterns. *Behaviour 49*: 1-61

— and Sweatman, H.P. (1974). Food searching behaviour of titmice in patchy environments. *Ecology 55*: 1216-32

Stanley, S.M. (1979). *Macroevolution: pattern and process*. San Francisco, Freeman

Thompson, D.B.A. and Barnard, C.J. (in press). Anti-predator responses in mixed species flocks of lapwings, golden plovers and black-headed gulls. *Anim. Behav.*

— and Barnard, C.J. (in press). Prey selection in plovers: optimal foraging in mixed-species groups. *Anim. Behav.*

Tinbergen, L. (1960). The natural control of insects in pinewoods I. Factors influencing the intensity of predation by songbirds. *Archs. neerl. Zool. 13*: 265-343

Vollrath, F. (1979). Behaviour of the kleptoparasitic spider *Argyrodes elevatus* (Araneae, Theridiidae). *Anim. Behav. 27*: 515-21

Willis, E.O. (1968). Studies of the behavior of lunulated and Salvin's antbirds. *Condor 70*: 128-48

Wyllie, I. (1981). *The Cuckoo*. London, B.T. Batsford

Young, A.M. (1971). Foraging for insects by a tropical hummingbird. *Condor 73*: 36-45

5 THE PRODUCER/SCROUNGER MODEL AND ITS RELEVANCE TO SEXUALITY

G.A. Parker

Some General Properties of the Producer/Scrounger Model

The central notion of the producer/scrounger (P/S) model (Barnard and Sibly, 1981) is akin to parasitism. Stated in selfish gene terminology (Dawkins, 1976, 1982), selection commonly favours replicators that build up resources such as energy reserves, tissues, bodies (or phenotypic extensions of bodies such as nests, food stores, etc.) which act as aids to replication. Such a strategy is a producer (*P*) strategy. A scrounger (*S*) strategy can be favoured if parasitism of the producer's resources is more profitable (in terms of replication rate) than gathering the equivalent resources directly.

An even more fundamental property of P/S systems relates to the fact that fitnesses will be frequency-dependent (Brockmann and Barnard, 1979; Dawkins, 1980). The *S* strategy does best when it is rare, and worst when it is common relative to the *P* strategy (Figure 5.1). As scroungers increase in frequency, competition for the limited resources held by the producers gets even more intense because there are relatively fewer *P* to exploit. This characteristic of P/S systems was discussed by Barnard and Sibly (1981). Provided that the mean replication rate of *S* ($\overline{W}_S$) exceeds that of *P* ($\overline{W}_P$) when *S* is rare, but is lower than that of *P* when *S* is common, the evolutionarily stable state (ESS; Maynard Smith, 1974; Maynard Smith and Price, 1973) will consist of a balanced mix of *P* and *S*. The equilibrium frequency p_* of *S* occurs when replication rates are equal (Barnard and Sibly, 1981); we can therefore write the equilibrium conditions:

$$\overline{W}_S(1) > \overline{W}_P(1); \overline{W}_S(0) < \overline{W}_P(0)$$
$$\overline{W}_S(p_*) = \overline{W}_P(p_*) \tag{1}$$

There are two important cases under which (1) may not apply if *P* and *S* are conspecific. These are:

1. Phenotypes are not determined parentally, but the 'decision' to be *P* or *S* depends on phenotype.

Figure 5.1: Fitness W (reproductive success) of producers and scroungers in relation to the frequency p of scroungers in the population

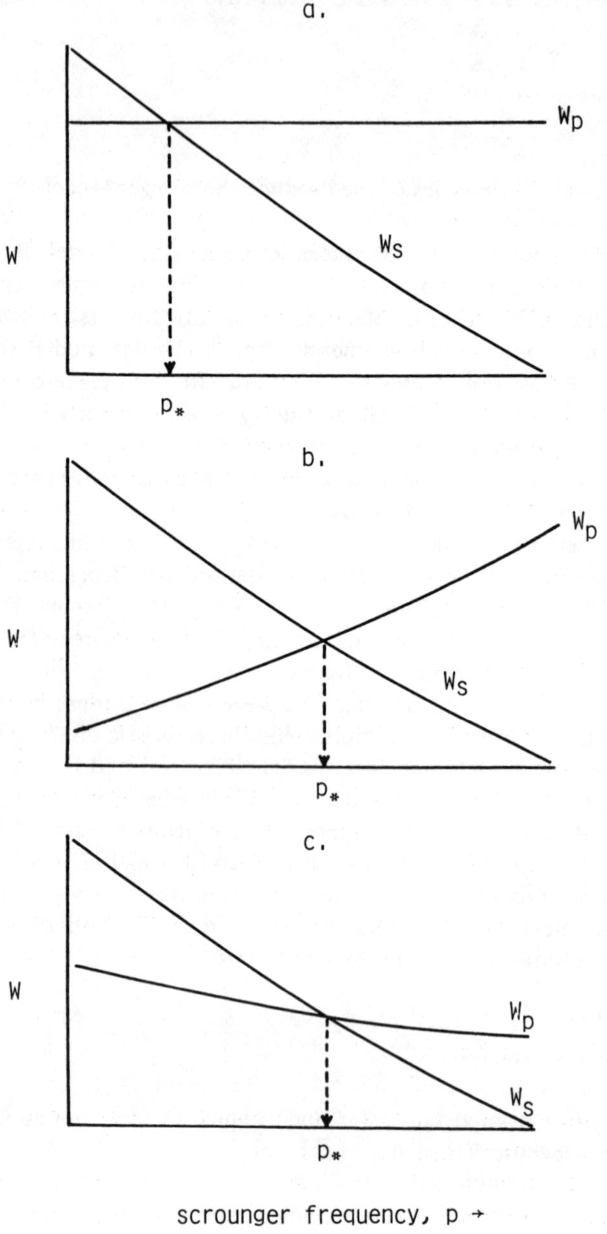

2. P and S are different phenotypes under parental control, and the costs to a parent of producing a P is not equal to that of an S.

These two cases will be discussed later.

Although $\overline{W}_S(p)$ is likely to be monotonic decreasing, it is less easy to make generalisations about $\overline{W}_P(p)$. In some cases, such as the sex ratio (discussed later), the replication rate of P may be independent of the frequency of S, at least over a wide range of P (Figure 5.1a). A second possibility is that scroungers actually *aid* producers, but are again affected by competitive effects between themselves. Here $\overline{W}_P(p)$ would be increasing (Figure 5.1b). However, it is parasitic cases of the sort shown in Figure 5.1c that best fit the P/S model. Here scroungers depress the replication rate of producers so that $W_P(p)$ is decreasing, though $W_S(p)$ must decrease at a higher rate than $W_P(p)$ if there is to be an equilibrium mix of P and S (Figure 5.1c).

The mutualistic case (Figure 5.1b) deserves further mention because it appears to have attracted little previous discussion. It is easy to imagine that the host (P) to some symbiont (such as a cleaner fish) might benefit increasingly with increasing densities of symbionts. However, the symbionts (S) will experience declining benefits as they become commoner because they must share the food available on a limited number of hosts (the host population size is likely to be only weakly dependent on the number of symbionts). Another instance of increasing $\overline{W}_P(p)$ that is again mathematically analogous to parasitic P/S is where there are 2 (or n) niches in which fitness gains are inversely proportional to population density. The familiar 'ideal free distribution' (Fretwell, 1972) occurs when $\overline{W}$ is equal for all individuals (for the 2 niche case, p_* prescribes the equilibrium proportion in one of the 2 niches).

The coexistence in a species of P and S at the equilibrium level p_* can be shown to be an ESS if it obeys condition (1), since deviations from p_* will always be pushed back towards p_*. The P/S model is an n-person game of the 'scramble' (Parker, 1982a, 1984) or 'playing the field' (Maynard Smith, 1982) type. If all competitors are functionally identical, then p_* would represent a mixed strategy in which individuals could *either* be consistently P or S (determined by genotype; genetic polymorphism), *or* could play P and S with probabilities $(1-p_*)$ or p_* (genetically monomorphic population). A third possibility is that the population consists of a mix of various 'genetic players' and 'probabilistic players' such that the global frequency of S is p_*.

One case where there is some real evidence for the stable coexistence of P and S at p_*, and where the P and S are apparently identical phenotypically, is the example of nest-building strategies in the digger wasp, *Sphex ichneumoneus* (Brockmann *et al.*, 1979; Brockmann and Dawkins, 1979). A female provisions underground chambers with several katydids that act as food for her single larva. Instead of digging a new burrow, a female may enter a burrow already dug by another female. If the two females meet, a fight occurs and the loser withdraws. It appears that digging (P) and entering (S) are a probabilistic, mixed ESS; the average payoff to each strategy is roughly equal (at least in one of the study populations) and wasps alternate between P and S. It is surprising that larger wasps do not tend to play S, and vice versa, for reasons we now discuss.

Phenotype-limited P/S Systems

Most P/S strategies are likely to be phenotype-limited in the sense of Parker (1982). A conditional strategy (Dawkins, 1980) is a strategy that prescribes playing (say) strategy X in condition x, strategy Y in condition y, etc. A phenotype-limited strategy is simply a form of conditional strategy in which the conditions, x, y, etc., are phenotypic conditions such as size, relative size, age, physiological state, etc.

Suppose that certain phenotypes (e.g. large individuals) have an intrinsically greater competitive ability in (say) P than in S. Selection will quickly favour a 'commonsense' form of phenotype limited strategy: 'play P when large, play S when small' (Parker, 1982a).

Continuous Phenotypes. First, suppose that the size (= phenotype) distribution of the population is continuous, and that competitive ability changes monotonically with size. At the ESS, there will be a threshold size (= phenotype) above which P will be played, and below which S will be played. An individual at the threshold will obtain equal fitness by playing either P or S (Parker, 1982).

This principle is explained in Figure 5.2 (see also Parker, 1982, 1983). The inherent competitive ability of a given phenotype is simply its fitness against other phenotypes playing the same strategy. Suppose that K_P, the inherent competitive ability of an individual, increases in the P strategy with age (and hence size) t. Competitive ability K_S in S may also be dependent on t – it may show a similar trend (Figure 5.2a (i)) or opposite trend (Figure 5.2a(ii)). Alternatively, K_S may be independent of t (Figure 5.2a(iii)) so that there are fitness differences *only* in the P strategy. Now, the ESS switch point t_* will be such that an individual at t_* experiences equal fitness in either strategy (Figure 5.2b). Thus the proportions of P and S in the population are adjusted so that an individual at t_* has equal expectation in either strategy. This adjustment comes about because of the frequency dependence of the P/S system. For instance, suppose that in a population the switch point from S to P occurs at $t_a < t_*$ (Figure 5.2c). Individuals *older* than t_a clearly do better to stay in S rather than switch to P; the switch age is therefore under selection to increase. In contrast, if the switch point is at $t_b > t_*$, then it is better to switch to P at an earlier age, thus reducing the switch age towards t_*.

Only when the switch age is at t_* is the system at ESS (Figure 5.2b), because no mutant strategy can profit by playing a different switch strategy with $t \neq t_*$. Whether a mutant can do better than the rest of the population is shown by projecting the solid line or dotted line beyond t_*; in all cases (i), (ii) and (iii) illustrated in Figure 5.2c, this deviation can be seen to be unprofitable, giving lower fitness than would be obtained by switching at t_*. However, it is clear that case (iv) in Figure 5.2c could not be an ESS; although a phenotype t_* does equally well in P and S, phenotypes with $t < t_*$ would do better to play P and those with $t > t_*$ would do better to play S. The system is unstable and will overturn in favour of one like case (i). This demonstrates that phenotype limitation must obey the rule that switches operate in a 'commensense' fashion with respect to competitive ability: if increased age (or size) is more favourable in P than S, the ESS is to play S when young and P when old. The commonsense rule is obvious for cases where competitive abilities in P and S show opposite trends in relation to phenotype (case (ii), Figure 5.2). If competitive abilities show the *same* trend in relation to phenotype (case (i), Figure 5.2) the commonsense rule is more subtle: the *dominant* competitive ability must be respected. In other words if competitive ability in P is more strongly affected by size than competitive ability in S, individuals will play S when small and P when large. As a formal definition of the phenotype-limited producer/scrounger model with continuous phenotypes and monotonic competitive abilities $K_S(t)$ and $K_P(t)$, we require, for a switch from S to P at t_* to be an ESS:

$$\left.\begin{array}{ll}\text{For all } t < t_* & W_{P,t}(p_*) < W_{S,t}(p_*) \\ \text{For all } t > t_* & W_{S,t}(p_*) < W_{P,t}(p_*)\end{array}\right\} \qquad (2a)$$

$$\text{For } t = t_* \qquad W_{P,t_*}(p_*) = W_{S,t_*}(p_*) \qquad (2b)$$

Figure 5.2: (a) Inherent competitive abilities, *K*, of phenotypes of age *t* against other individuals playing the same strategy (*P* or *S*). We assume that competitive ability increases with age (on size) *t* in one strategy (here *P*). Then competitive abilities in the other strategy (*S*) could be of various types: (i), (ii), (iii). (b) ESS for the phenotype-limited producer/scrounger model. At the ESS, an individual at t_* has equal fitness in either strategy (*P* or *S*) and no mutant can profit by having a different switch point. Case (iv) could *not* be an ESS because it would be profitable to play an opposite strategy (*P* at $t < t_*$, *S* at $t > t_*$). (c) Fitness W(t) of the range of phenotypes. If in the population the threshold for switching from *S* to *P* is at $t_b > t_a$, mutants that switch to *S* earlier in life do better

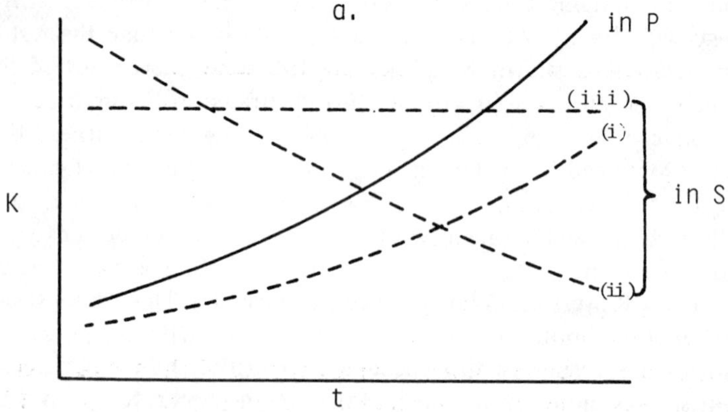

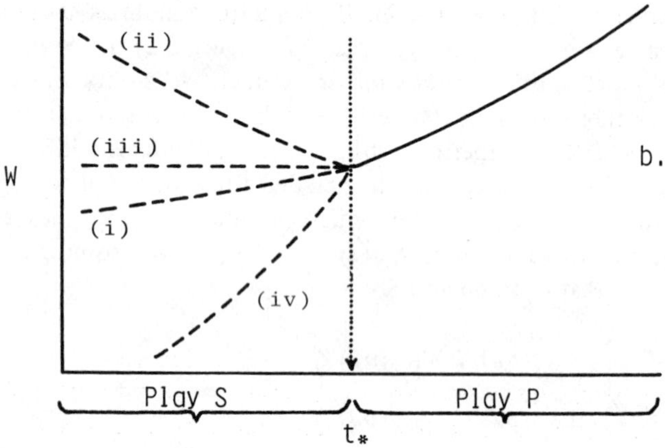

Figure 5.2(c)

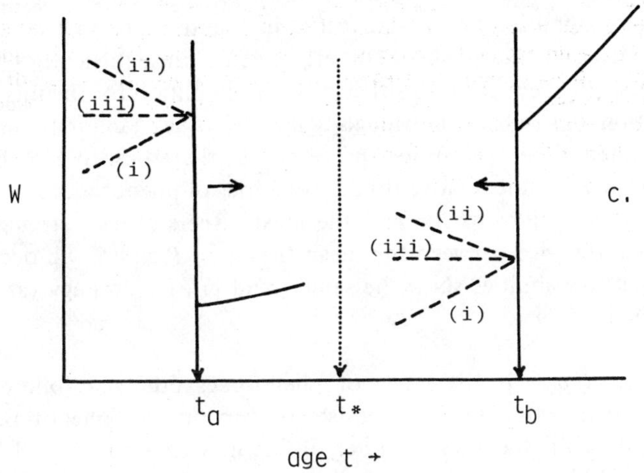

Rule (2a) states that a mutant with a deviant switch ($t \neq t_*$) cannot profit in a population at the equilibrium proportion of S, p_*; rule (2b) states that a phenotype at t_* does equally well in either P or S strategy in a population at the ESS, p_*. Since all individuals with $t > t_*$ play P, and those with $t < t_*$ play S, then where $\mu(t)$ is the frequency distribution of phenotypes:

$$\left.\begin{aligned} p_* &= \int_{t_{min}}^{t_*} \mu(t)\, dt \\ \text{and} \quad &\int_{t_{min}}^{t_{max}} \mu(t)\, dt = 1 \end{aligned}\right\} \tag{2c}$$

It is usually possible, using the sort of argument outlined here and in Parker (1982), to demonstrate that a population playing a mixed ESS (i.e. chooses P or S randomly with respect to phenotype) will evolve to a phenotype-limited system obeying equations (2a), (2b).

Figure 5.2b shows that the mean fitness of individuals playing S may well approximate to the mean fitness of those playing P, if competitive abilities show opposite trends (curve (ii) in Figure 5.2b) in the two strategies (see also Parker, 1984). This effect could cause difficulties for field-workers, since equal mean fitnesses is the condition for the mixed ESS, equation (1), where individuals select strategies randomly with respect to phenotype. The proper procedure is, therefore, to examine as closely as possible the relative fitness of a series of phenotypes *within* each strategy, as well as to compare the mean fitness of each strategy. The biggest difference between the mean fitness of P and S will occur when competitive abilities show the same trend in each strategy (as in curve (i), Figure 5.2).

Discrete Phenotypes. If the range of phenotypes does not follow a continuous distribution (e.g. it consists of two discrete phenotypes, 'large' and 'small'), the ESS is rather different. One strategy will be played uniquely by one phenotype, whilst the other phenotype will be forced (because of the frequency-dependence) to play a mix of P and S, such that its payoffs are equal in either strategy (Parker, 1982a).

Call the two phenotypes A and B. Depending on inherent competitive abilities, there will be a fitness difference between A and B phenotypes ($\bar{W}_A(p_*) \neq \bar{W}_B(p_*)$) but individuals that play the mixed strategy must by definition have equal fitness in either strategy. For the case where A plays the mixed strategy, and B is restricted to strategy P in which it has higher competitive ability, we need:

$$\bar{W}_{P,B}(p_*) > \bar{W}_{S,B}(p_*) \tag{3a}$$

$$\text{and} \quad \bar{W}_{P,A}(p_*) = \bar{W}_{S,A}(p_*) \tag{3b}$$

Rule (3a) asserts that a mutant B phenotype cannot profit by switching to S, and rule (3b) indicates that the numbers of A in the two strategies is balanced so that equal fitness expectations are achieved.

A detailed analysis of this sort of model is given by Parker (1982a); an outline of conclusions is given here for the sake of completeness. The ESS depends on the frequencies of A and B in relation to the 'niche space' for each strategy, P or S. Suppose that B phenotypes have higher competitive ability in P than A phenotypes. If there are rather few B phenotypes, all B will play P, but it will also pay some of the A to play P because of the frequency dependence. Thus A 'colonises' P until the fitness of A equilibrates in P and in S. This case is outlined in equations (3a) and (3b). The fitness difference between A and B is equivalent to the difference in competitive abilities in P. If there are rather few A phenotypes, a reverse solution is obtained. It is now B phenotypes that are forced out into the S strategy; B now plays a mixed strategy and A plays pure S. The fitness difference between phenotypes is now equivalent to the difference in competitive abilities in S. Hence the relative fitness of A and B depends upon the competitive abilities in the strategy in which they both occur together. The phenotype that does best is likely to be the one that is uncommon.

This model can be modified to cover the case where there are two non-overlapping distributions of phenotypes. Ultimately, it is much more likely that phenotypes are drawn from two continuous distributions than that all A or B are functionally identical. ESSs will now consist of various mixtures of equations (2) and (3). By way of demonstration, suppose that A phenotypes are very common and hence play *both* strategies, and also that Bs do better than As in P. For all B to remain in strategy P, we require:

$$\text{For all } B = B', \quad W_{P,B'}(p*) > W_{S,B'}(p*) \tag{4a}$$

Now consider the strategy of A phenotypes. Suppose that larger A individuals do better in strategy P than smaller or younger A individuals. Then by analogy with (2a) and (2b), A phenotypes should switch from S to P at age t_* such that:

$$\left.\begin{array}{ll} \text{For all } t < t_*, & W_{P,A,t}(p*) < W_{S,A,t}(p*) \\ \text{For all } t > t_* & W_{S,A,t}(p*) < W_{P,A,t}(p*) \end{array}\right\} \tag{4b}$$

$$\text{For } t = t_* \quad W_{P,A,t_*}(p*) = W_{S,A,t_*}(p*) \tag{4c}$$

Rule (4a) states that no B phenotype, B' can profit by shifting to S, rule (4b) states that no mutant A with $t \neq t_*$ can profit by its deviant switch point, and (4c) states that at t_*, an A individual should do equally well in P or in S. As before, the phenotype limitation must obey the 'commensense rule' concerning competitive abilities.

For cases where the distributions of A and B phenotypes overlap in competitive ability, the appropriate procedure is to rank all phenotypes in terms of competitive abilities, and then to follow the procedure in (2a) and (2b). This can lead theoretically to fragments of both A and B distributions included in each strategy P or S.

P/S systems involving two species are probably most likely to fit within the model suggested by equations (4a) to (4c); intraspecific

P/S systems are probably most likely to obey (2a) and (2b).

A Conspecific P/S Model with Parental Regulation of Phenotypes

The previous models examined P/S systems in which phenotype frequencies were fixed. This is a perfectly reasonable assumption, it is equivalent to assuming that the P/S game exerts a relatively weak influence on phenotype distributions, or that most phenotypic effects are due to chance environmental influences (e.g. age and hence size). In some cases, however, the frequencies of phenotypes in a population can be regulated by how a parent allocates its resources amongst its progeny. How does this influence P/S ESSs?

Consider the case where one phenotype is competitively best suited for the S strategy, and another is competitively best suited for P. Let the cost of a P individual to a parent (in terms of units of expenditure) by x, and the cost of an S individual = y. The proportion, r, of reproductive effort that is spent on producing S can be varied by parents; we seek an ESS level r_*. The following model is borrowed directly from sex ratio analysis (see Charnov, 1982), and makes a specific assumption about $\overline{W}_S(p_*)$.

Let the fitness (expected reproductive success) of a P individual be $\overline{W}_P(p^*)$, where $p_* = r_*y/[1-r_*)x + r_*y]$, the proportion of S in the ESS population. Let $E(S,P)$ be the expected payoffs of an S individual that successfully parasitises a P. Thus we assume that

$$\overline{W}_S(p_*) = \underset{\text{gain}}{E(S,P)}.\underbrace{\frac{(1-p_*)}{p_*}}_{\text{probability of gain}} = E(S,P).\frac{(1-r_*)x}{r_*y}$$

i.e. the S fitness is the product of the gain from a P and the ratio P:S, which indicates the probability of gain. (There are many possible formulations for $\overline{W}_S(p)$; this is the simplest. All are likely to involve the P:S ratio.) Then for r_* to be an ESS against any mutant strategy with $r \neq r_*$, we need

$$\overline{W}_P(p_*).(1-r)x + \overline{W}_S(p_*).ry < \overline{W}_P(p_*).(1-r_*)x + \overline{W}_S(p_*).r_*y$$

for all $r \neq r_*$, and the inequality must become equality at $r = r_*$. Thus

$$\frac{\partial}{\partial r}\left[\overline{W}_P(p_*).(1-r)x + \overline{W}_S(p_*).ry\right]_{r=r_*} = 0$$

Substituting $\bar{W}_S(p_*) = E(S,P).(1-r_*)x/r_*y$ we obtain

$$\frac{1-r_*}{r_*} = \frac{\bar{W}_P(p_*)}{E(S,P)} \qquad (5)$$

Thus the ESS, r_*, is independent of the costs x and y, and dependent only on the ratio of gains of P and S. At the ESS, the ratio of expenditures on P to S progeny must equal the ratio, fitness of a P/gain of an S from a P.

Suppose that $\bar{W}_p$ is independent of p, the scrounger frequency (Figure 1a). Then the RHS in (5) is constant = α. Thus if P and S have equal costs to parents ($x = y$), the ESS value $p_* = r_* = (\alpha+1)^{-1}$. If $x \neq y$, then

$$p_* = \left(\frac{\alpha x}{y} + 1\right)^{-1} \qquad (6)$$

and so if each P costs more than each S ($x > y$) then following intuition, the proportion of S will be increased ($x/y < 1$), and vice versa if S are more expensive ($x/y > 1$).

A possible example of unequal parental costs for the two phenotypes occurs in *Centris pallida* (Alcock *et al.*, 1977), in which males show two mating strategies. Large males tend to be 'patrollers' that search for females around pupation sites, where they mate with newly-emerged females. Fights usually occur between males and the bigger male tends to win and mate with the female. Smaller males tend to be 'hoverers' that search around trees and shrubs for females that had been missed by the patrollers. The system is likely to be frequency-dependent. Alcock *et al.* point out that the size of a male is maternally determined, and depends on the size of the brood cell and the amount of provisioning supplied to the larva. Thus large males are likely to be more costly than small males; the population would be in a stable state if r_* is devoted to one male size and $(1-r_*)$ to the other (see also Maynard Smith, 1982).

Males and Females as Producers and Scroungers

Intersexual Parasitism

Females are producers in the sense that they typically invest heavily in provisioning of their gametes, and may also expend extensively on the care of the zygote. Males parasitise this investment because their contribution to zygote survival is typically zero, the sperm contribution relating simply to the haploid set of chromosomes. It could be argued that males are not parasitic, they simply expend their reproductive effort in different ways from females (on inter-male competition rather than parental investment). Certainly all individuals of both sexes must be expected to expend to a maximum level in fitness-related activity, in the sense that it must always be favourable to do something, rather than nothing, that facilitates the spread of given replicators. The parasite analogy, however, relies on the fact that males do not (typically) invest in the welfare of zygotes, though females could benefit if they did. Males benefit equally to females from the expenditure made by females in the progeny, whilst making no investment themselves. This form of 'scrounging' leaves males free to expend against their competitors, in inter-male competition. The parasite analogy becomes even stronger when we consider the origin and maintenance of the two sexes (Parker *et al.*, 1972; also see later).

Intrasexual Parasitism

A second form of sexual parasitism occurs within a sex. I shall not discuss this case since it has already been treated at some length by Rubenstein (1980), Dunbar (1982) and Arak (Ch. 6 this volume). Also, the theory of intrasexual parasitism will follow the rules developed in the first section without presenting any of the special difficulties that I now consider for intersexual parasitism.

The Sex Ratio

So much has been written on the evolution of the sex ratio (see Charnov, 1982) since the original insights of Fisher (1930) and Shaw and Mohler (1953), that I shall deal with the topic only briefly, and for the case of autosomal sex ratio genes under diploidy.

Firstly, we can derive the sex ratio (females/males) very easily from equation (5). Since each offspring has a mother and a father with equal autosomal contributions, it is clear that $\overline{W}_P(p_*)$, the female fitness, is equal to $E(S,P)$, the expected gain to each male that is successful in mating with a female. Even for the case of extreme sperm competition (all the ejaculates of the male population compete for fusions with all the ova, as in the extreme external fertilisation), the ratio of $\overline{W}_P(p_*)/$

$E(S,P)$ must equal unity, since each zygote is the product of one sperm and one ovum. Thus from (5) we obtain Fisher's (1930) principle of equal expenditure, that

$$\underset{\text{expenditure on females}}{(1 - r_*)} = \underset{\text{expenditure on males}}{r_*}$$

$$\therefore r_* = \tfrac{1}{2}$$

The sex ratio is now obtained from:

$$\frac{\text{females}}{\text{males}} = \frac{(1-p_*)}{p_*} = \frac{(1-r_*)x}{r_* y} = \frac{x}{y}$$

i.e. the sex ratio is equal to the cost ratio (see Charnov, 1982).

It is important to stress that the ESS allocation, r_* is strictly a population equilibrium, not a mixed strategy in the sense that each individual parent must play r_*. Thus the population state can be constituted by any mix of individual strategies for expenditure, provided that the mean expenditure is r_*. However, there are good reasons why r_* is likely to become an individual's strategy as well as a stable population state (see Charnov, 1982). In a finite population, the population sex ratio will always deviate slightly from the equilibrium, p_*; in conditions of changing random deviations, p_* is uniquely the best ratio for an individual (Verner, 1965; Taylor and Sauer, 1980).

The preceding discussion relates the case where sexes are separate from birth. What about hermaphrodite animals?

There is now excellent evidence (Charnov, 1982) that sequential hermaphrodites (where, say, individuals are first male, then female) obey Ghiselin's (1974) 'size advantage' hypothesis. This hypothesis states that when reproductive success relates to age or size, and differently so for each sex, individuals are expected to change sex at a critical age or size, so that gains over a lifetime are maximised. The critical age (t_*) for a switch from one phenotypic strategy to the other should follow exactly the principle outlined by equations (2a) and (2c) and Figure 5.2. An individual should change sex at t_*, the age or size when it would do equally well in either sex. The sex ratio (females/males) is

$$\frac{(1-p_*)}{(p_*)} = \frac{1 - \int_{t_{min}}^{t_*} \mu(t)\,dt}{\int_{t_{min}}^{t_*} \mu(t)\,dt}$$

from equation (2c), where $\mu(t)$ is the frequency distribution of individuals of size (or age) t. It is clear from the arguments presented in section 1(ii), that the sex ratio will depend both on a) the distribution of size $\mu(t)$, and b) on the way that competitive ability alters with size in the two phenotypes (males and females). From Figure 5.2 it can be seen that there is no reason to expect the mean fitness of the two phenotypes to be equal at the ESS. Where size affects reproductive success similarly (but unequally) in the two sexes, mean reproductive success may be quite different, because of a distorted sex ratio.

Simultaneous hermaphrodites are both male and female within a single individual; the sex ratio is by definition unity. However, it seems unlikely that in any given circumstance it can be favourable to expend equally on maleness and femaleness; it is more likely to be favourable to expend on one or other strategy. Intuition suggests that a sequential hermaphrodite population might be composed of individuals ranged across circumstances, such that it pays to spend predominantly on maleness over one part of the range, and femaleness over the other. This 'functional' sex ratio might indeed then be ordered by similar principles to the sequential hermaphrodite case.

The Origin and Stability of Male/Female Systems

The rest of this chapter is devoted to two questions. The first asks how the male/female P/S system arose initially, and the second asks how it has remained stable throughout evolutionary time. It seems almost certain that the male/female system arose simultaneously with multicellularity, and may have several independent origins. In particular in animals, there have been many subsequent developments in the mode of reproduction that might be expected to affect the stability of the male/female system. Perhaps the most notable of such developments is the evolution of internal fertilisation.

The Origin of Anisogamy

A sex is best defined in terms of the type of gamete it produces; males produce microgametes and females produce megagametes. Sexuality in organisms that produce similar-sized gametes (*aniisogamous* organisms) concerns *mating types*; males and females are the mating types under anisogamy. The disruptive selection theory (Parker *et al.*, 1972) for the evolution of anisogamy proposes that parents that produce intermediate-sized gametes are lost due to selection in favour of microgamete producers (males) and megagamete producers (females). The disruptive effect arises because of an advantage on the one hand of producing gametes with high degree of provisioning for future survival and development, and on the other, of producing vast numbers of gametes and so gaining highest numbers of fusions. The theory has been the subject of several recent analyses (Bell, 1978, 1982; Charlesworth, 1978; Hoekstra, 1980; Maynard Smith, 1978, 1982; Parker, 1978).

Imagine an ancestral population in which the following assumptions hold:

(a) A large population of adults release gametes into an external medium such as sea water, so that gametes fuse randomly (independent of size).
(b) There is a fixed energy budget per parent, so that if an adult produces gametes of size m, the relative number of gametes released is proportional to m^{-1}.
(c) The viability, or success during development of a zygote increases with its size.

The first assumption (random fusion) can be relaxed in favour of the assumption that fusion is between different mating strains, without affecting the conclusions (Maynard Smith, 1982). From assumption (b) it is clear that there is an infinite series of possibilities for division of resources into gametes. We seek an ESS for gamete size strategy – and hence for the number (or rate) of gametes produced.

There are in fact two ESSs. The ESS is for stable anisogamy when the viability of a zygote is relatively strongly affected by its size, i.e. where cytoplasmic material is important in helping a zygote in its development to adulthood. This is especially likely for multicellular organisms, where a single cell must build up into a complex adult form. Alternatively, where cytoplasmic material is relatively unimportant to the development of the zygote (as might be expected for many unicellular organisms), the ESS is isogamy, with similar-sized, small gametes. Knowlton (1974) was indeed able to show a trend towards anisogamy in the Volvocidae with increasing degrees of coloniality (see also Bell, 1978).

Why is anisogamy the ESS for multicellular forms? An intuitive feel for the reason behind the anisogamy ESS can be gained as follows (see Figure 5.3). Imagine just three strategies for gamete size: large, medium and small. The fitness of a strategy is equivalent to the number of gametes produced (which is proportional to the number of fusions

Figure 5.3: Evolution of anisogamy. If there are three strategies for gamete production: large, medium and small, virtually all fusions (arrows) involve small gametes because of their numerical predominance. If surviving zygotes require substantial provisioning, only those fusions involving a large gamete survive. Small and large gamete-producing strategies therefore persist, whilst medium producing is selected out

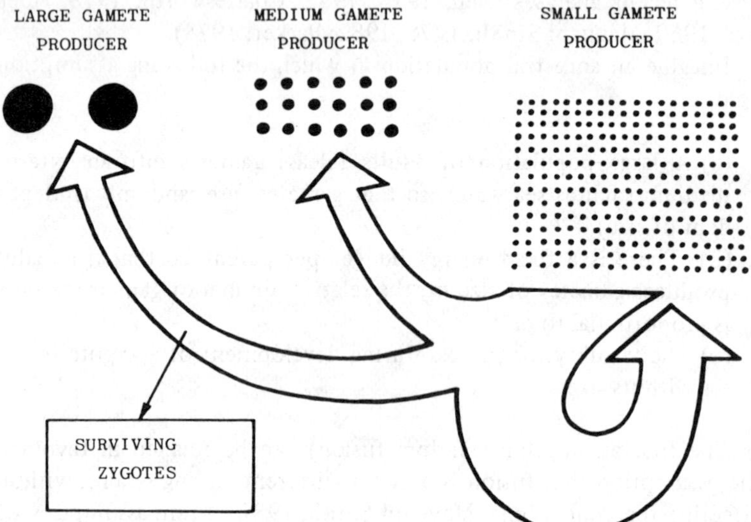

obtained) weighted by their viabilities as zygotes. Suppose that only fusions in which at least one gamete is 'large' have significant survival prospects as zygotes. Call these 'large zygotes' and the other fusion products 'small zygotes'. A population must consist of at least *some* large-gamete producers; large-gamete producers will always spread when rare (in a large population, they spread at a rate proportional to the ratio of viabilities of large and small zygotes). However, as they spread, small-gamete producers out-compete medium-gamete producers because they obtain more fusions with the valuable large gametes. As strategies for intermediate gamete sizes are lost, large-gamete producers will increase in frequency to the point where their fitness equals that of the small-gamete producers. If large-large fusions are very rare (high disparity in gamete sizes) and small-small fusions unviable, this will occur when the two strategies for gamete production (maleness and femaleness) occur with equal frequency, following the arguments pertaining to the sex ratio (see above). Further, selection will quickly favour the evolution of disassortative fusion (Parker,

1978) which will confirm the 1:1 sex ratio.

Sperm Competition and the Stability of Anisogamy

Thus males persisted by acting as scroungers on the investment-laden gametes from females. In the same way that parasites typically maximise infection rates by generative vast numbers of minute propagules, so males generated vast numbers of sperm in order to maximise fusions. It is, however, probably competition between males for fusions with the limited number of ova that *maintains* anisogamy as an ESS (Parker, 1982b). In the anisogamy model of the last section, the winning male strategy is that which maximises the number of fusions – males with fewer (and hence larger) gametes that make some contribution to zygote viability are selected out. Thus sperm competition (competition between the ejaculates of different males for fusions with ova; Parker, 1970) has an important role in the evolution and maintenance of the two sexes.

Without sperm competition, at least under the most simplistic assumptions, it would pay males to produce fewer, bigger gametes that make some contribution to zygote viability; i.e. anisogamy might therefore collapse. Let us define anisogamy as a state in which the following pair of strategies for gamete size coexist in stable equilibrium:

(1) m_{min} – males produce minimal-sized gametes, minimal in the sense that each contains the set of haploid chromosomes and just enough cellular machinery to ensure optimal survivorship *before* fusion, but no reserves that contribute to the subsequent survival of the zygote.

(2) m_{fopt} – females produce a gamete that contains an optimal amount of provisioning for the survival and subsequent development of the zygote; optimal on the assumption that the sperm contributes nothing to zygote survival (by playing m_{min}). The value of m_{fopt} can be calculated by the method of Smith and Fretwell (1974).

Call the ratio of sperm size/ovum size = A ('anisogamy ratio'); $A = m_{fopt}/m_{min}$. Imagine a population consisting of small, local groups in which males shed sperm into sea water to fertilise the ova produced by the females of the group. In order for the m_{min} strategy to be locally stable against mutant strategies that invest $m > m_{min}$ in each sperm, we need:

$$A > \frac{\text{number of males in group}}{\text{number of males} - 1} \tag{8}$$

(Parker, 1982b). In very large groups, it is easy to maintain anisogamy, since we require only that $A > 1$. In relatively small groups, the condition is also easy to satisfy – for instance, if on average there are just two males, then m_{min} is stable if $A > 2$. But if males spawn singly by females (number of males per group = 1) the RHS of (8) is infinity, and there is no value of A that can maintain the m_{min} strategy. It will pay to provision sperm with reserves to assist the zygote, and anisogamy will collapse. However, at most practical values for A it is clear that the average number of spawning males per group must exceed 1 only by a small amount in order to keep m_{min} stable. Thus provided that there is some minimal degree of sperm competition, anisogamy is 'safe'.

With the evolution of internal fertilisation, one might anticipate the disappearance of sperm competition, and hence the collapse of anisogamy. However, sperm competition is far from absent in internal fertilisers (see Smith, 1983). Groups with specialised sperm storage organs (notably insects) seem predisposed to quite high levels of sperm competition. Here ejaculates can often survive for long periods within the female, so that the probability of two matings having overlapping ejaculates is high, and much of male mating behaviour can be interpreted in terms of sperm competition (Parker, 1970).

In insects, sperm competition often appears to follow a pattern of sperm displacement (but see Gromko, 1983), i.e. the last male to mate appears to displace proportion z of previously-stored sperm on a volumetric basis, and replaces it with his own. Usually, $z > 0.5$ (Parker, 1970). In vertebrates, and other groups without sperm storage, sperm competition seems less likely, because of the shorter sperm life. However, there is a growing literature to suggest that even here, sperm competition can be significant (see Smith, 1983). Mammals, for instance, are highly mobile and often social, and chemical and visual cues of oestrus can attract several males quickly to a female. In groups without sperm storage, we would expect the displacement principle to be less prevalent (though sometimes possible); here the 'raffle principle' is the simplest hypothesis (Parker, 1982b). This asserts that a given male's success at fertilisation is proportional to the number of his sperm/total sperm in the female tract. There is some evidence for this (Martin *et al.*, 1974; Lanier *et al.*, 1979). The same principle should apply to groups of spawning external fertilisers, and was used in the derivation of equation (8).

Suppose that the probability that two males mate with the same female is p; there is no sperm competition on $(1-p)$ occasions. Where sperm competition follows the sperm displacement principle (insects), it can be shown that the m_{min} strategy is locally stable, if, for small p, approximately:

$$p > \frac{2}{z\,(1-z)A} \tag{9}$$

(Parker, 1982b). (6) shows that all or nothing displacement (z=1 or 0) would pose problems for anisogamy, but to my knowledge no insect (for which z is known) falls into this category. However, in the beetle *Ptinella aptera*, sperm are remarkably large (greater than the length of the adult insect) and very few are produced (the ejaculate contains some 1–4 sperm; see Taylor *et al.*, 1982), suggesting that (9) may not always be satisfied. The details and degree of sperm competition are not yet known for this species.

When sperm competition follows the raffle principle, we obtain, for small p, the approximation that

$$p > 4/A \tag{10}$$

in order for m_{min} to be stable. However infrequent the incidence of double mating, it is most improbable that it falls low enough in many species to pose a threat to anisogamy, since A is typically well above 10^4 in mammals and *many* orders of magnitude greater in reptiles and birds. Further, I earlier (Parker, 1982) omitted to note that (10) represents a *minimum* condition for the stability of m_{min}, since it assumes that female investment in a zygote is confined to the expenditure on the ovum. Arguably, we should include *all* the female investment. Thus in a species with zero male parental care, but with maternal care, we should possibly replace A with something closer to the investment ratio:

$$\mathrm{I} = \frac{\text{total energetic investment by female in each offspring}}{\text{energetic cost of sperm of size } m_{min}} \tag{11}$$

a quality which will be literally enormous in magnitude, allowing anisogamy to be robust even when the probability of sperm competition is close to zero. Such a modification could make anisogamy at least as ‘safe’ in mammals as in many birds and reptiles.

Increased Fertilising Capacity and the Stability of Anisogamy

For most multicellular species, the anisogamy ratio A is very high, which gives us circumstantial evidence that cytoplasmic provisioning of ova is indeed important for the survival and development of the zygote. It is therefore easy to see why the anisogamy ESS is so robust. Suppose that a mutant male doubled the sperm size, to 2 m_{min}. This would allow (roughly) one unit m_{min} to be contributed to zygote viability; the other m_{min} would be spent on chromosomes and on the apparatus of fusion and survival before fusion, as in a normal sperm. With an anisogamy ratio as low as 10^4, the contribution of one unit m_{min} shifts the zygote reserves from 10^4 to $10^4 + 1$; the change in zygote viability is marginal. (If it is appropriate to replace the anisogamy ratio by the investment ratio I (11), the change would be *entirely* marginal for mammals.) However, doubling the sperm size is assumed to halve the numbers of sperm produced, because of the fixed energy budget for gamete production. This has disastrous consequences. On each occasion that there is sperm competition, the mutant playing sperm size 2 m_{min} reduces its probable fertilisation gains by a half. In summary, because A (or I if more appropriate) is typically huge, a unit increase in sperm size will exert negligible increases in zygote viability, but reduces sperm numbers dramatically if there is a fixed energy budget for gamete production.

Are there any reasons other than sperm competition why decreases in sperm numbers could exert a major influence in male reproductive success? The classical explanation for large sperm numbers ('overproduction') is that this guarantees a high probability of conception. There is evidence from artificial insemination in cattle that increasing the dilution of the ejaculate does eventually decrease the probability of conception (Salisbury and Vandemark, 1961). However, the data indicate that the probability of conception for an undiluted ejaculate is near maximal – and close to the asymptotic value of the curve of conception probability, g, against increasing sperm numbers (proportional to m^{-1}).

It can be shown (Parker, 1982b) that the m_{min} strategy is stable if

$$g'(m_{min}^{-1}) > \frac{\text{probability of conception}}{A \times \text{sperm number}} \tag{12}$$

Again, it could be argued that A should be replaced by I (11) for species with female parental care and zero male parental care. It is possible to estimate for cattle that the gradient, $g'(m_{min}^{-1})$, must be *well below* 10^{-9} from the graph in Salisbury and Vandemark (1961), and probably much lower than this. From various sources of data (see Parker, 1982b), we can show that the RHS of (12) approximates to 0.75×10^{-14} (using A, not I). Thus stability of anisogamy could well be accounted for, in cattle, by the classical theory, especially if I is the appropriate ratio and not A. My guess is that for most species with internal fertilisation, both sperm competition *and* the classical theory (fertilisation efficiency) are important in ensuring the stability of anisogamy, though sperm competition is probably the more important effect.

An Alternative Hypothesis for the Stability of Anisogamy

All models considered so far have made the assumption that there is a fixed energy budget for gamete expenditure, so that the size of gametes is inversely proportional to the number produced. This assumption seems most reasonable for a sessile external fertiliser, which can achieve reproductive success only by expenditure on gametes. However, most animals could increase sperm size at the expense of some other reproductive allocation, and without any economy in the number of sperm produced. Thus sperm size could increase *either* by a reduction in sperm number, *or* by economies in other reproductive efforts. The present model examines this second possibility.

Suppose that each male has a total reproductive effort R, spends k on each ejaculate, and must spend effort j on searching for a female and obtaining each mating. The number of matings obtained is thus

$$\frac{R}{j+k}$$

A mutant male arises that increases his sperm size from m_{min} to $m > m_{min}$. He therefore has a reduced number of matings:

$$\frac{R}{j+k\,(m/m_{min})}$$

The relative number of matings obtained by the mutant is therefore

$$\frac{\text{number of matings by mutant}}{\text{number of matings by normal male}} = \frac{j+k}{j+k\,(m/m_{min})}$$

If we standardise $j = 1-k$, then k becomes the *proportion* of reproductive effort spent on sperm, and $1-k$ is the *proportion* of reproductive effort spent on obtaining matings. The relative number of mutant matings now is:

$$1/[1 - k + k\,(m/m_{min})]$$

By increasing sperm size to m, the mutant can make some contribution to the viability of the zygote. Let this additional benefit to zygote survival be $b(m)$, where an amount F is contributed by the female via the ovum. The zygote viability thus increases from F to $F + b(m)$. Let s = the sex ratio (females/males). For the m_{min} strategy to be stable, we need:

$$sF > s\,[1 - k + k\,(m/m_{min})]^{-1}\,[F + b(m)] \qquad (13)$$

Now m_{min} will be locally stable if the gradient with respect to m of the RHS of (13), evaluated at $m = m_{min}$, is negative. Hence, for m_{min} to be stable:

$$\frac{d}{dm}\left\{[1 - k + k\,(m/m_{min})]^{-1}\,[F + b(m)]\right\} < 0, \text{ for } m = m_{min}$$

Differentiating, setting $m = m_{min}$, and remembering that $b(m_{min}) = 0$ (since at m_{min} the male contributes nothing to zygote survival) we obtain

$$b'\,(m_{min})m_{min} < k\,F$$

The gradient $b'\,(m_{min})$ is the gradient of the curve of zygote viability at the optimum provisioning m_{fopt} by the female, when the male contributes nothing. It is therefore the gradient of the curve where $b = F$, if we assume that maternal and paternal contributions to zygote survival would be equivalent. From the tangent method (Smith and Fretwell, 1974; see Parker, 1982b) we can calculate that $b'\,(m_{min}) = F/m_{fopt}$. Substituting:

$$k > \frac{m_{min}}{m_{fopt}}$$

or

$$k > A^{-1} \tag{14}$$

in order for m_{min} to be stable. Once again, it may be more appropriate to substitute I (11) for A in equation (14). Either way, it is clear that the fraction k of reproductive effort devoted to sperm production need only be minutely small in order for it to be too costly to increase sperm size at the expense of other reproductive efforts such as mate searching. Again the negligible benefit to zygote survival achieved by a unit increase in sperm size is offset by severe costs, in this instance through reduced numbers of matings.

A Summary on the Stability of Anisogamy

In the last few sections, I have asked the question why sperm stay small, since this helps to explain why males persist as scroungers rather than evolving towards producers. There are two ways in which males can be parasitic. One is by providing no parental care in species where the female does. Reasons why this form of parasitism persists (and sometimes does not persist) are explained in ESS terms by Maynard Smith (1977). A second way in which males are parasitic, and, by definition, are *always* parasitic, is by providing apparently no cellular reserves towards the success of the zygote, when the female does. It is this part of male-female parasitism, and its exceptional robustness in multicellular organisms, that I have attempted to explain.

An important index in all the models has been the anisogamy ratio A or the investment ratio I, which tells us how many minimal-sized gametes can be produced for the same cost as the female investment. Both A and I are likely to be enormous, because high levels of zygote provisioning (and sometimes also subsequent parental investment) are important to the viability and success of the progeny. This makes anisogamy highly stable, because an enormous number of minimal-sized gametes can be produced at equivalent expense to the investment paid by a female in a single progeny. The greater A or I, the more marginal the value of unit increase in sperm size as a contribution to zygote viability. But whilst benefits are marginal, costs of unit increases in sperm size are potentially severe. These costs depend on how the increase in sperm size could be achieved:

(1) If increases in sperm size are achieved by reduction in sperm numbers (gamete expenditure is fixed) costs arise

a) by decreased competitiveness under sperm competition;

b) by decreased efficiency in fertilisation (reduced probability of conception).

(2) If increase in sperm size is achieved by reduction in expenditure on achieving matings, the costs in terms of lost mating opportunities are likely to be severe unless the proportion of reproductive effort spent on sperm is infinitesimal.

How Much Reproductive Effort Should a Male Spend on Sperm?

Finally, I shall consider how much of its reproductive effort a male should expend on its parasitic propagules, the sperm. Since sperm function to obtain fusions with ova, then following previous arguments, high sperm production could yield advantages through sperm competition and/or through increased probability of conception. For both cases, we seek an ESS level of expenditure, k_*, which is the stable proportion of effort to be expended on sperm.

First, consider the effect of sperm competition. Imagine an internal fertilising species in which females experience double matings with frequency p, on which occasions sperm competition follows the raffle principle. Following the previous model, a mutant male that expends $k \neq k^*$ will gain $1/(1 - k_* + k)$ matings relative to each mating by a normal, k_*, male. However, the mutant produces a different sperm number (sperm size stays constant) and gains $k/(k_* + k)$ of the conceptions under sperm competition.

For k_* to be an ESS, all mutants with $k \neq k_*$ must have lower fitness, except a mutant with $k = k_*$, which will have equal fitness. Hence

$$\frac{d}{dk}\left[\underbrace{\left(\frac{1}{1 - k_* + k}\right)}_{\text{number of matings}} \left(\underbrace{1 - p}_{\substack{\text{no sperm}\\ \text{competition}}} + \underbrace{\frac{pk}{k_* + k}}_{\substack{\text{sperm}\\ \text{competition}}}\right)\right]_{k=k_*} = 0$$

which gives

$$k_* = \frac{p}{4 - 2p} \tag{15}$$

or, for small p,

$$k_* \simeq \frac{p}{4} \tag{16}$$

The same approximation was suggested earlier (Parker, 1982b) but (15) is not identical to my former treatment which included a less rigorous formulation for the relative number of mutant matings.

I have been unable to find data on the natural frequency of double mating in a vertebrate. However, there is nothing about (16) to suggest that this level of sperm expenditure is implausible. For example, if double matings occur on 10 per cent of occasions, some 2½ per cent of a male's total reproductive effort should be expended on sperm. These figures seem entirely possible, but judgement must be suspended until data are available.

Now consider the second possibility – that large numbers of sperm are produced to increase the probability of conception. A male expending k on sperm has $R/(j+k)$ matings, and probability $g(k)$ of achieving conception, where $g(k)$ is the fertilisation probability with increasing ejaculate expenditure k. To solve for an optimum the product of number of matings and conception probability must be maximised. Hence

$$\frac{d}{dk}\left[\frac{(R)\,g(k)}{j+k}\right]=0$$

$$\therefore g'(k)=g(k)/(j+k)$$

and setting $j = 1 - k$, so that k is a proportion of reproductive effort spent on sperm, we obtain

$$g'(k_*)=g(k_*)/1 \qquad (17)$$

The evidence from artificial insemination in cattle is that $g(k)$ is monotonic increasing, to an asymptotic value around 0.75 (see previous section). The gradient of the curve $g(k)$ at the value of k for a normal ejaculate is exceedingly small; a normal ejaculate gives a probability of conception very close to the asymptotic value and halving or doubling k would hardly alter the probability. Hence cattle cannot possibly obey rule (17). Thus whilst k_* could be optimised with respect to sperm competition, it is unlikely to be optimised with respect to conception probability.

References

Alcock, J., Jones, C.E. and Buchmann, S.L. (1977). Male mating strategies in the bee *Centris pallida*, Fox (Anthophoridae, Hymenoptera). *Am. Nat.* *111*: 145-55

Barnard, C.J. and Sibly, R.M. (1981). Producers and scroungers: a general model and its application to captive flocks of house sparrows. *Anim. Behav. 29*: 543-50

Bell, G. (1978). The evolution of anisogamy. *J. theor. Biol. 73*: 247-70

— (1982). *The Masterpiece of Nature: The Evolution and Genetics of Sexuality*. Croom Helm, London/University of California Press, Berkeley and Los Angeles

Brockmann, H.J. and Barnard, C.J. (1979). Kleptoparasitism in birds. *Anim. Behav. 27*: 487-514

— and Dawkins, R. (1979). Joint meeting in a digger wasp as an evolutionary stable preadaptation to social life. *Behaviour 71*: 203-45

— Grafen, A. and Dawkins, R. (1979). Evolutionarily stable nesting strategy in a digger wasp. *J. theor. Biol. 77*: 473-96

Charlesworth, B. (1978). The population genetics of anisogamy. *J. theor. Biol. 73*: 347-57

Charnov. E.L. (1982). *The Theory of Sex Allocation*. Princeton, New Jersey, Princeton University Press

Dawkins, R. (1976). *The Selfish Gene*. Oxford, Oxford University Press

— (1980). Good strategy or evolutionarily stable strategy? In G.W. Barlow and J. Silverberg (eds.), *Sociobiology: Beyond Nature/Nurture*. Boulder, Colorado, Westview Press

— (1982). *The Extended Phenotype*. New York, Freeman

Dunbar, R.I.M. (1982). Intraspecific variations in mating strategy. In P.H. Klopfer (ed.), *Perspectives in Ethology:* 385-431. Plenum Publishing Co.

Fisher, R.A. (1930). *The Genetical Theory of Natural Selection*. Oxford, Clarendon Press

Fretwell, S.D. (1972). *Populations in a Seasonal Environment*. Princeton, New Jersey, Princeton University Press

Ghiselin, M.T. (1974). *The Economy of Nature and the Evolution of Sex*. Berkeley, California, University of California Press

Gromko, M.G., Gilbert, D.G. and Richmond, R.C. (1983). Sperm transfer and use in the repeat mating system of *Drosophila*. In R.L. Smith (ed.), *Sperm Competition and the Evolution of Animal Mating Systems*, New York, Academic Press

Hoekstra, R. (1980). Why do organisms produce gametes of only two different sizes? Some theoretical aspects of the evolution of anisogamy. *J. theor. Biol. 87*: 785-93

Knowlton, N. (1974). A note on the evolution of gamete dimorphism. *J. theor. Biol. 46*: 283-5

Lanier, D.L., Estep, D.Q. and Dewsbury, D.A. (1979). Role of prolonged copulatory behaviour in facilitating reproductive success in a competitive mating situation in laboratory rats. *J. Comp. Physiol. Psychol. 93*: 781-92

Martin, P.A., Reimens, T.J., Lodge, J.R. and Dzink, P.J. (1974). The effect of ratios and numbers of spermatozoa mixed from two males on the proportions of offspring. *J. Reprod. Fert. 39*: 251-8

Maynard Smith, J. (1974). The theory of games and the evolution of animal conflict. *J. theor. Biol. 47*: 209-21

— (1977). Parental investment: a prospective analysis. *Anim. Behav. 25*: 1-9

— (1978). *The Evolution of Sex*. Cambridge, Cambridge University Press

— (1982). *Evolution and the Theory of Games*. Cambridge, Cambridge University Press

— and Price, G.R. (1973). The logic of animal conflict. *Nature. Lond. 26*: 15-18

Parker, G.A. (1970). Sperm competition and its evolutionary consequences in the insects. *Biol. Rev. 45*: 525-67

Parker, G.A. (1978). Selection on non-random fusion of gametes during the evolution of anisogamy. *J. theor. Biol. 73*: 1-28

— (1982a). Phenotype limited evolutionarily stable strategies. In King's College Sociobiology Group (eds.), *Current Problems in Sociobiology*. Cambridge, Cambridge University Press

— (1982b). Why are there so many tiny sperm? Sperm competition and the maintenance of two sexes. *J. theor. Biol. 96*: 281-94

— (1984). Evolutionarily stable strategies. In: J.R. Krebs and N.B. Davies (eds.), *Behavioural Ecology: an Evolutionary Approach*, 2nd ed. Oxford, Blackwells.

—, Baker, R.R. and Smith, V.G.F. (1972). The origin and evolution of gamete dimorphism and the male-female phenomenon. *J. theor. Biol. 36*: 529-53

Rubenstein, D.I. (1980). On the evolution of alternative mating strategies. In J.E.R. Staddon (ed.), *Limits to Action: the Allocation of Individual Behavior*: 65-100. New York, Academic Press

Salisbury, G.W. and Vandemark, N.L. (1961). *Physiology of Reproduction and Artificial Insemination of Cattle.* San Francisco, Freeman

Shaw, R.F. and Mohler, J.D. (1953). The selective advantage of the sex ratio. *Amer. Natur. 87*: 337-42

Smith, C.C. and Fretwell, S.D. (1974). The optimal balance between size and number of offspring. *Amer. Natur. 108*: 499-506

Smith, R.L. (ed.) (1983). *Sperm Competition and the Evolution of Animal Mating Systems.* New York, Academic Press

Taylor, P.D. and Sauer, A. (1980). The selective advantage of sex ratio homeostasis. *Amer. Natur. 116*: 305-10

Taylor, V.A., Luke, B.M. and Lomas, M.B. (1982). The giant sperm of a minute beetle. *Tissue & Cell 14*: 113-23

Verner, J. (1965). Selection for the sex ratio. *Amer. Natur. 99*: 419-21

6 SNEAKY BREEDERS

Anthony Arak

The central theorem of modern ethology is that animals are expected to behave in such a way as to maximise their gene contribution to future generations. In sexually reproducing species, selection usually favours adaptations in males for obtaining as many mates as possible or increasing fertilisation success. These adaptations are often as bizarre as they are diverse: male mites which mate with their sisters before they are born (Hamilton, 1967), male dragonflies which scoop out of the female any sperm left by previous mates (Waage, 1979), male cephalopods which use detachable, self-propelled limbs to inseminate females (Lane, 1957).

Much of the early work in ethology was primarily directed towards the goal of explaining differences in social behaviour between species, but over the last few years there has been an upsurge of interest in the variation of behaviour between individuals *within* a species. Reproduction is one area where examples of individual variability in behavioural tactics are legion (reviewed by Rubenstein, 1980; Davies, 1982; Dunbar, 1982). In an increasing number of species, males are found to adopt 'alternative strategies' of mate-searching. While some males invest time and energy in obtaining and defending resources such as territories or females, other males forgo this investment and instead adopt sneaky strategies in an attempt to obtain clandestine copulations. For example, some male scorpionflies (*Mecoptera*) catch small insects and offer them to females as nuptial gifts but other males steal these gifts and use them as their own offerings (Thornhill, 1979); some male fish defend nests in which females spawn while other males pose as females and steal fertilisations (Van Den Assem, 1967; Loiselle and Barlow, 1978); and some male birds defend territories during the spring while other males sneak into these territories and sneak copulations with their mates (Alatalo *et al.*, 1983)

In this chapter I shall attempt to explain the evolutionary basis and consequences of this variability in male mating behaviour. Although, strictly speaking, the word 'strategy' should be used to describe a genetically determined behavioural alternative (Dawkins, 1980), I use the term here in a much looser sense to describe how an animal behaves in a particular situation (i.e. a behaviour pattern). I begin by discussing how alternative reproductive strategies may be maintained in a population, even when one strategy appears to be much less successful than the other. I then go on to discuss conflicts of interest within and between sexes. Not only will there be conflicts of interest between males adopting different mating tactics, but in many cases the aim of the female will differ from that of the male resulting in a 'battle of the sexes' (*sensu* Dawkins, 1976). A complex set of adaptations may evolve in response to sexual conflicts and the concept of the evolutionary 'arms race' (Dawkins and Krebs, 1979) will be used here to explore some of these adaptations in detail and to discuss how such conflicts may be resolved.

Evolutionary Explanations of Behavioural Variability

At one time individual variation in social behaviour was often considered maladaptive or aberrant behaviour of little evolutionary interest. Opinions have now changed, and whenever an animal is seen to be doing something unusual a specific function is ascribed to the new behaviour. The application of games theory to the study of animal behaviour (see Maynard Smith, 1982) and the realisation that evolutionary arguments must be framed in terms of advantage to the individual or gene rather than for the good of the species (see Williams, 1966; Dawkins, 1976), are the main factors responsible for this change of interpretation. There are currently two principal categories of explanation concerning the maintenance of alternative strategies in a population:

(1) The alternative strategies have equal payoffs and are maintained in equilibrium because their benefits are negatively related to their frequency in the population (a mixed evolutionarily stable strategy, or mixed ESS).
(2) The alternatives have unequal payoffs but selection favours different strategies in individuals of different competitive ability or in different environmental circumstances (a pure conditional strategy).

To clarify this distinction, let us consider some examples (see Table 6.1). Some male frogs call to attract females ('callers') while others ('satellites') sit silently next to a caller and attempt to intercept approaching females. It is easy to see how the relative success of each strategy

Table 6.1: A classification of alternative mating strategies

Mechanism	Possible examples
1. Mixed evolutionarily stable strategy	
a) Genetic polymorphism	Resident and satellite male ruffs (van Rhijn, 1973) Parental and cuckolders in blue-gill sunfish (Gross, 1982)
b) Strategy varies from contest to contest	Callers and satellites in the green treefrog (Perrill *et al.*, 1978, 1982)
2. Pure conditional strategy	
a) Strategy depends on phenotype or competitive ability	Harem-holders and sneaks in red deer (Clutton-Brock *et al.*, 1979) Patrollers and hoverers in bees (Alcock *et al.*, 1977a)
b) Strategy depends on circumstances	
Ownership conventions	Speckled wood butterfly (Davies, 1978) Hamadryas baboons (Bachmann and Kummer, 1980)
Side-payment strategy	Opportunistic cuckoldry by male pied flycatchers (Bjorkland and Westman, in press) Sexual interference by salamanders and newts (Arnold, 1976; Verrell, 1982)
Conditional on environment	Male sticklebacks are dull in shallow water but have red throats in deep water (Moodie, 1972; Semler, 1971) Male natterjack toads switch from calling to searching as chorus size increases (Arak, 1983)

may depend on its frequency in the population. When most males are calling many females arrive so it pays to be a satellite; conversely, when most males are satellites few females arrive so it is better to call. Thus we would expect the population to evolve to consist of a mixture of callers and satellites, such that the success of each strategy is equal.

To demonstrate a mixed ESS in nature it would be necessary to show that the success of the two strategies calling and satellite are equal and that each strategy does better when rare. Perrill *et al.* (1978) studied callers and satellites in the green treefrog (*Hyla cinerea*) and found that the two strategies were about equally successful in capturing females released near caller-satellite associations: the calling male mated with the female on 17 occasions while the satellite did so on 13 occasions. Although the result of this experiment is consistent with a *mixed ESS*, the evidence is not conclusive. First, even if the benefits of the two strategies are equal the costs may be different. Calling probably involves considerable energetic expenditure and predation risk (Bucher *et al.*, 1982; Ryan *et al.*, 1982) and these costs would have to be taken into account to estimate the *net benefits* of each strategy. Also, the average payoff to a male playing each strategy would depend critically on the proportion of males in the population adopting each strategy, not just on the relative success of callers and satellites in the experimental situation.

Theoretically, there are two mechanisms by which a mixed ESS could be achieved. First, a genetic polymorphism may give rise to alternative strategies with individuals adopting different pure strategies, or second, the population may be genetically monomorphic with individuals adopting both strategies in the proportions of the mixed ESS. In the latter case, individuals may either play each strategy at random in accordance with an unconscious behaviour programme ('play *A* with probability *p*, play *B* with probability (1–*p*)'), or they may use a simple rule of behavioural assessment to decide which strategy to adopt depending on what other individuals in the population are doing. Perrill *et al.* (1982) showed that male green treefrogs frequently switched strategies and appeared to be assessing what other males in the population were doing. When calling males were removed from caller-satellite associations, satellites often began calling; also, calling males switched to satellite behaviour when a loudspeaker emitting male calls was placed near them.

Examples of conditional strategies are provided by bullfrogs (*Rana catesbeiana*; Howard, 1978a) and natterjack toads (*Bufo calamita*; Arak, 1982). In these species a male's strategy depends on his body size: large males mainly call and smaller males adopt the satellite strategy. In both species large males obtained many more mates than smaller males, indicating that the smaller individuals were probably parasitising the large males' ability to attract females. In natterjack toads it is unlikely that satellite males could have achieved greater reproductive success by calling since their calls are not as loud as those of larger males and they are not preferred by females (Arak, 1982). Clearly males adopt a conditional strategy dependent on their phenotype, i.e. 'if big call, if small adopt satellite behaviour'.

Conditional strategies of reproduction are common where natural asymmetries, such as age or size differences, exist between individuals. Differences in competitive ability between individuals may be fixed for life, as exemplified by male dungflies (*Scatophaga stercoraria*; Borgia, 1979), and bees (*Centris pallida*; Alcock *et al.*, 1977a) where adult size depends on nutrient conditions during the larval stages. Alternatively, competitive ability may change with age as an animal grows e.g. elephant seals (*Mirounga angustirostris*; Le Boeuf, 1972, 1974) and red deer (*Cervus elephus*; Clutton-Brock *et al.*, 1979). Whatever the proximate cause of these differences, smaller or younger males adopt the less successful strategy because they are unsuccessful playing the dominant strategy. Poor competitors must make the best of a bad job by using an alternative strategy with fewer rewards.

But even 'best of a bad job' strategies can be stable against invasion by alternative strategies. Parker's (1982) models predict that in a population consisting of a range of phenotypes differing in competitive ability, each phenotype will have its own ESS, such that no alternative strategy can do better when played by an individual of the same phenotype. This prediction is supported by the experimental work of Perrins and Moss (1975) on clutch size in great tits (*Parus major*). They showed that birds producing smaller than average clutch sizes were doing the best they could given their phenotypic constraints; when their brood size was increased artificially they fledged fewer young than birds with naturally large broods. Similarly, males who opt out of intense competition for females by adopting sneaky strategies are probably doing the best thing possible given their constraints.

Conditional strategies may sometimes occur in the absence of any obvious or permanent asymmetries between individuals. Davies (1978) showed that male speckled wood butterflies (*Pararge aegeria*) competing for sunspots where females arrive play to the rule 'resident always wins'. This rule appears to be evolutionarily stable in the sense that it cannot be beaten by individuals who play to a different rule. For example, the high risk of injury or high energy cost would militate against a male who always escalates on encountering the 'owner' of a sunspot. In some cases, individuals may adopt 'side-payment strategies' to collect opportunistically whatever small gains can be obtained on the side while pursuing the dominant strategy (Dunbar, 1982). Frogs may call most of the time but adopt the satellite strategy when tired; thus while resting they have a chance of obtaining a female using the sneaky tactic. Finally, the best strategy to adopt may depend on the environmental conditions, as, for example, in the populations of three-spined sticklebacks mentioned in Chapter 1, where different colour morphs are found at different depths in the water owing to the trade-off between female attraction and vulnerability to predators (Moodie, 1972; Semler, 1973).

The ESS approach, although enlightening, is not without its difficulties. First, there are some practical problems involved in measuring costs and benefits of alternative strategies (Davies, 1982). Often short-term measurements of reproductive success may not accurately reflect consistent differences between individuals and therefore it will be necessary to measure lifetime reproductive success to compare different strategies. There is also the problem of finding a suitable common currency to measure such diverse costs as time, energy, predation risk and injury. Finally, even if we are certain that we have accurately measured all the costs and benefits of each strategy, we may find that the strategies have unequal payoffs. If there are no obvious morphological or age differences between individuals adopting different strategies we are left with the possibility that one strategy may be non-adaptive. This may be because evolution has not yet caught up with some recent environmental change or because one strategy merely represents non-heritable variation. A possible example of such non-heritable variation is the adverse response shown by some mammals to high population densities. Individuals often develop gastric ulcers, grow tumours in the pituitary gland, suffer a decrease in fertility and behave unusually aggressively or withdraw from conflict situations (Christian, 1963; von Holst, 1974). Most biologists would not seriously consider such pathological responses as *adaptations* (but see Wynne-Edwards, 1962), but merely the unfortunate *consequences* of crowding. However, we should be careful to avoid the ready acceptance of such non-adaptive explanations of behaviour for they can easily become a cloak for ignorance, discouraging the researcher from seeking out and testing further adaptive explanations. The ESS approach, on the other hand, encourages us to measure the costs and benefits of different forms of behaviour. In the example above, for instance, it might be worth while to ask whether an animal showing symptoms of social stress performs better or worse in conflict situations compared with an unstressed individual with little previous experience of intense competition.

Sneaks as Scroungers

Recently, Barnard and Sibly (1981) suggested that the coexistence of some alternative mating strategies can be viewed as relationships between 'producers' and 'scroungers'. In a reproductive context, while producers invest energy in attracting females, defending territories or harems, scroungers attempt to exploit this investment by stealing fertilisations. We shall now look in detail at some examples where scroungers exploit the reproductive efforts of producers without providing any benefits to the producer. The question of interest is, how are these alternative strategies maintained in the population?

Bluegill Sunfish: Parentals and Cuckolders

The bluegill (*Lepomis macrochirus*) is a common freshwater fish in North America and Canada where it breeds colonially in ponds and lakes. Gross (1982) discovered two quite distinct male breeding strategies which he referred to as 'parental' and 'cuckolder'. Parentals are large males which construct nests on the lake floor, then attract females to the nests and provide sole parental care of the developing eggs (Figure 6.1a). Parental males remain reproductively inactive until they are large, at about seven years old. In contrast, cuckolder males become sexually mature when small, at about two years old. At first they behave as 'sneakers'; lurking in vegetation near a nest tended by a parental male, they rush into it and attempt to fertilise some of the female's eggs as she spawns there (Figure 6.1b). Sneakers later become 'satellites' when they reach a size similar to sexually mature females. When a female enters a parental male's nest the satellite moves *slowly* into the nest, between the spawning female and the parental male, and releases sperm onto the eggs. Satellites are able to achieve fertilisations in this way because they mimic females in their size, colour, pattern and behaviour (Figure 6.1c).

Gross (1982) showed that parentals and cuckolders represent discrete reproductive pathways and individuals cannot switch between them. When parental males were removed from their nests cuckolders

Figure 6.1: Mating tactics of male bluegill sunfish. (a) A parental male constructs a nest by sweeping the substrate with his caudal fin; then a female visits the nest. (b) Sneaker males hide in vegetation and dart into the nest releasing sperm as the female spawns. (c) A satellite male, who resembles a female, slowly descends between the parental male and the female and releases sperm

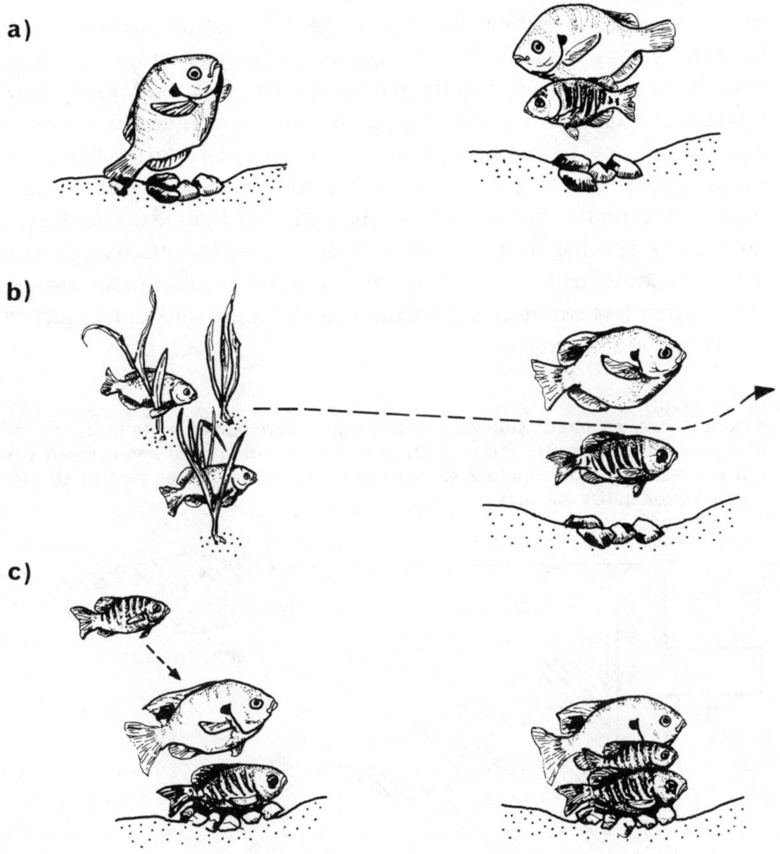

Source: redrawn from Gross (1982)

did not attempt to defend them but continued adopting sneaky tactics. Switching from cuckolder to parental is probably not possible because once individuals mature as cuckolders they grow much more slowly than males in the parental pathway and never become large enough to compete effectively for nest sites (see also Drager and Chiszar, 1982).

Approximately one-fifth of all males enter the cuckolder pathway and the remainder become parentals (Figure 6.2). Gross (1982) estimated that by using sneaky tactics cuckolders fertilise about one-fifth of all the eggs, suggesting that the average fertilisation success of males in each strategy is equal. Thus the population may be adopting a mixed ESS determined by a genetic polymorphism such that each male's strategy is fixed for life. The population will be maintained at the ESS equilibrium due to negative frequency-dependent payoffs. When most males are constructing nests and caring for young there will be abundant opportunities for cuckolders to steal fertilisations. However, as cuckolders become more common their relative success will decrease because simultaneous nest intrusions will cause intense sperm competition, higher levels of aggression from parental males and and an increase in the number of females deserting nests.

Figure 6.2: A schematic summary of alternative developmental pathways of male bluegill sunfish. Parental males delay sexual maturation until about seven years old whereas cuckolders mature at two years old. About 21 per cent of all males enter the cuckolder pathway

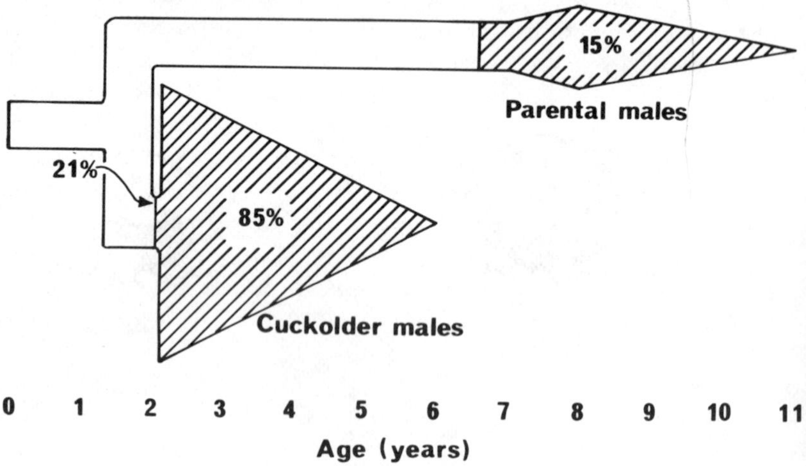

Source: redrawn from Gross (1982)

Pied Flycatchers: Faithful Males and Philanderers

Male pied flycatchers (*Ficedula hypoleuca*) settle on breeding territories during the spring and sing to attract females. Some males in the population are faithful: they form a monogamous pair-bond with a female and help to feed the young. Other males, however, are philanderers: after mating with a 'primary' female they move on to another territory and attempt to attract a 'secondary' female. Approximately 15 per cent of all males at a study site in Sweden succeeded in attracting and mating with more than one female (Alatalo *et al.*, 1981; Lundberg, personal communication).

Polygynous males help to feed the young of both females, but provide less food to the brood of the secondary female and consequently some young birds starve in the nest. Whereas monogamous and primary females fledge on average 5.4 young, secondary females fledge only 3.4. Thus faithful males, at first sight, appear to be less successful than philanderers, raising on average only 5.4 young compared with the philanderers' total of 8.8 young. Faithful males tend to be smaller, younger males who arrive on territories late in the breeding season and so may not have time to attract a secondary female (Alatalo *et al.*, 1981, 1982). It would seem, then, that faithful males are poor-quality males who are making the best of a bad job by mating with only one female. However, the story is complicated by the fact that males have the opportunity to obtain some offspring through cuckoldry. By estimating heritability of tarsus length to assess the incidence of multiple paternity within broods. Alatalo *et al.* (1983) found that, on average, one quarter of the offspring in a given brood were fathered by a male other than the one associated with the nest. This value of one quarter is an average estimate for all males, but it seems reasonable to suppose that faithful males who remain in close association with their female on a single territory are less susceptible to cuckoldry than philanderers who move between several territories. Also, faithful males are in a better position to monitor the presence or absence of males in neighbouring territories and sneak copulations with their neighbour's mate. In experimental removals of territorial males, monogamous males from neighbouring territories were seen sneaking in and copulating with the female (Bjorkland and Westman, in press). Alatalo *et al.* (1983) found a positive correlation between tarsus length of fledglings and tarsus length of the nearest neighbouring male indicating that these extra-pair copulations certainly resulted in successful fertilisations on some occasions.

These observations suggest that the disparity in reproductive success

Table 6.2: Estimates of reproductive success for polygynous and monogamous male pied flycatchers (data from Alatalo *et al.*, 1981, 1983)

Polygynous male (philanderer)			Monogamous male (faithful)		
Primary female (1°♀)	=	5.4 young	Monogamous female	=	5.4 young
Secondary female (2°♀)	=	3.4 young			
Paternity loss from 1°♀	=	–1.3 young	Gain by cuckoldry	=	1.3 young
Paternity loss from 2°♀	=	–0.8 young			
Total[a]	=	6.7 young	Total[a]	=	6.7 young

Note: a. These estimates give the impression that the net reproductive success of polygynous and monogamous males is equal. However, there may be great individual variation in paternity gains and losses through cuckoldry, so the estimates shown here are speculative. The estimates assume a polygynous male loses 25% of his young from each nest by cuckoldry and a monogamous male gains an additional 25% (see Alatalo *et al.*, 1983).

between faithful males and philanderers may be smaller than estimated by measures of fledgling success alone (Table 6.2). The success of the two strategies may in fact be equal: by adopting the side-payment strategy of cuckoldry, faithful males might be able to 'make up' for the number of young they lose by failing to mate with a secondary female. Until the costs and benefits of all possible strategies are measured taking cuckoldry into account, it will not be possible to say for certain whether male strategies are conditional on phenotype or whether they are maintained as a mixed ESS through frequency-dependence. However, this study serves as a useful example illustrating some of the problems faced by the field-worker in interpreting data and devising new experiments to test alternative hypotheses.

Sneaks as Helpers

Although the majority of alternative male mating strategies involve subordinate males scrounging matings from dominants, there are a few cases where sneaks offer something in return for the matings they steal. In some species, sneaks may 'pay for permission' to mate with females by helping a dominant male to attract mates or defend a territory. To understand how such associations evolve and remain stable, two crucial questions must be asked: do helpers really help, and does helping benefit the helper?

Displays on Leks: Residents and Satellites

On early mornings during the spring, male ruffs (*Philomachus pugnax*) gather together at traditional arenas or 'leks' where they display vigorously to attract females (Hogan-Warburg, 1966; van Rhijn, 1973). There are two distinct male mating strategies that are thought to be maintained by a genetic polymorphism. Some males, called 'residents', defend small territories on the lek, and others, called 'satellites', behave non-aggressively and associate with a resident on his territory. This behavioural difference is correlated with a marked difference in plumage colour. Resident males have dark head tufts and dark ruffs whereas satellite males have white tufts and ruffs.

Two kinds of observations made at ruff leks are of interest to our discussion here. First, it is clear that, as in other lekking species, almost all of the copulations are performed by a small proportion of males who occupy central positions on the lek. Males compete vigorously for these central positions by displaying to each other and fighting.

Secondly, it has been noted that under some circumstances resident males tolerate one or more satellites on the territory but at other times the resident chases satellites away (Hogan-Warburg, 1966; van Rhijn, 1973). These observations pose some intriguing questions about lek behaviour. In view of the high level of competition between males on a lek, why do males bother to display in groups at all? Are interactions between males on a lek exploitative or co-operative in nature?

Several different (though not mutually exclusive) explanations of lek behaviour have been suggested (Table 6.3). Lack (1968) saw lek behaviour as an adaptation designed to reduce the risk of predation because groups of displaying males would be more likely to spot approaching predators. This hypothesis is supported by much recent work on the effect of group size on predator-spotting and escape in flocks of foraging birds (Pulliam, 1973; Kenward, 1978; Lazarus, 1979; Bertram, 1980; Elgar and Catterall, 1981).

The most fashionable explanation of lek behaviour is that it has evolved through female choice (e.g. Alexander, 1975; Borgia, 1979; Bradbury, 1981). In effect males are forced to display in groups because females will only mate in situations where they can compare several males. Despite the general appeal of this idea, there is little evidence to support it, except that in many species there is a strong tendency for females to mate with males occupying territories at the centre of the lek. In some species central males tend to be older, more experienced individuals, characteristics that may be indicative of high survival ability. In black grouse (*Lyrurus tetrix*), the most successful males are those which perform the 'rookooing' display at the highest rate (Kruijt, personal communication), but in other species such as manakins (*Manacus manacus*) and sage grouse (*Centrocercus urophasianus*), no correlations have been found between any aspects of male morphology or display and mating success (Lill, 1974; Wiley, 1973). Most of the evidence for *active* female choice can be criticised because similar results would be expected as a result of *passive* responses by females to high levels of courtship display. The tendency of females to mate with central males, for example, may be the consequence of a superstimulus or facilitation effect rather than a function of assessment of these males as suitable mating partners. The female, once arrived at the centre of the lek, is subject to stimuli from all sides, hastening arousal to a maximum (Davison, 1982). Furthermore, in some lekking species females mate with more than one male during a visit to a lek (see later) which suggests that females are not choosing between males.

Otte (1974, 1977) suggested that a possible reason for communal

Table 6.3: Hypotheses concerning the evolution of lek behaviour (adapted from Bradbury and Gibson, 1983)

Hypothesis	Proponents
Male-initiated reasons for clumping:	
1. Males clump to reduce the risk of predation	Lack, 1968; Wiley, 1974
2. Males clump to increase signal range or amount of time signals are being emitted	Otte, 1974, 1977
3. Males require specific display habitats which are limited and patchily distributed	Snow, 1974
4. Males clump at places through which females are likely to pass	Payne and Payne, 1977
5. Unattractive males associate with and parasitise the display efforts of more attractive males and attempt to steal copulations when the females arrive	Arak, 1983, and see below
Female-initiated reasons for clumping:	
6. Females prefer clumped males because this reduces predation on females	Wittenberger, 1978
7. Females prefer males who clump in less-desirable habitats thus reducing competition between the sexes for food	Wrangham, 1980
8. Females prefer clumps of males because this facilitates mate choice	Alexander, 1975; Borgia, 1979; Bradbury, 1981

display may be that males 'co-operate' to attract females from a wider area. The essence of this idea is that males accrue a mutual benefit by displaying in large groups. Bradbury (1981), however, demonstrated that for most types of signal – visual, auditory and olfactory – the *absolute* number of females attracted would be expected to increase as the number of displaying males increases, but the number of females attracted *per male* would remain the same or actually decrease. This is because signal strength does not increase in a simple additive way as the number of emitters increases. Bradbury's theoretical conclusions are supported by the results of some studies but not by others. For example, Cade (1981) found that the number of female crickets (*Gryllus integer*) attracted to loudspeakers emitting male calls was relatively independent of the size of the artificial chorus, and Koivisto (cited in Lack, 1968) found no difference in the number of females attracted to small and large leks in black grouse. In contrast, Ryan *et al.* (1981) found that larger choruses of frogs (*Physalaemus pustulosus*) attracted more females per calling male than smaller choruses.

Bradbury (1981) only considered how communal signalling would affect the long-range attraction of females to groups but it is possible that there are advantages in combining signalling efforts once females have arrived at the lek. Very high levels of display afforded by a group of males may facilitate the sexual arousal of individual females and make them more willing to mate. For example, in the ruff, the probability of a resident male attracting a female is increased by tolerating a satellite male on the territory (van Rhijn, 1973). Presumably the satellite male also benefits by associating with a resident because he can sometimes steal copulations when the resident is involved in territorial defence.

In the rather special case where males associate with their relatives, benefits of group display may accrue through kin selection. In wild turkeys (*Meleagris gallopavo*), males display together on leks in sibling groups of one to four brothers (Watts and Stokes, 1971; but see Balph *et al.*, 1980). Although all the males within a group display vigorously to attract females and fight together against intruding sibling groups, usually only one or two dominant males in each group copulate with the hens. Watts and Stokes remark that out of 170 males on four different leks, no more than six males accounted for all 59 observed copulations. Despite the close relatedness of the males in each display group, there does appear to be some conflict between brothers for mating opportunities, and the variance in male reproductive success is no greater than in other species where displaying males are unrelated.

Thus, there is no suggestion that subordinate males are actively forgoing reproduction to aid their dominant brothers.

Although I have so far only emphasised the benefits gained by associating with other displaying males, lek behaviour almost certainly involves exploitative interactions as well. Earlier in this chapter I suggested that small male toads with quiet calls may adopt the satellite strategy to exploit the superior calling abilities of larger males. This form of exploitative interaction in frog choruses may also occur in lek displays of other species. Peripheral or satellite males on leks may be unable to produce attractive displays to compete with their more experienced rivals; their best option may be to associate with these males and attempt to steal females as they arrive on the territory. An attractive male, although he would do better by displaying alone, may be unable to prevent satellites clustering around him because he is outnumbered by them. In accord with this idea is an observation made by De Vos and co-workers (1983) on black grouse. They found that one of the most successful males during their 15-year study of grouse leks was an individual who displayed solitarily and avoided the high levels of competition on nearby leks.

In conclusion, group display probably involves both co-operative and exploitative interactions between males. While benefits of displaying in groups may arise through co-operative female attraction and predator avoidance, costs of competition for mating opportunities may impose an upper limit on group size. Individuals would be expected to trade off these costs and benefits in making decisions to join groups and tolerate other males in accordance with their own attractiveness or competitive ability. For an attractive or dominant male, the optimal trade-off may be realised at quite small lek sizes, when the benefits of tolerating satellite males (in terms of co-operative attraction of females) are not offset by the costs of being cuckolded by these satellites. In contrast, satellites may prefer larger leks than do dominants, because the chance of stealing copulations probably increases as the number of arriving females surpasses the rate at which dominant males can 'service' or defend them. On ruff leks observed by Hogan-Warburg (1966), the tolerance of satellites by resident males varied depending on the size of the lek. On one small lek satellites were tolerated by residents and they achieved about the same rate of copulations as residents; on a larger lek, however, residents were very aggressive towards satellites and continually attempted to drive them away so they obtained very few copulations. These observations suggest that resident ruffs prefer to display on small leks but satellites are still prepared to join leks larger than the optimum size for residents.

Waterbuck: Owners and Satellites

In some species, subordinate males may help a dominant male to defend a territory without any immediate prospect of reward. The return on investment in helping may occur at a later time when the dominant vacates his territory and the subordinate male takes it over. An example of this is provided by satellite male helpers in the waterbuck (*Kobus ellipsiprymnus*). Wirtz (1981, 1982) studied the social behaviour of waterbuck at Lake Nakuru National Park in Kenya. Some males defended territories bordering the lake all year round, and females were attracted to these territories to gain access to high-quality grazing and water. The population of waterbuck at Lake Nakuru was extremely dense and only 7 per cent of adult males held territories. Of the remaining 93 per cent, 84 per cent were bachelor males and 9 per cent were satellites on the territories of dominant males.

Satellites were probably tolerated on territories because they helped in territorial defence. In 64 per cent of intrusions onto shared territories the satellite male, rather than the territory owner, challenged the intruder. Owners who tolerate satellite males clearly may benefit by reducing the high energetic costs involved in aggressive defence and by decreasing the risk of injury due to fighting. Also, owners may be able to hold their territories for a longer period of time with a satellite than they can without one.

How does the satellite male benefit from the association? When the owner was challenging an intruder, satellites sometimes managed to sneak copulations with females, but usually territory owners displaced satellite males from oestrous females during mounting attempts. The relative frequency of copulations by territory owners, satellites and bachelor males was 67:4:1. This indicates that satellite males achieved a slightly, though not significantly, higher rate of copulation than bachelor males. It seems likely that the benefit of occasional copulations with a female is a 'side-payment' while the major strategy is gaining possession of the territory. Of eleven observed cases of changes in territory ownership, satellites took over seven territories and bachelor males took over four. However, there were nearly seven times as many bachelors as satellites in the population, which means that the probability of territory takeover was about twelve times greater for a satellite male than for a bachelor.

It is possible that the alternative strategies of male waterbuck, like those of red deer and elephant seals, are age-dependent. As bachelor males grow older and stronger they may settle on a territory as a satellite male or attempt to take over the territory outright. However, because relatively few territories are available and takeovers are fairly infrequent, it seems unlikely that all bachelor males eventually become territory owners: some may remain lifelong bachelors.

A similar case of alternative strategies involving helpers occurs in the anemone fish (*Amphiprion akallopisos*). Pairs of fish defend anemones in which they breed and sometimes non-breeding fish help the pair to chase off intruders. Fricke (1979) suggests that these helpers, like satellite male waterbuck, may be 'hopeful reproductives' waiting for a chance to take over the territory when one of the breeders leaves or dies.

Conflict and Compromise

It is clear from the examples discussed above that alternative male mating tactics will cause conflicts of interest between dominant and subordinate males. Where sneaks act as scroungers, providing no help in return for the benefits they obtain by interacting with a dominant male, these conflicts may be settled by overt aggression. Sneaky male red deer who enter harems and attempt to steal copulations are driven away by the stags (Clutton-Brock *et al.*, 1979); young male elephant seals who use similar tactics may be viciously attacked by dominant bulls (Le Boeuf, 1972, 1974). Various tactics may evolve in males to get the better of their rivals and evolutionary escalations of counter-adaptations or 'arms races' (Dawkins and Krebs, 1979) would be expected to occur between competing males. In cases where sneaks act as scroungers, we should expect natural selection to increase dominant males' abilities to detect and repel sneaks while concurrently improving sneaks' abilities to escape detection by dominant males (Barnard, Ch. 1 this volume).

Where subordinate males help dominant individuals to attract females or defend territories, both males may benefit by maintaining the association. However, because natural selection acts on individuals, we would expect to find undercurrents of conflict and selfishness even in apparently co-operative interactions (Vehrencamp, Ch. 8 this volume). We would also expect any conflict to take a more subtle form in these interactions, compared with situations where benefits are entirely one-sided. For instance, satellite male waterbuck might be expected to undermine the interests of territory owners by mating with more females than they 'pay' for in terms of the assistance they provide in territory defence. Territory owners, on the other hand, will be selected to be exacting task-masters and coerce satellite males into providing greater help in defence for fewer rewards. In effect, each male faces a difficult dilemma in deciding whether to maintain the relationship with the concomitant risk of being exploited by his partner, or whether to break off the association and settle for smaller but more assured benefits. Hence, a seemingly co-operative partnership may be inherently unstable and a compromise of the divergent interests of the partners must be reached if the association is to endure. Given the delicate nature of the interaction between an owner and a satellite, selection may favour the development of complex negotiating procedures by which an individual regulates his altruistic and selfish tendencies in response to the same tendencies in the other individual (see Axelrod and Hamilton, 1981). Breakdowns in the negotiating procedure are perhaps less likely to occur if individual recognition is possible and the males have experienced previous interactions with the same partner. It is interesting to note on this point, that in both ruffs and waterbuck, satellite males tend to form fairly stable associations with the same resident on different occasions (van Rhijn, 1973; Wirtz, 1982).

The question of which male wins in arms races such as these is an intriguing one. Four factors may be important. Firstly, the male adopting the dominant strategy may have an inherent advantage in size or strength which can be used to force the outcome in his favour. Secondly, there may be an asymmetry in the costs associated with winning or losing a contest. For example, the cost of failing to achieve any copulations at all will be greater for a sneaky strategist than the cost of a slight chance of cuckoldry for a dominant strategist. Therefore, selection pressure may act more strongly on improving sneaks' abilities to escape detection than on improving dominants' abilities to detect sneaks. Thirdly, different rates of reproduction between the strategies may affect which strategy is likely to win the arms race. If the strategies are maintained by a mixed ESS such that both achieve equal reproductive success then counter-adaptations can potentially evolve at equal rates in each strategy. If, however, one strategy is the 'best of a bad job' with males siring fewer progeny compared with males playing the dominant alternative strategy, then the potential rate of evolution of counter-adaptations may be slower in subordinates. However, the fact that males adopting the strategy with the lower payoff have more to lose in conflict with dominant males (see above), will tend to counterbalance the effect of different rates of reproduction between the strategies. Finally, which strategy 'wins' may vary unpredictably from time to time as conditions change. In ruff leks, satellite males obtained as many copulations as did residents on a small lek, but virtually none at all on a large lek.

Escaping Detection

The problem faced by sneaks in escaping detection varies in difficulty according to the species and the ecological circumstances in which mating takes place. For example, many species of frogs and toads are nocturnal breeders and satellite males can easily escape detection simply by sitting still and remaining silent. In natterjack toads (Arak, 1982) satellite males adopt a characteristic 'low posture' in the water to avoid being detected by callers. When attempting to intercept females, satellites are often attacked and driven away by callers. The problem of escaping detection is obviously more difficult if breeding occurs during the day. However, one tactic adopted by sneaky males in some insects (Thornhill, 1979), fish (Loiselle and Barlow, 1978) amphibians (Verrell, 1982), reptiles (Trivers, 1976) and birds (Rohwer *et al.*, 1980) is to mimic females while attempting to steal food items or matings from other males. Female mimicry takes a particularly interesting form in many species of passerine birds. In both monogamous and polygamous species where sexual dimorphism in plumage colour occurs, often males do not achieve plumage maturity until one year after their first potential breeding season. Rohwer *et al.* suggest that these young males with female-like plumage may be attempting to breed by deceiving older males. When young males do not breed, female mimicry may be a tactic used to explore territories defended by older males and determine the best places to hold territories as adults in the following year. It is also possible that these sub-adult males may be able to trespass on older males' territories to steal occasional copulations (but Rohwer *et al.* discount this possibility, suggesting that it would not be to females' advantage to copulate with non-territorial males). The employment of such 'transvestite' strategies by some males in the population doubtless imposes selection pressures on individuals defending females or resources to sharpen up their perceptual abilities in order to discriminate between pseudo- and real females.

Another method by which sneaks steal fertilisations is to wait until the dominant male has left the territory or is otherwise occupied. Satellite males on leks often copulate with females when the dominant male is copulating or defending his territory against another rival male (Waltz, 1982). Cuckoldry by male pied flycatchers often occurred on neighbours' territories when the territory owner was temporarily away or after experimental removals of males from their territories (Bjorklund and Westman, in press). In the dunnock (*Prunella modularis*), two males sometimes defend the same territory and share the same female (Davies, 1983). One of the males (the 'alpha' male), who is dominant over the other male (the 'beta' male), persistently tries to prevent the beta male copulating with the female. However, the beta male sometimes succeeds in mating when the alpha male is feeding or defending the territory and cannot remain close to the female (Figure 6.3a). As a counter-measure against cuckoldry, the alpha male attempts to assure his paternity of the offspring by copulating with the female more frequently the greater the proportion of time the beta male has spent close to her (Figure 6.3b).

Coalitions Between Sneaks

If it is impossible for sneaks to avoid detection by a dominant male

Figure 6.3: Conflict between male dunnocks sharing the same territory and mate. (a) The subordinate (beta) male is more likely to copulate as the distance between the dominant (alpha) male and the female increases. (b) The alpha male copulates more frequently the longer another male has spent near her

(a)

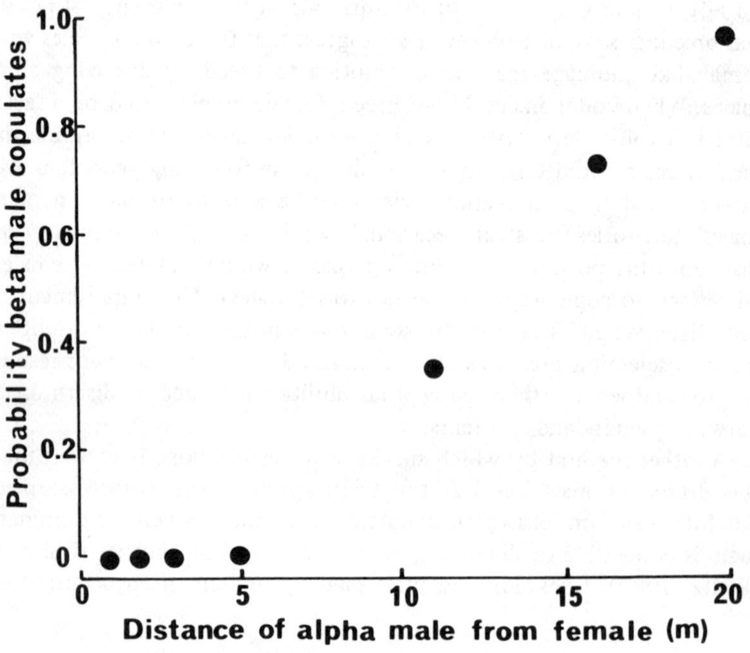

(b)

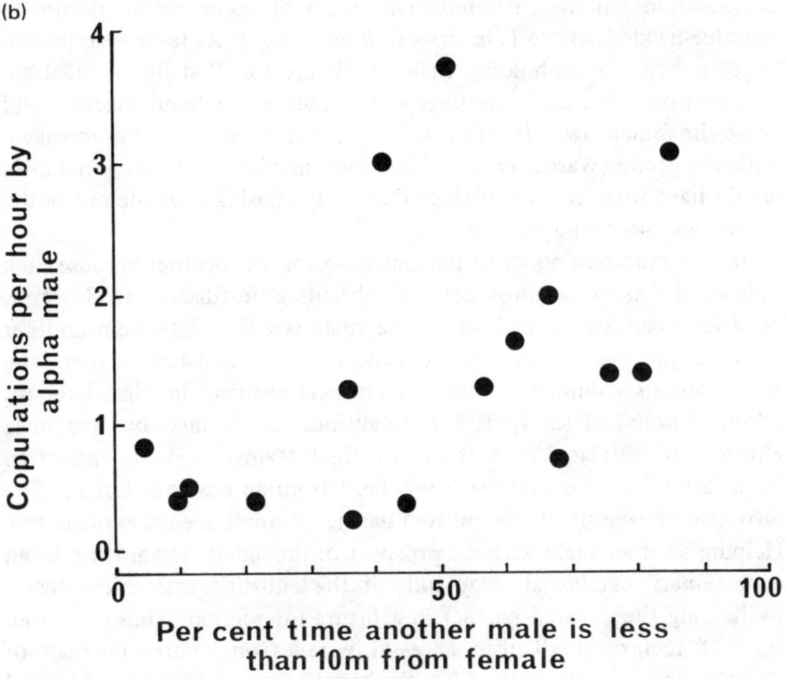

Source: redrawn from Davies, 1983 (in press)

then they may profit by banding together to steal fertilisations. By forming coalitions, sneaks may be able to distract a dominant male or dilute his aggression, and hence increase their individual chances of successfully usurping females. In the bluegill sunfish, for example, several sneaky males sometimes entered a nest simultaneously causing confusion in the defensive behaviour of the parental male (Gross, 1982). In this case, apparent co-operation between sneaks may be a purely coincidental effect of simultaneous nest intrusion, but even if this is true, it is easy to see how this could be the starting point favouring deliberate coalitions between sneaks. Imagine, for instance, that two sneaks arrive simultaneously at a parental male's nest just as a female is entering and is about to spawn. If, when this occurs, one sneak is able to fertilise the female's eggs while the other is driven away by the parental, the probability of stealing fertilisations will be 0.5 for each male (assuming that both sneaks are equally as likely to be driven away by the parental male). As long as this value is greater than the probability of achieving fertilisation when alone, say 0.2, co-operation will be favoured. This form of co-operation occurs in the blue-headed wrasse (*Thalassoma bifasciatum*). As females approach large, gaudy station-holding males they are solicited by small, drab, non-territorial males. Sometimes these small males band together and chase the female away from the colony and then stimulate her to spawn with the group (Warner *et al.*, 1975). Presumably, a non-territorial male would have little chance of single-handedly chasing a female out of the colony and spawning with her.

In the examples above, it pays sneaks to work together because each individual has some prospects of obtaining fertilisations. However, coalitions can be formed when one male is enlisted to help another with no prospect of *immediate* benefit to the enlisted partner. The most famous example of this is reciprocal altruism in olive baboons (*Papio anubis*) (Packer, 1977). Coalitions are formed by one male enlisting an unrelated partner to help fight against an opponent, often in an attempt to separate the opponent from an oestrous female. The altruistic behaviour of the enlisted partner requires special explanation. Helping another male with no prospect of immediate reward will be an evolutionarily stable strategy only if the enlisting male reciprocates by helping the enlisted partner in a future interaction. Thus the evolution of reciprocal altruism necessarily relies on a large element of 'trust' between individuals. Also, the benefits to the male who is helped must be large compared with the costs to the helper (Trivers, 1971).

The major problem faced by the theory of reciprocal altruism is to explain how an evolutionary trend towards co-operative behaviour could have got started in the first place in a population of entirely selfish individuals. Axelrod and Hamilton (1981) suggest that altruism between close relatives may be a starting point from which reciprocal altruism between non-relatives could evolve. Alternatively, reciprocal altruism could evolve if a small group of clustered individuals co-operate with each other but not with other individuals outside the group. Since clustering is often associated with kinship, the two mechanisms would probably reinforce each other in promoting the initial spread of altruism (Axelrod and Hamilton, 1981).

Another question is, what can dominant males do to prevent being exploited by coalitions of sneaks? One possibility is that dominant males could bear a grudge against sneaks who have previously co-operated in coalitions by punishing them when they are encountered alone (assuming individual recognition is possible). This grudging strategy does not occur in baboons, possibly because the dominant male cannot afford to leave his oestrous female unguarded while punishing sneaks. However, some consorting males were able to withstand the efforts of coalitions, either by moving the female away from the vicinity of the coalition, or by manipulating the coalition partners into threatening someone else. Further, those males with the most experience in consorting were best able to avoid potentially harzardous situations which could result in takeovers (Packer, 1979).

Sexual Interference

What is important in natural selection is not an individual's fitness in absolute terms but its fitness relative to that of other individuals (Arnold, 1976; Halliday, 1978). Hence if a subordinate male is unable to obtain a mate using sneaky tactics it may pay him to interfere with the sexual exploits of rival males. Male millipedes, collembola, pseudoscorpions and oribatid mites are reported to eat the spermatophores of rival males when they come across them (Schaller, 1971). Male scorpionflies (*Panorpa* sp) eat their rivals' salivary secretions which are offered to females, and male hanging flies (*Hylobittacus apicalis*) pose as females and steal nuptial food gifts from other males (Thornhill, 1976, 1979). North American salamanders (*Ambystoma maculatum* and *A. tigrinum*) deposit spermatophores on top of those of other males making it impossible for their rivals' sperm to be picked up by females (Arnold, 1976). These examples of sexual interference all fall into the realm of *selfish* behaviour because in each case males detract from the reproductive success of others while obtaining a personal benefit in the form of nutrients or fertilisations.

At first glance, some examples of sexual interference are more difficult to explain. In salamanders of the genera *Plethodon* and *Ensatina* homosexual courtships between males are commonly observed. The courting male appears to induce the other male to waste spermatophores by depositing them in places where they will never be picked up by a female. Male satin bowerbirds (*Ptilinorhynchus violaceus*) spend considerable time and effort wrecking the bowers of other males (Marshall, 1954), and in a similar manner, male sticklebacks sometimes destroy the nests of their rivals (Van Den Assem, 1967) or steal eggs from them and place them in their own nests (see Rohwer, 1978). These examples seem to represent truly *spiteful* behaviour because males appear to cause harm to their rivals at some cost, and no apparent absolute benefit, to themselves. The mathematical models of Hamilton (1970), Rothstein (1979) and Knowlton and Parker (1979) show that such spiteful behaviour is unlikely to be favoured except in very small populations because the benefits gained through spite will be usually outweighed by the cost of performing the spiteful behaviour. Also, as spiteful individuals become more common in the population they will play mainly against each other, thus preventing the fixation of spiteful behaviour.

How, then, do we explain the occurrence of apparently spiteful behaviour in nature? The crucial question is whether the harm delivered to a rival is adaptive in itself because it increases the individual's relative fitness (*spite*), or whether the harm is an adaptation for reproduction which increases the individual's absolute fitness (*selfishness*). The examples mentioned above can be easily interpreted to fit in with the latter explanation and so avoid the theoretical difficulties concerning the evolution of spite. For example, the supposedly 'spiteful' wrecking practised by male bowerbirds may actually increase the absolute number of matings that a male achieves. Bowerbirds are polygamous so a male who wrecks his neighbour's bower probably experiences an increase in the number of females visiting his own. Wrecking behaviour is often accompanied by stealing of adornments which are used to elaborate the wrecker's own nest. Similarly, male sticklebacks who destroy neighbours' nests may increase the attractiveness of their own nest to females. Females have been shown to prefer spawning in nests which contain eggs compared with empty nests, so egg-stealing can also be regarded as an adaptive behaviour designed to increase the attractiveness of the egg-stealer's nest (Rohwer, 1978).

Sexual interference naturally leads to the evolution of counter-adaptations for sexual defence. Male salamanders may steer females away from rivals or deposit such a large number of spermatophores that the rival has little chance of covering them all with his own spermatophores (Arnold, 1976). Male gladiator frogs (*Hyla rosenbergi*) defend their eggs in clay nests and attack rival males who attempt to cannibalise or destroy them. Males possess very sharp prepollical spines which are used in prolonged fights to stab at the eyes and eardrums of rival males (Kluge, 1981), and such armaments almost certainly discourage sexual interference.

Conflict Between Males and Females

So far we have considered alternative male mating strategies only from the point of view of competition and conflict between males for females, or as Darwin (1871) called it, the 'law of battle' between males. However, because males and females invest differentially in each offspring, individual selection often leads to a battle of the sexes (Trivers, 1972; Dawkins, 1976; Parker, Ch. 5 this volume). In many polygynous species, males attempt to maximise the number of matings they achieve and bypass any choice that females attempt to exercise. But females are selected to maintain their options, not to be misled or have their interests subverted (Lloyd, 1979).

Reluctant Females

In species where males invest nothing but sperm in their offspring, females may have little to gain as a result of mating repeatedly with one or several males; females are expected to be choosy, to maximise the quality rather than the quantity of their mates (Bateman, 1948; Trivers, 1972).

Females of many species appear actively to discriminate against certain males when choosing a mate. Female frogs and toads avoid non-calling satellite males by swimming or running away, and if a satellite male achieves successful amplexus they can sometimes shake him off (Arak, 1983). In general, there are several advantages that females might gain by avoiding satellite males. In species where males offer material resources to females such as food, territories or nest sites, there is an obvious advantage gained by avoiding males who do not possess these resources. In bullfrogs, for example, the largest calling males occupy the best oviposition sites where spawn predation by leeches is the lowest (Howard, 1978b). Therefore it pays female bullfrogs to avoid mating with satellite males because they are excluded from spawning in the best places.

In species where males normally contribute towards rearing the young it may pay females to reject males who are likely to desert the offspring. For example, male pied flycatchers differ in their propensity to feed the young depending on the order in which females are mated. Polygynous males with two females do not help to feed the second brood as much as the first brood. Thus female flycatchers who mate with already-mated males suffer a brood reduction because some young starve in the nest. In this case the question immediately arises: why should the female accept an already-mated male? Alatalo *et al.* (1982) suggest that male pied flycatchers deceive females about their true marital status. After a male has attracted his first mate he begins to sing from another territory some distance from his first one. Because of the male's polyterritorial behaviour, only a prolonged courtship period would enable a female to assess his true marital status, but she is forced into accepting him immediately because the prospects for successful breeding decline rapidly as the season progresses.

Sometimes there may be good proximate reasons why females should avoid mating with sneaky males. Borgia (1981) suggests that female dungflies may prefer to mate with large males because large males protect females from attack by other males. In addition, females paired with large males are less susceptible to takeovers resulting in more rapid completion of copulation and oviposition (Parker, 1970; Borgia, 1981). In nocturnally breeding frogs, the avoidance of heterospecific matings may be the main advantage gained by females who pair with calling males. Calls provide the most important cue for species recognition and females who pair with non-calling satellites cannot be sure that these males are conspecifics (Gerhardt, 1982; Arak, 1983).

Females may prefer to mate with dominant males for genetic reasons. Often dominant males are larger or older individuals and these traits may be indicative of high 'genetic quality'. However, the question of whether there is sufficient genetic variance in traits directly associated with fitness (in the sense of survival ability) is a somewhat contentious issue (see Lande, 1976; Maynard Smith, 1978; Parker, 1982; Andersson, 1982; Taylor and Williams, 1982), although there is now considerable evidence building up to suggest that the heritability of these traits may be higher than is usually assumed (e.g. birds: Boag and Grant, 1978; van Noordwijk *et al.*, 1980; fruit flies: Partridge, 1980).

There will be a conflict of interest between non-preferred males and females in all situations where females prefer to mate with some males and not with others (Table 6.4). In some species, sneaky breeders successfully bypass the process of prolonged courtship during which a female could potentially assess male quality. A common way of achieving this is by forced copulation. Male scorpionflies sometimes forcibly capture and inseminate females when they are unwilling to mate (Thornhill, 1980). Coercion of females by males takes a more subtle form in the red-spotted newt (*Notophthalmus viridescens*). In this species there are two different modes of courtship (Verrell, 1982). If the female is initially responsive to the male when approached, the male performs a 'hula' display, in which he waves his body and tail. If the female then approaches him, he turns round and begins the spermatophore transfer phase of courtship. This mode of courtship is quick, taking only a few minutes to complete the display and transfer spermatophores. A much longer courtship sequence occurs, however, when the female is initially unresponsive to the male. If she swims away the

Table 6.4: Examples of conflict between the sexes in species with alternative male mating strategies

Species	Interests of the male	Interests of the female	Reference
Elephant seal	Young subordinate male attempts to copulate with females	Female resists copulation attempts by subordinate males	Cox and Le Boeuf, 1977
Pied flycatcher	Already-mated males attempt to attract additional females	Female prefers to mate with a male who will remain faithful and help feed her offspring	Alatalo *et al.*, 1982
Scorpionfly	Males without nuptial offerings attempt forced copulations with unreceptive females	Female prefers to mate with males who offer salivary gland secretions	Thornhill, 1980
Dunnock	Dominant male attempts to monopolise female to ensure his paternity	Female solicits to both males sharing territory to encourage both to feed offspring	Davies, 1983
Baboon Chimpanzee	Dominant male consorts oestrous females and attempts to monopolise matings	Female solicits to low-ranking males or mates promiscuously to ensure successful fertilisation or to solicit future help in offspring care	Rasmussen, 1980; Tutin, 1979

Figure 6.4: The influence of female behaviour on mating success of male elephant seals. (a) Males of low status are most likely to be interrupted by other males during mounting attempts, especially if the female protests. (b) Females protest more frequently when mounted by a male of low rank

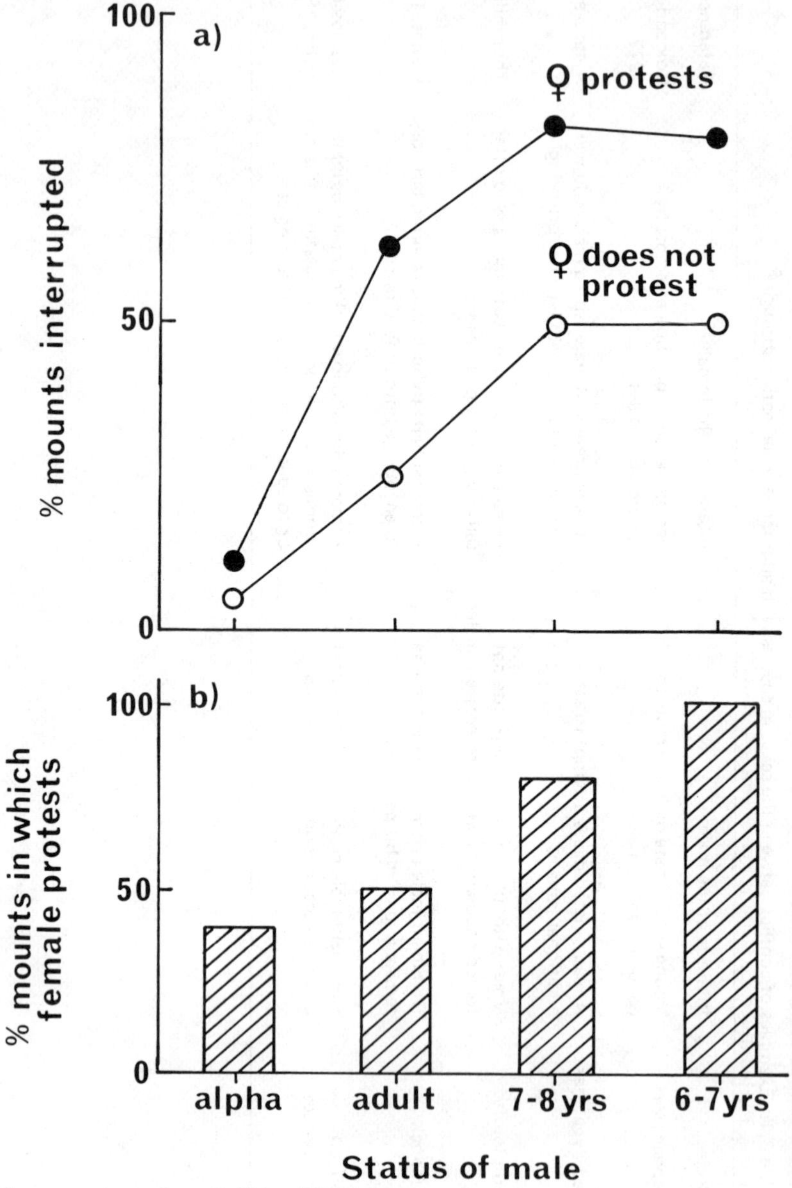

Source: redrawn from Halliday (1978), based on data in Cox and Le Boeuf (1977)

male pursues her and attempts to clasp her round the neck with his hindlimbs. If he succeeds, a prolonged period of amplexus follows, which may last for three hours. During amplexus the male stimulates the female by rubbing his cheek glands against her snout and by beating his tail. This behaviour appears to increase the sexual receptivity of the female since 71 per cent of females who are initially unreceptive have become responsive after capture and amplexus. Assessment of males by females during the amplexus period cannot be discounted here because females have the opportunity to escape when released by the male after amplexus. However, this study demonstrates the potential for 'persuasion' (*sensu* Tinbergen, 1953; Halliday, 1983) of females that can be exercised by males in some species.

Counter-adaptations against forced copulation may evolve in females of some species if there are disadvantages associated with mating with certain males. The behaviour of females of many species appears to intensify competition making it more difficult for subordinate males to mate. For example, female elephant seals seem to incite competition between males by giving loud vocal protests when they are mounted by a low-ranking male (Figure 6.4). These protest calls often attract the attention of the dominant bull nearby who drives away the subordinate male (Cox and Le Boeuf, 1977). Similar tactics are used by female natterjack toads who emit protest calls when they are captured by a satellite male; these calls sometimes attract the calling male who then attempts to dislodge the satellite (Arak, 1982).

Promiscuous Females

Despite the general prediction of sexual selection theory that females will be coy and unwilling to mate with several males, female promiscuity is probably more widespread in nature than is usually assumed.

Multiple mating by females has been recorded in many species belonging to a wide variety of taxonomic groups. In some species, the female may not obtain any benefits by mating more than once, but extra copulations may be forced on females by physically superior males and the cost of avoiding male harassment may exceed the cost of wasting time in supernumerary copulations (e.g. dung flies: Parker, 1970; bees: Alcock *et al.*, 1977b). However, the evidence suggests that in at least some species females benefit directly by multiple mating.

Sometimes it may pay a female to mate with additional males in order to gain access to a male-controlled resource such as an oviposition site (e.g. dragonflies: Campanella and Wolf, 1974) or a food source (e.g. nuptial feeding by male insects and birds). If males are likely to assist in parental care of the young, then a female may benefit by mating with more than one male in order to increase the number of offspring that she can raise. On breeding territories of the dunnock, where two males share the same female, the female actively solicits copulations from *both* males. This behaviour is favoured because when both males copulate then both provide help in feeding the young and the female can raise a larger brood (Davies, 1983). In this case it is clear that the female benefits by copulating with more than one male, but why should both males help to feed the brood? After copulating with the female why doesn't one male desert the nest and leave the other one lumbered with the task of caring for the offspring? The answer to this question probably lies in the fact that each male, having copulated with the female, has a genetic stake in the offspring so it pays him to feed them. However, sperm competition between males is likely to reduce the chance that a given male will fertilise all the eggs in a clutch, and therefore may favour desertion by males. A curious feature of courtship in this species is that hen dunnocks have evolved a fascinating behaviour which may encourage both males to stay and feed the young by increasing each male's confidence of paternity. Prior to copulating, the male pecks the female's cloaca and she ejects some sperm from previous matings. Sperm ejection by the hen dunnock prior to copulation can thus be seen as an adaptation designed to persuade each male in turn to 'accept responsibility' for the offspring (Davies, 1983).

Another condition that will favour multiple mating by females is when sperm concentrations fall below the level at which successful fertilisation is possible. There are several situations where this may be true. In externally-fertilising organisms, especially those shedding sperm pelagically or in fast currents, there is a risk that the sperm will be washed away or fail to come into contact with the eggs due to a slight asynchrony in gamete release between male and female. Female salmon (*Salmo salar*), which spawn on the bed of fast-flowing streams, indicate the moment of egg release by pouting and pushing the anal fin downwards, and this behaviour attracts one or several younger males who join the spawning pair and release sperm (Jones and King, 1952). Extra males are also involved in spawning by northern longear sunfish (*Lepomis megalotus*). In this species, a gravid female enters the nest of a territorial male and adopts a conspicuous tilting movement which attracts neighbouring territorial males. These males rush into the nest and shed sperm at the moment of spawning (Keenleyside, 1972). We cannot be certain that these signals given by females just before spawning have evolved specifically to solicit extra males, but in principle, it is easy to see why it would be advantageous for females to do so if sperm is likely to be quickly dispersed. Keenleyside (1972) remarked that a high proportion of eggs released during intrusions by accessory males were found to be fertile, but he did not compare fertilisation success in these matings with success in single-pair spawnings.

Multiple mating by internally-fertilised females may be favoured if sperm concentrations in the male testis drop below a critical limit due to repeated mating by males. Sperm is often assumed to be cheap and in unlimited supply, but recently a significant decline in sperm concentrations or fertilisation rates with successive ejaculations has been noted in both external and internal fertilisers (e.g. Nakatsuru and Kramer, 1982; Gibson and Jewell, 1982). It follows that if the amount of sperm is limited due to high levels of polygyny or high rates of copulation with single females, sperm may become a limited resource for which females compete. In some species females do this by exercising choice between males to ensure they obtain a fertile mate (e.g. fish: Nakatsuru and Kramer, 1982), whereas in other species females may mate sequentially with many males (e.g. insects: Parker, 1970; Walker, 1980; birds: Hogan-Warburg, 1966; Stacey, 1982; domestic sheep: Gibson and Jewell, 1982; antelope: Buechner and Schloeth, 1965; primates: Goodall, 1965; Stacey, 1982).

Sivinski (1980) has suggested that by encouraging sperm competition females may be selecting for males with the 'best' sperm. In a sense, the sperm reservoir acquired from several males may serve as a kind of lek from which the egg chooses its mate. Sperm phenotypes are generated by genes in the diploid genotype of the father, not in the haploid genotype of the sperm itself (Beatty and Gluecksohn-Waelsch, 1972) so the outcome of sperm competition may be to produce offspring with the ability to manufacture efficient sperm. A possible evolutionary outcome of multiple mating by females is an arms race between males to produce large quantities of highly efficient sperm. An interesting point here is that females may actually benefit by actively encouraging sperm competition between males. If the female receives a number of ejaculates, and there is not priority to any given ejaculate, then statistically speaking, the male with the most efficient sperm or the largest ejaculate has the greatest chance of fertilising the eggs. A female who is fertilised by a highly potent male may increase her genetic contribution to future generations by being the bearer of potent progeny, as long as there is some initial genetic variation in sperm efficiency or ejaculate size between males. Female promiscuity and male potency may evolve hand in hand and spread through the population in an analagous way to the spread of female choice and a male trait by sexual selection (Fisher, 1930).

There is some evidence that this process of selection for potent progeny has occurred in primates. In many species of Old World monkeys, females show pronounced sexual swellings during oestrus which may serve to attract the attention of males during the period when the female is fertile (Clutton-Brock and Harvey, 1976). Female sexual swellings are most common in genera containing many species which typically live in multi-male groups (i.e. *Pan*, *Papio*, *Cercocebus*, *Macaca*). In these groups females mate with several males during the oestrous period and males have large testes. In comparison, monkeys which live in multi-male groups in the New World (most *Cebids*) show neither pronounced sexual swellings in females nor large testis size in males (data on testis size in Harcourt *et al.*, 1981). Although it is possible that these differences between the Old World and New World species are caused by phylogenetic or ecological differences, the correlation between the presence of female sexual swellings and large testis size in males suggests that these traits, rather than representing indeppendent adaptations, have coevolved (see also Short, 1979). In conclusion, it is important to realise that sperm quality, the behaviour of females as well as competition between males may have influenced the selection and survival of the present-day phenotypes.

Promiscuous tendencies in females, whatever their cause, will tend to intensify competition between males. While subordinate males, who would not normally have access to receptive females, may benefit from these tendencies, dominant males will lose out unless they can control females and prevent them from mating with other males. Here the sexual conflict will be between the dominant male and the female (Table 6.4). Again, who wins this conflict may depend on initial asymmetries between male and female. In chimpanzees (*Pan trogolydytes*), for example, sexual dimorphism in body size is not very marked and under some circumstances the female seems to get away with being highly promiscuous (Tutin, 1979). Goodall (1965) noted on one occasion that seven males mounted one female, with less than two minutes between the first five copulations! In contrast, male Hamadryas baboons (*Papio hamadryas*) are much larger than females and have extremely large canine teeth. Dominant males herd groups of females together and behave aggressively towards females who wander from the group, often biting them on the neck as a punishment (Kummer and

Kurt, 1963). Like the female chimp, the female Hamadryas baboon has well-developed sexual swellings suggesting that it is to her advantage to attract the attentions of several males. The fact that she is rarely allowed to be promiscuous without being punished suggests, perhaps, that this may represent a case where the male has won the evolutionary arms race.

Summary

In this chapter I have attempted to provide evolutionary explanations of how alternative male reproductive strategies may evolve and be maintained in a population. The contemporary view of evolution leads us to predict that whenever we find some individuals investing time or energy in attracting mates, defending territories or caring for offspring, then other individuals will be found exploiting this investment. In the face of a deluge of terms used in the literature to describe such alternative strategies, I have, for simplicity, referred to males who exploit the investment of others to obtain matings as 'sneaky breeders'.

In general, alternative strategies may be maintained in a population either by frequency dependence, such that each strategy achieves equal reproductive success (a mixed evolutionarily stable strategy), or because some individuals are poor competitors and must opt for a strategy with a lower payoff (a conditional strategy). In nature, we are perhaps more likely to find the latter type of explanation more commonly applicable, because in most species there are significant differences in competitive ability between individuals. Although in many cases, inferior competitors appear to minimise their losses by attempting to steal matings, there are a few examples where subordinate males form co-operative associations with dominant males. By co-operating with the dominant male in attracting females or defending a territory, subordinate individuals 'pay for permission' to remain in close association with the dominant and reap some of the rewards gained through their mutual efforts.

Conflicts of interest between males will always occur when alternative male mating strategies exist in a population. Where sneaks steal matings and offer no help in return we may see 'nature red in tooth and claw'. Complex adaptations may evolve in sneaks to obtain matings by escaping detection, by forming coalitions with other sneaks or by interfering with the sexual exploits of dominant males. At the same time, however, dominant males will evolve counter-measures to repel sneaks, punish them or bear a grudge against them so that it is not worth their while attempting to steal fertilisations. In cases where subordinate males help dominant males, sexual conflict may be more subtle. Each party in such an association faces a difficult dilemma: while it is better to co-operate than 'go it alone', in order to obtain maximum benefits from the interaction each male faces the risk of causing the other male to desert. Adaptations and arms races in co-operative associations such as these must involve far more subtle forms of psychological manipulation than will be selected for in interactions that are only exploitative. Although nothing much is known of the process of negotiation and compromise which must occur between males in such co-operative associations, it is a promising area for future research. One of the important questions which would be worth examining is, how does the pattern of threats or concessions by the dominant male affect the subordinate male's subsequent behaviour, and how does the level of help offered by the subordinate affect the dominant male's tolerance of him?

Sexual conflict is not restricted to a battle between males for females but there also will be a battle of the sexes. Where sneaks offer neither material nor genetic resources to the female, then an arms race may develop between them in the classical sense of sexual selection theory. Males will be selected to be eager and willing to mate, resorting to forced copulations with females if possible, whereas females will be selected to be coy and resistant, sometimes deliberately inciting rivalry between males to ensure they mate with the 'best' male. But in certain circumstances it may pay females to be promiscuous, against the best interests of dominant males. This may be true if sperm is in short supply because of the environmental circumstances in which fertilisation takes place, or in polygynous species, because the sperm count of the most dominant or favoured males drops below the level for successful fertilisation. Promiscuity by females (resulting in sperm competition) may be the behavioural agent that selects for increased testis size and sperm efficiency in males, and if these traits are heritable it may be to the female's advantage to encourage sperm competition to increase her chances of obtaining potent sons. The outcome of arms races within and between the sexes may in some cases depend on the relative costs of losing the conflict, or on the initial ability of each opponent to win the dispute by brute force.

Acknowledgements

I thank Chris Barnard, Nick Davies, Mark Elgar, David Gibbons, David Harper, Arne Lundberg and Paul Verrell for valuable comments on an earlier version of this chapter.

References

Alatalo, R.V., Carlson, A., Lundberg, A. and Ulfstrand, S. (1981). The conflict between male polygamy and female monogamy: the case of the pied flycatcher *Ficedula hypoleuca*. *Amer. Nat*. *117*: 738-53

—Lundberg, A. and Stahlbrandt, K. (1982). Why do pied flycatcher females mate with already-mated males? *Anim Behav*. *30*: 585-93

—Gustafsson, L. and Lundberg, A. (1983). High frequency of cuckoldry in pied and collared flycatchers. *Oikos*, in press

Alcock, J., Jones, C.E. and Buckman, S.L. (1977a). Male nesting strategies in the bee *Centris pallida* Fox (Anthophoridae: Hymenoptera). *Amer. Nat. 111*: 145-55

—Eickwort, G.C. and Eickwort, K.R. (1977b). The reproductive behaviour of *Anthidium maculosum* (Hymenoptera: Megachilidae) and the evolutionary significance of multiple copulations by females. *Behav. Ecol. Sociobiol. 2*: 385-96

Alexander, R.D. (1975). Natural selection and specialized chorusing behavior in acoustical insects. In D. Pimentel (ed.), *Insects, Science and Society*: 37-77. New York, Academic Press

Andersson, M. (1982). Sexual selection, natural selection and quality advertisement. *Biol. J. Linn. Soc. 17*: 375-93

Arak, A. (1982). Male-male competition and mate choice in frogs and toads. Unpublished Ph.D thesis, University of Cambridge

—(1983). Male-male competition and mate-choice in anuran amphibians. In P.P.G. Bateson (ed.), *Mate Choice*: 181-210. Cambridge, Cambridge University Press

Arnold, S.J. (1976). Sexual behaviour, sexual interference and sexual defence in the salamanders *Ambystoma maculatum, A. tigrinum* and *Plethodon jordani. Z. Tierpsychol. 42*: 247-300

Axelrod, R. and Hamilton, W.D. (1981). The evolution of cooperation. *Science, 211*: 1390-6

Bachmann, C. and Kummer, H. (1980). Male assessment of female choice in hamadryas baboons. *Behav. Ecol. Sociobiol. 6*: 315-21

Balph, D.F., Innis, G.S. and Balph, M.H. (1980). Kin selection in Rio Grande turkeys: a critical assessment. *Auk 97*: 854-60

Barnard, C.J. and Sibly, R.M. (1981). Producers and scroungers: a general model and its applicability to captive flocks of house sparrows. *Anim. Behav. 29*: 543-50

Bateman, A.J. (1948). Intra-sexual selection in *Drosophila. Heredity 2*: 349-68

Beatty, R.A. and Gluecksohn-Waelsch, S. (1972). *The Genetics of the Spermatozoon*. Edinburgh, Department of Genetics of the University

Bertram, B.C.R. (1980). Vigilance and group size in ostriches. *Anim. Behav. 28*: 278-86

Bjorklund, M. and Westman, B. (in preparation). The importance of mate-guarding in the pied flycatcher (*Ficedula hypoleuca*): a removal experiment

Boag, P.T. and Grant, P.R. (1978). Heritability of external morphology in Darwin's finches. *Nature 274*: 793-4

Borgia, G. (1979). Sexual selection and the evolution of mating systems. In M.S. Blum and N.A. Blum (eds.), *Sexual Selection and Reproductive Competition in Insects*: 19-80. London, Academic Press

—(1981). Mate selection in the fly *Scatophaga stercoraria*: female-choice in a male-controlled system. *Anim. Behav. 29*: 71-80

Bradbury, J.W. (1981). The evolution of leks. In R.D. Alexander and D.W. Tinkle (eds.), *Natural Selection and Social Behavior: Recent Research and New Theory*: 138-69. New York, Chiron Press

—and Gibson, R.M. (1983). Leks and mate choice. In P.P.G. Bateson (ed.), *Mate Choice*: 109-40. Cambridge, Cambridge University Press

Bucher, T.L., Ryan, M.J. and Bartholomew, G.A. (1982). Oxygen consumption during resting, calling and nest-building in the frog *Physalaemus pustulosus. Physiol. Zool. 55*: 10-22

Buechner, H.K. and Schloeth, R. (1965). Ceremonial mating behaviour in Uganda Kob (*Adenota kob thomasi*, Neumann). *Z. Tierpsychol. 22*: 209-25

Cade, W.H. (1981). Field cricket spacing, and the phonotaxis of crickets and parasitoid flies to clumped and isolated cricket songs. *Z. Tierpsychol. 55*: 365-75

Campanella, P.J. and Wolf, L.L. (1974). Temporal leks as a mating system in a temperate zone dragonfly (Odonata: Anisoptera) I. *Plathemis lydia* (Drury). *Behaviour 51*: 49-87

Christian, J.J. (1963). Endocrine adaptive mechanisms and the physiological regulation of population growth. In W.V. Mayer and R.G. van Gelder (eds.), *Physiological Mammalogy*, vol. 1. New York, Academic Press

Clutton-Brock, T.H. and Harvey, P.H. (1976). Evolutionary rules and primate societies. In P.P.G. Bateson and R.A. Hinde (eds.), *Growing Points in Ethology*: 195-237. Cambridge, Cambridge University Press

— Albon, S.D., Gibson, R.M. and Guinness, F.E. (1979). The logical stag: adaptive aspects of fighting in red deer (*Cervus elephas* L.). *Anim. Behav. 27*: 211-25

Cox, C.R. and Le Boeuf, B.J. (1977). Female incitation of male competition: a mechanism of mate selection. *Amer. Nat. 111*: 317-35

Darwin, C.R. (1871). *The Descent of Man and Selection in Relation to Sex*. London, John Murray

Davies, N.B. (1978). Territorial defence in the speckled wood butterfly *Pararge aegeria*: the resident always wins. *Anim. Behav. 26*: 138-47

—(1982). Alternative strategies and competition for scarce resources. In Kings College Sociobiology Group (eds.), *Current Problems in Sociobiology*: 363-80. Cambridge, Cambridge University Press

—(1983). Polyandry, cloaca pecking and sperm competition in dunnocks *Prunella modularis. Nature 302*: 334-6

—(in press). Sexual conflict and a variable mating system in dunnocks, *Prunella modularis*. In R. Smith and R. Sibly (eds.), *Behavioural Ecology: The Ecological Consequences of Adaptive Behaviour. Symposium of the British Ecological Society*. Oxford, Blackwells Scientific Press

Davison, G.W.H. (1982). Sexual displays of the great argus pheasant *Argusianus argus. Z. Tierpsychol. 58*: 185-202

Dawkins, R. (1976). *The Selfish Gene*. Oxford, Oxford University Press

—(1980). Good strategy or evolutionarily stable strategy? In G.W. Barlow and J. Silverberg (eds.), *Sociobiology: Beyond Nature/Nurture*: 331-67, Boulder, Colorado, Westview Press

—and Krebs, J.R. (1979). Arms races between and within species. *Proc. Roy. Soc. Lond. B 205*: 489-511

De Vos, G.J. (1983). Social behaviour of black grouse, an observational and experimental field study. *Ardea 71*: 1-103

Drager, B. and Chiszar, D. (1982). Growth of bluegill sunfish (*Lepomis macrochirus*) maintained in groups and in isolation. *Bull. Psychonom. Soc. 20*: 284-6

Dunbar, R.I.M. (1982). Intraspecific variations in mating strategy. In P.P.G. Bateson and P. Klopfer (eds.), *Perspectives in Ethology*, vol. 5: 385-431. New York, Plenum Press

Elgar, M.A. and Catterall, C.P. (1981). Flocking and predator surveillance in house sparrows: test of an hypothesis. *Anim. Behav. 29*: 868-72

Fisher, R.A. (1930). *The Genetical Theory of Natural Selection*. Oxford, Clarendon Press

Fricke, H.W. (1979). Mating system, resource defence and sex change in the anemone fish *Amphiprion akallopisos. Z. Tierpsychol. 50*: 313-26

Gerhardt, H.C. (1982). Sound pattern recognition in some North American tree-frogs (Anura: Hylidae): implications for mate choice. *Amer. Zool. 22*: 581-95

Gibson, R.M. and Jewell, P.A. (1982). Semen quality, female choice and multiple mating in domestic sheep: A test of Trivers' sexual competence hypothesis. *Behaviour 80*: 9-31

Goodall, J. (1965). Chimpanzees of the Gombe Stream Reserve. In I. De Vore (ed.), *Primate Behavior. Field Studies of Monkeys and Apes*: 425-73. New York, Holt, Rinehart & Winston

Gross, M.R. (1982). Sneakers, satellites and parentals: polymorphic mating strategies in North American sunfishes. *Z. Tierpsychol. 60*: 1-26

Halliday, T.R. (1978). Sexual selection and mate-choice. In J.R. Krebs and N.B. Davies (eds.), *Behavioural Ecology: An Evolutionary Approach*: 180-213. Oxford, Blackwell

—(1983). The study of mate choice. In P.P.G. Bateson (ed.), *Mate Choice*: 3-32. Cambridge, Cambridge University Press

Hamilton, W.D. (1967). Extraordinary sex ratios. *Science 156*: 477-88

—(1970). Selfish and spiteful behaviour in an evolutionary model. *Nature 228*: 1218-20

Harcourt, A.H., Harvey, P.H., Larson, S.G. and Short, R.V. (1981). Testis weight, body weight and breeding systems in primates. *Nature 293*: 55-7

Hogan-Warburg, A.J. (1966). Social behaviour of the ruff, *Philomachus pugnax* (L). *Ardea 54*: 109-229

Howard, R.D, (1978a). The evolution of mating strategies in bullfrogs, *Rana catesbeiana. Evolution 32*: 850-71

—(1978b). Factors influencing early embryo mortality in bullfrogs. *Ecology 59*: 789-98

Jones, J.W. and King, G.M. (1952). The spawning of the male salmon parr (*Salmo salar*, Linn., juv.). *Proc. Zool. Soc. of Lond. 122*: 615-19

Keenleyside, M.H.A. (1972). Intraspecific intrusions into nests of spawning longear sunfish (Pisces: Centrarchidae). *Copeia 1972*: 272-8

Kenward, R.E. (1978). Hawks and doves. Factors affecting success and selection in goshawk attacks on wood-pigeons. *J. Anim. Ecol. 47*: 449-60

Kluge, A.G. (1981). The life-history, social organization, and parental behaviour of *Hyla rosenbergi*, Boulenger, a nest-building gladiator frog. *Misc. Publ. Mus. Zool. Univ. Mich. 160*: 1-170

Knowlton, N. and Parker, G.A. (1979). An evolutionarily stable strategy approach to indiscriminate spite. *Nature 279*: 419-21

Kummer, H. and Kurt, F. (1963). Social units of a free-living population of Hamadryas baboons. *Folia Primatologica 1*: 4-19

Lack, D. (1968). *Ecological Adaptations for Breeding in Birds*. London, Methuen

Lande, R. (1976). The maintenance of genetic variability by mutation in polygenic characters with linked loci. *Gen. Res. Camb. 26*: 221-35

Lane, F.W. (1957). *Kingdom of the Octopus: The Life-history of the Cephalopoda.* London, Jarrolds
Lazarus, J. (1979). The early warning function of flocking in birds: an experimental study with captive quelea. *Anim. Behav. 27*: 855-65
Le Boeuf, B.J. (1972). Sexual behaviour in the northern elephant seal, *Mirounga anqustirostris. Behaviour 41*: 1-26
—(1974). Male-male competition and reproductive success in elephant seals. *Amer. Zool. 14*: 163-76
Lill, A. (1974). Sexual behaviour of the lek-forming white-bearded manakin (*Manacus manacus trinitatus*). *Z. Tierpsychol. 45*: 225-55
Lloyd, J.E. (1979). Mating behavior and natural selection. *Florida Entomol. 62*: 17-23
Loiselle, P.V. and Barlow, G.W. (1978). Do fishes lek like birds? In E.S. Reese and F.J. Lighter (eds.), *Contrasts in Behaviour. Adaptations in the Aquatic and Terrestrial Environments*: 31-75. Chichester, John Wiley
Marshall, A.J. (1954). *Bowerbirds*. Oxford, Oxford University Press
Maynard Smith, J. (1978). *The Evolution of Sex*. Cambridge, Cambridge University Press
—(1982). *Evolution and the Theory of Games*. Cambridge, Cambridge University Press
Moodie, G.E.E. (1972). Predation, natural selection and adaptation in an unusual three-spined stickleback. *Heredity 28*: 155-67
Nakatsuru, K. and Kramer, D.L. (1982). Is sperm cheap? Limited male fertility and female choice in the lemon tetra (Pisces, Characidae). *Science 216*: 753-5
Otte, D. (1974). Effects and functions in the evolution of signalling systems. *Ann. Rev. Ecol. Syst. 5*: 385-418
—(1977). Communication in Orthoptera. In T.A. Sebeok (ed.), *How Animals Communicate*. Bloomington, Indiana University Press
Packer, C. (1977). Reciprocal altruism in *Papio anubis. Nature 265*: 441-3
—(1979). Male dominance and reproductive activity in *Papio anubis. Anim. Behav. 27*: 37-45
Parker, G.A. (1970). The reproductive behaviour and the nature of sexual selection in *Scatophaga stercoraria* L. (Diptera: Scatophagidae). V. The female's behaviour at the oviposition site. *Behaviour 37*: 140-68
—(1982). Phenotype-limited evolutionarily stable strategies and fitness in three types of intrasexual competition. In Kings College Sociobiology Group (eds.), *Current Problems in Sociobiology*: 173-202. Cambridge, Cambridge University Press
Partridge, L. (1980). Mate-choice increases a component of offspring fitness in fruit flies. *Nature 283*: 290-1
Payne, R.B. and Payne, K. (1977). Social organization and mating success in local song populations of village indigo birds *Vidua chalybeata. Z. Tierpsychol. 57*: 209-14
Perrill, S.A., Gerhardt, H.C. and Daniel, R. (1978). Sexual parasitism in the green treefrog, *Hyla cinerea. Science 200*: 1179-80
—Gerhardt, H.C. and Daniel, R. (1982). Mating strategy shifts in male green treefrogs (*Hyla cinerea*): an experimental study. *Anim. Behav. 30*: 43-8
Perrins, C.M. and Moss, D. (1975). Reproductive rates in the great tit. *J. Anim. Ecol. 44*: 695-706
Pulliam, H.R. (1973). On the advantages of flocking. *J. theor. Biol. 38*: 419-22
Rasmussen, K.L.R. (1980). Consort behaviour and mate selection in yellow baboons (*Papio cynocephalus*). Unpublished Ph.D. thesis, University of Cambridge
Rohwer, S. (1978). Parent cannibalism of offspring and egg raiding as a courtship strategy. *Amer. Nat. 112*: 429-40

Rohwer, S. and Fretwell, S.D. and Niles, D.M. (1980). Delayed maturation in passerine plumages and the deceptive acquisition of resources. *Amer. Nat. 115*: 400-37

Rothstein, S.I. (1979). Gene frequencies and selection for inhibitory traits, with special emphasis on the adaptiveness of territoriality. *Amer. Nat. 113*: 317-33

Rubenstein, D.I. (1980). On the evolution of alternative mating strategies. In J.E.R. Staddon (ed.), *Limits to Action: The Allocation of Individual Behavior*: 65-100. New York, Academic Press

Ryan, M.J., Tuttle, M.D. and Taft, L.K. (1981). The costs and benefits of frog chorusing behaviour. *Behav. Ecol. Sociobiol. 8*: 273-8

—Tuttle, M.D. and Taft, L.K. (1982). Bat predation and sexual advertisement in a neotropical frog. *Amer. Nat. 119*: 136-9

Schaller, F. (1971). Indirect sperm transfer by soil arthropods. *Ann. Rev. Entomol. 16*: 407-46

Semler, D.E. (1971). Some aspects of adaptation in a polymorphism for breeding colours in the three-spined stickleback (*Gasterosteus aculeatus*). *J. Zool. Lond. 165*: 295-302

Short, R.V. (1979). Sexual selection and its component parts, somatic and genital selection, as illustrated by man and the great apes. *Adv. Study. Behav. 9*: 131-58

Sivinski, J. (1980). Sexual selection and insect sperm. *Florida Entomol. 63*: 99-111

Snow, B. (1974). Lek behaviour and breeding of Guy's hermit hummingbird *Phaethornis guy*. *Ibis 116*: 278-97

Stacey, P.B. (1982). Female promiscuity and male reproductive success in social birds and mammals. *Amer. Nat. 120*: 51-64

Taylor, P.D. and Williams, G.C. (1982). The lek paradox is not resolved. *Theor. Popul. Biol. 22*: 392-409

Thornhill, R. (1976). Sexual selection and paternal investment in insects. *Amer. Nat. 110*: 153-63

—(1979). Adaptive female-mimicking behaviour in a scorpion fly. *Science 205*: 412-14

—(1980). Rape in *Panorpa* scorpion flies and a general rape hypothesis. *Anim. Behav. 28*: 52-9

Tinbergen, N. (1953). *Social Behaviour in Animals*. London, Methuen

Trivers, R.L. (1971). The evolution of reciprocal altruism. *Q. Rev. Biol. 46*: 35-57

—(1972). Parental investment and sexual selection. In B. Campbell (ed.), *Sexual Selection and the Descent of Man*: 139-79. Chicago, Aldine

—(1976). Sexual selection and resource accruing abilities in *Anolis garmani*. *Evolution 30*: 253-69

Tutin, E.G. (1979). Mating patterns and reproductive strategies in a community of wild chimpanzees (*Pan troglodytes schweinfurthi*). *Behav. Ecol. Sociobiol. 6*: 29-38

Van Den Assem, J. (1967). Territory in the three-spined stickleback *Gasterosteus aculeatus* L., an experimental study in intraspecific competition. *Behav. Suppl. 16*: 1-164

van Noordwijk, A.J., van Balen, J.H. and Scharloo, W. (1980). Heritability of ecologically important traits in the great tit. *Ardea 68*: 193-203

van Rhijn, J.G. (1973). Behavioural dimorphism in male ruffs *Philomachus pugnax* (L). *Behaviour 47*: 153-229

Verrell, P. (1982). The sexual behaviour of the red-spotted newt *Notophthalmus viridescens* (Amphibia: Urodela: Salamandridae). *Anim. Behav. 30*: 1224-36

von Holst, D. (1974). Social stress in the tree shrew: its causes and physiological and ethological consequences. In R.D. Doyle and A.C. Walker (eds.), *Prosimian Biology*: 389-411. London, Duckworth

Waage, J.K. (1979). Dual function of the damselfly penis: sperm removal and

transfer. *Science 203*: 916-18
Walker, W.F. (1980). Sperm utilization strategies in non-social insects. *Amer. Nat. 115*: 780-99
Waltz, E.C. (1982). Alternative mating tactics and the law of diminishing returns: the satellite threshold model. *Behav. Ecol. Sociobiol. 10*: 75-83
Warner, R.R., Robertson, D.R. and Leigh, E.G. (1975). Sex change and sexual selection. *Science 190*: 633-8
Watts, C.R. and Stokes, A.W. (1971). The social order of turkeys. *Sci. Am. 224*: 112-18
Wiley, R.H. (1973). Territoriality and non-random mating in the sage grouse, *Centrocercus urophasianus. Anim. Behav. Monogr. 6*: 87-169
—(1974). Evolution of social organization and life history patterns among grouse (Aves: Tetraonidae). *Q. Rev. Biol. 49*: 201-27
Williams, G.C. (1966). *Adaptation and Natural Selection*. Princeton, N.J., Princeton University Press
Wirtz, P. (1981). Territorial defence and territory take-over by satellite males in the waterbuck *Kobus ellipsiprymnus* (Bovidae). *Behav. Ecol. Sociobiol. 8*: 161-2
—(1982). Territory holders, satellite males and bachelor males in a high density population of waterbuck (*Kobus ellipsiprymnus*) and their associations with conspecifics. *Z. Tierpsychol. 58*: 277-300
Wittenberger, J.F. (1978). The evolution of mating systems in grouse. *Condor 80*: 126-37
Wrangham, R.W. (1980). Female choice of least costly males: a possible factor in the evolution of leks. *Z. Tierpsychol. 54*: 357-67
Wynne-Edwards, V.C. (1962). *Animal Dispersion in Relation to Social Behaviour*. Edinburgh, Oliver & Boyd

7 BROOD PARASITISM WITHIN SPECIES

Malte Andersson

Parental care such as nest building, brooding, food provisioning and defence of offspring takes time and energy, and may reduce the survival of the parent. Animals which manage to have their progeny raised by other adults through 'brood parasitism' might therefore be more successful than parents which raise their offspring the normal way. Brood parasitism between different species is well known in birds and insects, and it is gradually becoming clear that parasitism within species is also important in some groups, particularly birds. This chapter reviews the tactics and evolution of intraspecific brood parasitism (see also Arak, Ch. 6, Vehrencamp, Ch. 8 this volume, for parasitism and other exploitation among co-operative breeders).

About 1 per cent of all bird species are interspecific brood parasites (Payne 1977a), and so are many wasps and bees (Wilson, 1971; Michener, 1974). Interspecific brood care also occurs in certain fish, but the relationship appears mutualistic rather than parasitic (McKaye, 1981). Brood parasitism between species is often conspicuous as parasite and host are easy to distinguish, and it has been analysed at depth in certain insects and birds (Wilson, 1971; Eickwort, 1981; Rothstein, 1975, 1982; Payne, 1977a).

Brood parasitism within species is often more elusive, since host and parasite are difficult to distinguish, as, in birds and other ovipositors, are their eggs. Detailed observations of marked adults are therefore required to find out exactly what happens. Such studies are still rare and intraspecific parasitism is poorly known, but there are reasons to expect it to occur in many animals with parental care. No complicated adaptations need evolve: the parasitic offspring receive proper species-specific care if the host accepts them, which is likely except in animals that recognise their own progeny. Brood parasitism within species therefore might evolve more easily than between species. On the other hand, in territorial animals, defence is strongest against conspecifics, which might make intraspecific parasitism difficult in such species (Hamilton and Orians, 1965).

Most studies of brood parasitism within bird species have relied on indirect means such as differences in shape or colour among the eggs in a clutch, addition of more than one egg per day, and appearance of new eggs after the start of incubation (Yom-Tov, 1980). Intraspecific brood parasitism may turn out to be widespread as detailed studies of egg-laying by marked females become more common, particularly if combined with biochemical analyses of parentage (Manwell and Baker 1975; Evans, in preparation; Gowaty and Karlin, in preparation).

Intraspecific brood parasitism has mainly been reported for birds and a few insects; possibly it also occurs in certain salamanders (Blanchard, 1934). In birds, it has been recorded regularly in fewer than 100 species (Yom-Tov, 1980); it is common among ratites, quail and a few passerines, and particularly among waterfowl (Anatidae). Surprising numbers of eggs are laid by several females in 'dump nests' (or 'kolchoz nests', as they are sometimes termed (Fredriksson, 1968)). The record is held by a black-bellied whistling duck *Dendrocygna autumnalis* which incubated 101 eggs; 38 of them hatched (Delnicki *et al.*, 1976). In some populations of ducks between half and two-thirds of the nests are parasitised (Morse and Wight, 1969; McCamant and Bolen, 1979). So far, however, no population has been reported to reproduce only through intraspecific brood parasitism.

After a brief treatment of insects this review necessarily concentrates on birds, as less is known about intraspecific brood parasitism in other groups. However, it is probably more important than presently realised in certain other animals with parental care.

Insects

Usurpation of nest burrows is common among solitary bees and wasps (Hamilton, 1964; Alcock, 1975; Brockmann *et al.*, 1979; Eickwort, 1981), and parasitic laying in nests built and provisioned with food by another female occurs in some species. It has been observed in the genera *Hoplitis*, *Chalicodoma*, *Osmia* and *Dianthidium* among megachilid bees (Eickwort, 1975) and in the sphecid wasp *Trypoxylon politum* (Brockmann, 1980). In some cases parasitism is common: Eickwort (1975) observed intraspecific parasitism by over half the number of females in aggregations of nesting *Hoplitis anthocopoides*.

The female in these species builds a nest which requires considerable labour: a mud nest with cells on rocks (*Hoplitis*, *Chalicodoma*, *Trypoxylon*), or nest chambers in cavities (*Osmia*, *Heriades*). In these groups a common form of parasitism is to chew open a recently finished and food-provisioned cell, eat or remove the owner's egg, lay a new egg in the cell, and close it. Such behaviour qualifies as brood parasitism since the parasite lays her egg in the nest of another female, and the parasitic larva is raised on food gathered by the host; the parasite probably saves considerable time and energy. If the host is present at the nest, she usually repels the parasite, but there are sometimes long fights, and the host may lose (Eickwort, 1975).

It is not known whether parasitism represents a genetic polymorphism or a phenotypic option; Eickwort (1975) suggested the latter. All parasites observed by him and by Fabre (1914) also built their own nests, so parasitism was not genetically fixed in these females. Parasitism occurred even when unused nest sites were abundant, so it is not being a response to lack of sites.

Obligate parasitism, where the usurper always lays eggs in the nest of another species, is the ultimate step in the direction taken by intraspecific brood parasites (Eickwort, 1975). Such obligate 'cleptoparasitism' (as it is termed by entomologists) has apparently evolved several times independently in megachilid bees (Michenor, 1974; Eickwort, 1975). In *Stelis bilineolata*, the female substitutes her own egg for that of the host bee, *Euglossa cordata* (Bennet, 1966). The parasite hence acts as the intraspecific parasites just described, and it seems likely that the interspecific relation may evolve from parasitism within species. Interspecific usurpation has been observed among normally non-parasitic megachilid bees (Bohart, 1955).

Altricial birds

The Starling Sturnus vulgaris

It has recently become evident that intraspecific brood parasitism is important in the starling (Yom-Tov *et al.*, 1974; Yom-Tov, 1980; Power *et al.*, 1981; Evans, 1980; Feare, 1983; Karlsson, 1983). It is the most thoroughly studied altricial bird in this respect; for details of its breeding biology, see Feare (1983) and Karlsson (1983).

The male builds most of the nest cup in a tree hole or nest box. It is mainly the female which incubates, but parental foraging is more evenly shared. The unparasitised clutch size is usually 4-6 eggs; some males are serially polygamous. Like most passerines, starlings lay one egg per day, usually between 08.00 and 10.00 hours, but parasites tend to lay at other times, when the host is unlikely to be present (Feare *et al.*, 1982).

Many consequences of parasitism depend on how offspring survival relates to clutch size. Chick survival over the first three months is inversely related to clutch size in starlings (Lack, 1948), and food delivered per young decreases with increasing brood size, as does fledging success (Tinbergen, 1981). When brood size increases, fledglings weigh less and parents lose more weight while nesting (Tinbergen, 1981; Westerterp *et al.*, 1982). Therefore, survival probably decreases with increasing brood size in parents as well as young.

Figure 7.1: Number of host eggs and parasitic eggs in parasitised nests of the starling *Sturnus vulgaris* in the UK. There is evidence that starlings often remove eggs from nests; this is possibly a tactic used by parasites

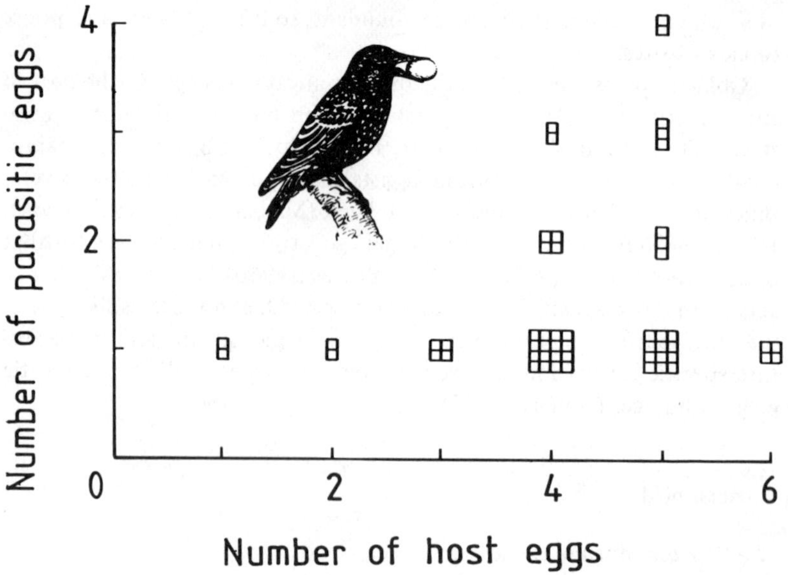

Source: modified after Evans (1980 and in preparation)

For these reasons, it should be disadvantageous for a starling to become parasitised. If the host does not respond by laying fewer eggs of her own, the total clutch becomes larger, with the negative consequences discussed. If the host does lay fewer eggs, her reproduction is reduced correspondingly. Selection for traits which reduce the risk of parasitism should therefore be strong.

How common is brood parasitism among starlings? Yom-Tov *et al.* (1974) estimated that 40 per cent of the nests in their study area were parasitised. Power *et al.* (1981) recorded parasitism in 46 per cent of 24 nests. Electrophoretic and other evidence indicated that at least

12 to 32 per cent of all nests were parasitised in two British colonies during the three years of study (Evans, 1980). Parasitised nests usually contained one parasitic egg, but up to four parasitic eggs were recorded (Figure 7.1). In the most extreme season and site, an average of about 2.3 foreign eggs occurred in each parasitised nest (Figure 7.2). Karlsson (1983) recorded parasitism in 30 per cent of dispersed nests, but in only 5 per cent of the nests in a starling colony.

Figure 7.2: Mean clutch size for normal nests (n) and for hosts (h) and parasites (p) in two populations of starlings, UK, 1977-79. Total clutch size is larger in parasitised nests, but the host clutch (h) is smaller than the clutch of non-parasitised females (n)

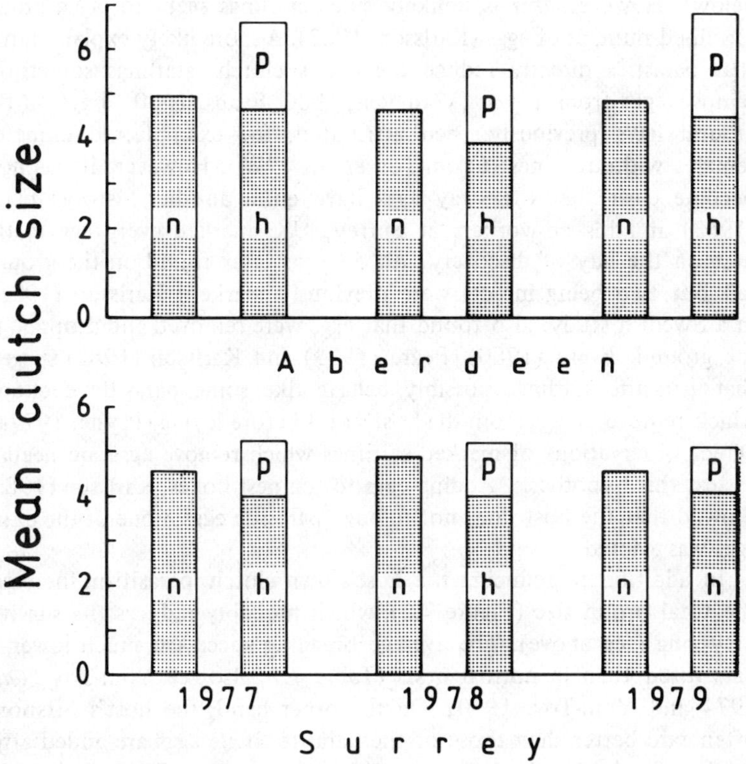

Source: based on results in Evans (1980)

Which are the parasitic females? Evans (1980) suggested that they were mainly of three types: (a) mated females from pairs contesting nests occupied by other pairs; (b) secondary females of polygynous males; (c) 'professional' parasites which lay in other nests in addition to their own, or possibly without making any nest at all. However, the identity and detailed tactics of parasitic females remain to be clarified.

What are the consequences of being parasitised? Although the total clutch became larger, hosts laid fewer eggs than in normal nests (Figure 7.2), so parasitism probably reduces host reproduction.

Why are host clutches smaller in parasitised nests? One possibility is that the host responds to parasitic eggs as to her own, and tends to stop laying when the total clutch reaches normal size, as do parasitised goldeneyes *Bucephala clangula* (Figure 7.4) and other ducks (see below). However, this is unlikely since starlings seem to lay a predetermined number of eggs (Karlsson, 1983). A more likely explanation is that parasites directly reduce the host's clutch: starlings sometimes remove eggs from a nest (Verheyen, 1980; Evans, 1980). Eggs on the ground have previously been attributed, for example, to unmated females without a nest (Yom-Tov *et al.*, 1974). However, it has now become clear that the stray eggs have quite another history: Feare (1983) and his co-workers at Surrey, UK, marked every egg in the nest on the day of discovery. All 58 eggs later found on the ground (all but two being intact) were previously marked! Karlsson (1983), in a Swedish study, also found that eggs were removed and dropped to the ground. Evans (1980), Feare (1983) and Karlsson (1983) suggest that parasitic starlings possibly behave like some parasitic cuckoos, which remove an egg from the host's nest before laying (Payne, 1977a). Direct observations of marked starlings which remove eggs are needed to test this hypothesis. Adding eggs to ten nest boxes, Karlsson (1983) showed that the host does not remove parasitic eggs; none of the extra eggs was evicted.

In addition to reducing the host's own clutch, parasitism increases the total brood size (Figure 7.2), which probably reduces the survival of young (see above). The average breeding success is much lower in parasitised than in normal nests (Table 7.1; also see Yom-Tov *et al.*, 1974 and Yom-Tov, 1980). On the other hand, the host's offspring often fare better than those of the parasite. If its eggs are added after the host begins incubation they will have low success (Table 7.2). However, when added during the host's laying sequence, parasitic eggs for unknown reasons hatched less well than those of the host, resulted in lighter young, and were only half as successful (Table 7.2) as normal eggs in nests which escaped parasitism (Table 7.1). Parasitic eggs usually seem to be added towards the end of the host's laying sequence (Evans, in preparation). Since embryonic development in starlings starts before the clutch is full, hatching is often spread over a day (Karlsson, personal communication). Parasitic eggs may then be among the latest to hatch in a brood, which could partly explain their low success compared to host eggs.

Table 7.1: Average breeding success (% eggs resulting in fledged young) in normal and parasitised nests of starlings

	Normal nests	Parasitised nests
Aberdeen	70	45
Surrey	44	21

Source: Based on first clutches from Evans' (1980) Table 5; the result is similar if renests and second clutches are included.

Table 7.2: Success of parasitic eggs (% eggs resulting in fledged young) laid during and after the host starling's laying sequence

	Eggs laid during the host's laying sequence	Eggs laid after the host's laying sequence
Aberdeen	35	9
Surrey	22	7

Source: Based on Evans' (1980)

Yom-Tov (1980) concluded that the lower fledging success in parasitised nests was mainly due to higher mortality among parasitic young, and the overall egg success for parasitic females seems to be less than half that of a normal female. Hence a purely parasitic hen must produce at least twice as many eggs to achieve normal production. A parasitic female might be favoured by increased survival since she does not raise a brood, but this possibility has not been tested.

Are there any host adaptations which reduce the risk or disadvantage of being parasitised? The remarkable synchrony of egg-laying may be such a trait (Yom-Tov, 1980); about 80 per cent of the females start laying within five days of each other (Dunnet, 1955), and 95 per cent within seven days (Karlsson, 1983). Evans (1980) found some support for Yom-Tov's hypothesis. However, the 'hypersynchrony' may also have evolved for other reasons, such as reduced predation or enhanced foraging efficiency for young which leave the nest during the peak of the fledgling period (Karlsson, 1983).

Other Altricial Birds

The high incidence of parasitism among starlings is probably not typical of most altricial birds. Odd cases of intraspecific brood parasitism have been reported for certain other such species (Yom-Tov, 1980). However, intraspecific parasitism with one or a few eggs can be hard to detect without biochemical analysis. Such studies, which are still exceptions, may reveal surprising frequencies of parasitism. A recently discovered example is the hole-nesting eastern bluebird *Sialia sialis*, a North American epitome of apparently harmonious family life. Gowaty and Karlin (in press) from electrophoretic evidence concluded that at least 15 per cent of the females were caring for at least one offspring not their own. Intraspecific brood parasitism, which previously was unknown for the eastern bluebird, apparently is common. Further similar surprises may emerge as biochemical analyses of parentage become more widespread. For conclusive demonstration of parasitism, electrophoresis seems to need supplementation by other evidence showing that the biochemical differences within the brood are not a result of matings with several males by the female.

Tactics of Parasites and Hosts

Although recent studies of brood parasitism have yielded new and surprising results, they raise even more questions than they answer. Neither the identity nor the tactics of parasitic females have been clarified in detail for any species. This section discusses tactics of hosts and parasites which are not close kin; parasitism involving relatives is treated later on.

How Flexible is the Parasitic Behaviour?

The simplest possibility is that a female is either entirely parasitic, or not at all so. There is no firm evidence for such a fixed difference between females. A second possibility is that a parasite lays in the nest of another female, and in addition rears a brood of her own, as do some redheads *Aythya americana* (Weller, 1959). Since none of 42 trapped parasitising redheads had a brood patch, Weller inferred that part of the population was strictly parasitic, whereas a third cohort seemed to be normal nesters (a complicating feature is that redheads often parasitise other ducks). The situation is similar among wood ducks *Aix sponsa* (Clawson *et al*., 1979; Heusmann *et al*., 1980), in which parasitism is phenotypically flexible: some females parasitised in one year, but nested normally the next season, whereas other females did the opposite (Clawson *et al*., 1979).

A third possibility is that parasites are mainly young birds (e.g. Weller, 1959; Grenquist, 1963; Grice and Rogers, 1965). It might be particularly advantageous for young females not to nest normally if raising a brood severely reduces the survival of inexperienced birds (Evans, 1980). Parasitism seems more prevalent among yearling red-heads than among adults, but some adults are clearly also parasitic (Weller, 1959). In the wood duck, parasites occur among all reproductive age classes (Clawson *et al*., 1979; Heusmann *et al*., 1980).

Success of Hosts and Parasites

Parasitism may evolve to some frequency in a population if a parasitic mutant has higher Darwinian fitness than normal females. Frequency-dependent reproduction in hosts and parasites might equalise the success of the two tactics (Hamilton, 1971; Maynard Smith, 1982). Parasites should be most successful when their proportion in the population is low and host nests are easily found, and least successful when parasites make up a large proportion of the population, and the few host nests are hard to find. At some frequency, parasitism could then be an evolutionarily stable strategy yielding the same fitness as normal nesting. (For a detailed example, see Brockmann *et al*., 1979.)

Another possibility is that parasites always have lower fitness, and turn to parasitism because it is at least better than non-breeding, or nesting at an inferior site. Parasitism can be the best tactic for females of young age or poor condition if correlated with lower egg-laying capacity. The low production from a small clutch might not repay nest building, defence and incubation, which require almost as much time and energy as with a larger brood.

The possibility seems small that there has never been genetic variation in the tendency for parasitism. It is probably a genetically evolved trait: for example, among ducks in the same area and habitat, there are marked and consistent differences between species in the occurrence of parasitism (e.g. Weller, 1959).

In starlings, the breeding success for a pure parasite may be lower than for normally nesting females; a possibly higher survival of parasites can probably not compensate for this disadvantage (Yom-Tov *et al*., 1974; Yom-Tov, 1980). Parasitism in starlings may be an inferior alternative, enforced by circumstances such as scarcity of nest sites. Parasitism increased with the relative occupancy of nest boxes between years and areas (Evans, 1980; Karlsson, 1983). Similar results have been reported for many other species (see below). However, some wood ducks laid parasitic eggs after starting their own nests, so parasitism was clearly not the only alternative open to these females (Clawson *et al*., 1979).

Optimal Clutch Size in Hosts

For birds subjected to brood parasitism, traditional models of optimal clutch size may not apply if mortality of parent or offspring depend on the total clutch size. This was shown in a model by Andersson and Eriksson (1982), where a parasite added a fixed number of eggs b_P to the nest, independent of the number of eggs b_H laid by the host. For simplicity, the proportion of the total brood surviving, $P(b)$, was assumed to decrease linearly with slope a as total brood size b ($=b_H + b_P$) increases, so that $P(b) = 1 - a\,b$. (A power function $1 - ab^d$, with $d > 1$, might often be more realistic, at least for species with parental feeding of young.)

With these assumptions, the number of host fledglings $F(b_H)$ produced is

$$F(b_H) = b_H P(b) = b_H[1 - a(b_H + b_P)] \qquad (1)$$

Examples of how $F(b_H)$ varies with b_H and b_P are given in Figure 7.3a. Seeking the maximum of F by differentiation with respect to b_H and setting the derivative = 0 yields the optimal brood size for the host

$$b_H{}^* = \frac{1}{2a} - \frac{b_P}{2} \qquad (2)$$

To produce the maximum number of own surviving offspring, the host should therefore lay half an egg less for each egg added by the parasite (Figure 7.3a).

What happens if the parasite does not add a fixed number of eggs, but starts to lay when the host has produced k eggs, and then adds one for each new host egg? The host controls the onset of incubation, after which the parasite cannot profitably lay more than perhaps one additional egg; later eggs may not hatch. The host could then start incubation when the clutch size is most favourable for her; if she begins after laying b_H eggs and the parasite later adds another one, the clutch will contain $b_P = b_H - (k - 1)$ parasite eggs.

Figure 7.3: Expected number of host fledglings $F(b_H)$ as a function of host clutch size b_H (curves based on equation (1), with $a = 0.05$). In (a) the parasite adds a fixed number of eggs bp. The optimal host clutch (*) then decreases with half an egg for each parastitic egg added (see equation (2)). In (b) the parasite starts to lay when the host has produced k eggs, the parasite then lays another egg for each additional one produced by the host. The optimal host clutch (*) in this case increases with 1/4 for each egg it already contains when the parasite starts to lay (equation (3))

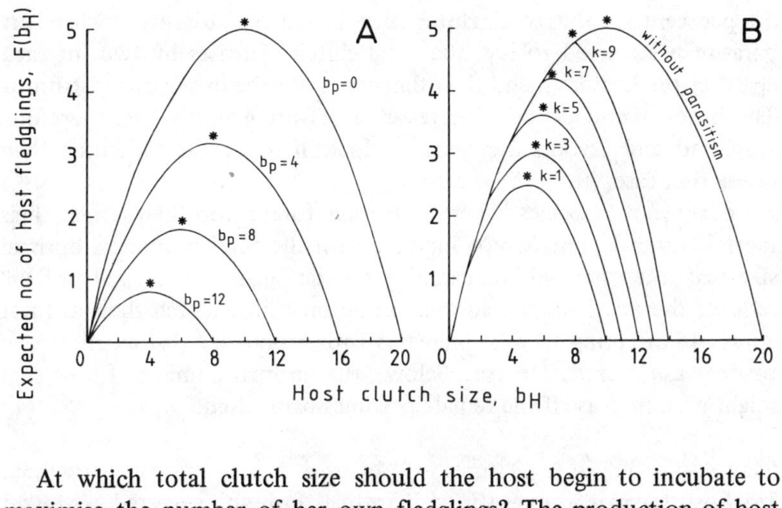

At which total clutch size should the host begin to incubate to maximise the number of her own fledglings? The production of host fledglings in this case is also given by equation (1), the final number of parasitic eggs now being $b_H - (k - 1)$. The optimal host clutch then becomes

$$b_H{}^* = \frac{1}{4a} + \frac{k-1}{4} \tag{3}$$

Hence the optimal host clutch decreases strongly with parasitism, particularly if it starts early, when the host has laid few eggs (k small). The production of host fledglings declines rapidly with k, and the optimal host clutch if parasitism starts after the first host egg is only half as large as without parasitism (Figure 7.3b).

If the host lays the optimal number of eggs, the total clutch in this case becomes $b_H{}^* + b_H{}^* - (k-1) = 1/2a - (k-1)/2$. Hence the total clutch becomes smaller than the optimal size without parasitism, which is $1/2a$ (with $b_P = 0$ in equation (1)). On the other hand, in the first case where the parasite adds a fixed number of eggs b_P and the host responds optimally (equation (2)), the total clutch becomes $b_H{}^* + b_P = 1/2a + b_P/2$, i.e. larger than the optimal non-parasitised clutch.

The difference in optimal host response arises because, in the first case where the parasite adds a fixed number of eggs, the host can increase its own relative contribution to the total clutch by laying more eggs. In the second case, however, the proportion of host eggs always decreases after parasitism has begun, and declines towards 50 per cent as the total clutch size increases. Moreover, when the parasite has started to lay, the total clutch increases by two for each egg that the host lays, and the other egg is not the host's own offspring. The brood size-dependent decrease in offspring survival is, therefore, amplified and favours a stronger reduction of the host's clutch than in the first case.

In precocial species where offspring forage for themselves, their mortality need perhaps not increase with the total brood. Its optimal size may then depend for example on the egg-laying capacity of the female; the previous conclusions on optimal host clutch then will not apply. If offspring survival increases with brood size, for example due to decreasing predation (see below), the optimal number of host eggs might even increase if the female becomes parasitised.

Host Responses

Do host females respond in the optimal fashion suggested by either of these two models? Redheads, wood ducks and several other ducks lay fewer eggs if parasitised (Weller 1959; Clawson *et al*., 1979. Heusmann *et al*., 1980). Goldeneyes supplied with extra eggs responded to early added eggs as to their own, and laid as many fewer as had been added (Andersson and Eriksson, 1982). Ducks in general are indeterminate layers (Sowls, 1950; Weller, 1959) which start incubation when a clutch of normal size is present in the nest, even if some eggs are parasitic. However, if extra eggs were supplied late in a goldeneye's laying sequence, the host did not lay fewer eggs (Figure 7.4); probably she already carried several well-developed ova which could not be resorbed.

These results suggest that ducks cannot discriminate eggs added by a conspecific female, and there is no evidence that they respond in either of the two ways suggested by the optimality models. Evolution of such precise responses may not be feasible. The reproductive results for host and parasite are usually interdependent; to respond optimally the host must therefore correctly anticipate the laying tactic of the parasite, and

Figure 7.4: Host clutch sizes (mean ± s.e.) in goldeneye *Bucephala clangula* nests experimentally 'parasitised' with one or four eggs, early (< 5 host eggs) and late (5-8 host eggs) during the host's laying sequence. Hosts laid significantly fewer own eggs if parasitised with four early eggs than if parasitised with one early egg or four late eggs. With four eggs added early, the host laid correspondingly fewer own eggs

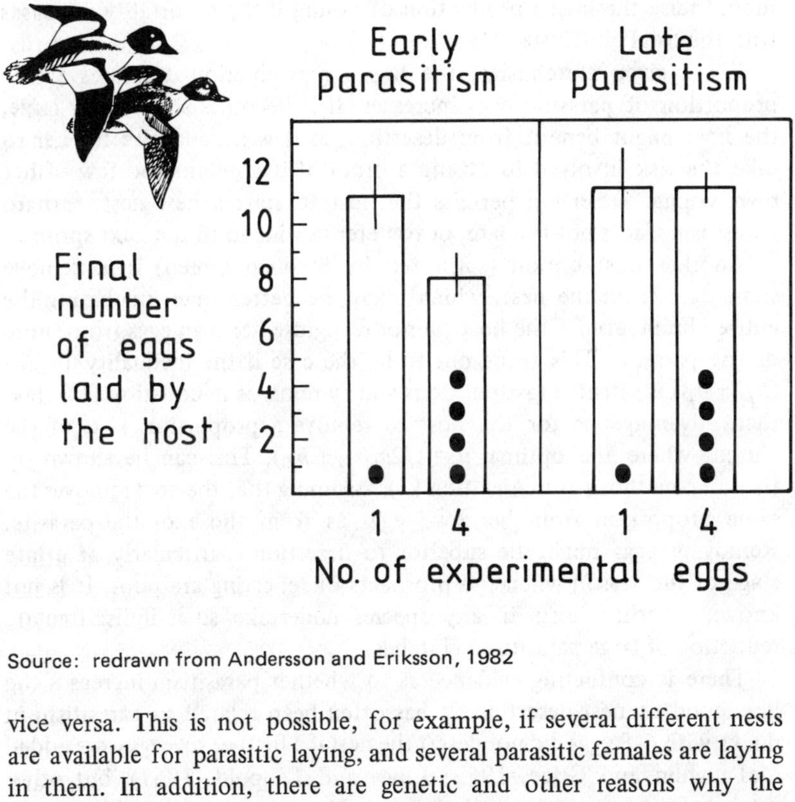

Source: redrawn from Andersson and Eriksson, 1982

vice versa. This is not possible, for example, if several different nests are available for parasitic laying, and several parasitic females are laying in them. In addition, there are genetic and other reasons why the optimal responses suggested by equations (2) and (3) may not evolve (Maynard Smith, 1978; Gould and Lewontin, 1979). For example, Rothstein (1982) suggested that the reason why adaptive rejection of eggs of parasitic cowbirds (*Molothrus ater*) is lacking in some (but not all) victimised species is that the appropriate mutations have not occurred. For this and other reasons, host adaptations are probably simple and robust rather than finely attuned to any particular situation, and models of optimal clutch size may be useful mainly as standards with which to compare and evaluate actual host responses, not as realistic descriptions of such responses.

One robust response for a host is to reduce its own final clutch by one or a few eggs if the nest receives several parasitic eggs. If the host is parasitised both early and late during the laying sequence, a response such as that of goldeneyes (Andersson and Eriksson, 1982) leads to a smaller host clutch than without parasitism. Compared with laying a fixed number of eggs regardless of parasitism, a moderate reduction should raise the host's production of young if their mortality increases with the total clutch size.

For a given clutch size, the host's reproduction decreases as the proportion of parasitic eggs increases. If it becomes sufficiently large, the host might benefit from desertion, as it is maladaptive for her to take the risk involved in raising a brood if it contains too few of her own young. Desertion permits the host to start a new nest, turn to parasitism if it is not too late, or to defer nesting until the next spring.

Another host option (suggested by Stephen Emlen) is to remove some eggs from the nest, which might be better than incubating the entire clutch, even if the host cannot recognise her own eggs from those of the parasite. This turns out to be the case if the inequality $1 < 2a (b_H + b_P)$ is satisfied (assumptions and symbols as in equation (1)). It is then advantageous for the host to remove a proportion $1-u$ of the clutch, where the optimal $u = 1/2a(b_H + b_P)$. This can be shown by the same methods as in equation (2), assuming that the host removes the same proportion from her own eggs as from those of the parasite. Removing eggs might be superior to desertion, particularly at a late stage of the season when the prospects of renesting are poor. It is not known whether hosts in any species undertake such indiscriminate reduction of large parasitised clutches.

There is conflicting evidence as to whether parasitism increases the likelihood of nest desertion. It has often been related to parasitism in ducks with 6-8 eggs did not desert the nest if a further six eggs were added McLaughlin and Grice, 1952; Jones and Leopold, 1967), but other workers found no such association (e.g. Morse and Wight, 1969; Joyner, 1976; Clawson *et al.*, 1979). Further, where an association has been shown, cause and effect are not clear. Does parasitism lead to desertion, or does desertion make later 'parasitism' more likely as there is no owner to deter other females? The latter alternative is usually not excluded; it may be particularly relevant where the nest has been parasitised and deserted during the incubation period (Andersson and Eriksson, 1982).

Experimental addition of eggs sheds some light on desertion. Wood ducks with 6-8 eggs did not desert the nest if a further six eggs were added at one time (Heusmann *et al.*, 1980). Neither did goldeneyes which received two extra eggs on each of two successive days, but goldeneyes with only 1-3 eggs which received seven extra eggs simultaneously did desert. However, the first case with 2 + 2 eggs is more realistic since usually only one parasitic egg is added per day.

Since kin selection probably plays a role in brood parasitism among ducks (see below), one may ask whether hosts are prone to resist or desert if parasitised by unrelated females, but not if the parasite is close kin. This would require individual recognition and parasitism when the host is present, as often appears to occur (see wood duck below). Such discrimination should be possible to study among marked females of known relatedness. Kin recognition has recently been demonstrated in a variety of animals, reviewed by Holmes and Sherman (1983).

Improved Reproductive Success for the Host?

As precocial young forage for themselves, extra offspring need perhaps not raise brood mortality (but it did so in one well-studied case; Safriel, 1975). The survival of young might even increase with brood size (Nudds, 1980; Eadie and Lumsden, in preparation). This could occur because of (a) more rapid discovery of an approaching predator; (b) 'dilution' of risk as the predator selects from more prey; or (c) Hamilton's (1971) 'selfish herd' effect, if the host's own young concentrate in the centre of the brood, where risk is lowest. Eadie and Lumsden (in preparation) found that parasitic chicks of the hooded merganser *Mergus cucullatus* disappeared faster than did the goldeneye host's own offspring, which apparently survived better in parasitised than in normal broods. McKaye (1981) reviewed similar cases of interspecific brood care in fish.

Such effects could also be important within species. McKaye and McKaye (1977) discuss communal care and even 'kidnapping' of conspecific fry in cichlids (also see Ribbink *et al.*, 1981). These authors suggest that the behaviour is adaptive for the adopting parents in 'diluting' the risk of predation for their own progeny, and probably also for the adopted fry. Merging of several broods into a 'crêche' sometimes reduces duckling predation in eiders (*Somateria mollissima*) (Munro and Bedard, 1977) and possibly goldeneyes (Eadie and Lumsden, personal communication). Other things being equal, it might then be adaptive to become parasitised. However, if the host lays considerably fewer eggs under parasitism (see above), the advantage vanishes.

Whether increased brood size reduces predation probably depends on predator species and other aspects of the environment. Duckling mortality might increase with their number if they must often travel between small bodies of water due to rapid food depletion; such overland movements are hazardous (Ball *et al.*, 1975; Eriksson, 1979). Smaller broods may deplete their food less rapidly, travel less often overland, and have lower mortality. In large lakes where many broods share the feeding area, there may be no corresponding advantage for a small brood: its share of the total food consumption is small, and there are few overland travels. Mortality is then less likely to increase with brood size; it might even decrease for the reasons discussed. If so, selection should favour acceptance of being parasitised, unless the advantage is offset by reduced clutch size or hatching success for the host.

Parasitism of One or Several Host Nests?

If several nests are available, a parasite can choose between 'multiple parasitism' in one of them, and 'scattered laying' among several nests (Payne, 1977a, b). Which tactic is the best?

Scattered laying may be adaptive in raising the chances that at least some of the eggs escape predation and other hazards (Payne, 1977b; Clawson *et al.*, 1979). This is a version of the argument that natural selection among reproductive tactics with equal expected success often favours the one with lowest variance (Gillespie, 1977; Rubenstein, 1982).

The importance of this principle for brood parasitism is not clear. Scattered laying should enhance the initial spread of parasitism by reducing the probability of extinction of a new allele for parasitism (Rubenstein, 1982). But once the trait is widespread, chance extinction is unlikely and it probably matters little whether parasites use scattered laying or multiple parasitism, as long as the two tactics have equal expected success. Scattered laying requires searching for several nests, but higher mean success may make it superior. As laying is never perfectly synchronised among females, scatter-laying should give the parasite longer access to clutches not yet incubated, and hence let her lay more eggs. Also, if the parasite scatters her eggs among several nests, her young may be raised in smaller broods than if she lays in only one host nest. Scattered laying should therefore be favoured if offspring mortality increases with brood size.

If the parasite removes host eggs, scattered laying may be particularly favourable, since the parasite can remove eggs from each nest before laying in it. A parasite which benefits from egg eviction should therefore also benefit from scattered laying. In addition, if several parasites lay in the same nests, a scatter-layer is more likely to evict eggs of competing multiple-layers than vice versa. Scattered laying might then outcompete multiple laying.

With few exceptions, it is not known whether intraspecific brood parasites use scattered laying. At least some wood ducks spread their eggs among several hosts (Clawson *et al.*, 1979; Heusmann *et al.*, 1980). Since egg laying occurs over more than one month in this and other ducks (e.g. Bellrose, 1976), a parasitic wood duck may produce a large total clutch. One female is known to have laid at least 20 eggs among four nest boxes, almost twice the normal clutch (Heusmann *et al.*, 1980).

Optimal Clutch Size in Parasites

How many eggs should a parasite lay in the nest of one host to produce as many own fledglings as possible? If the host starts incubation at the total clutch yielding the maximum number of surviving young, and lays as many fewer eggs as added by the parasite, then the parasite should simply lay as many as possible before the host starts to incubate. On the other hand, if the host lays a fixed number of eggs b_H, then it can be shown that the optimal parasite clutch is $1/2a - b_H/2$, analogous to equation (2) for the host.

Probably a more realistic assumption is that the host cannot discriminate parasite eggs, but lays one less for each parasite egg added early in the sequence. The host cannot cease laying immediately, but will lay its last m eggs even if the parasite keeps adding eggs (Figure 7.4). Assume further that the host aims at a total clutch of $1/2a$ eggs (equation (2)) and that the parasite starts laying when the host has produced k eggs. Should the parasite then add one egg for each host egg until incubation starts, or can the parasite obtain more fledglings by laying fewer eggs? With these assumptions, for a parasite to benefit from laying fewer eggs it is necessary that $m > (k + 1/2a)/3$. Hence it requires that the host has a 'braking distance' m which is at least one-third of its normal clutch size. Since this seems unlikely at least in birds with large clutch sizes, such as ducks, the parasite should usually lay as many eggs as possible until the host starts to incubate.

For the parasite, removal of one or more host eggs before laying the first of its own eggs would seem adaptive. This will either reduce the final number of host offspring in the nest, which reduces competition from them, or prolong the laying period of the host, which permits the parasite to add more eggs. Removal of an egg should also reduce the host's chances of detecting parasitism and counteracting it. A female before laying sometimes removes eggs produced by other females in communally nesting anis (*Crotophaga sulcirostris*) (Vehrencamp, 1977) and acorn woodpeckers (*Melanerpes formicivorus*) (Mumme *et al.*, in press); in the latter species such behaviour even occurs between sisters (see Vehrencamp, Ch. 8 this volume).

As is evident from clutches of over 50 or even 100 eggs, some parasitic ducks continue to lay even if the total clutch becomes much larger than normal. However, such clutches are often deserted and do not hatch (e.g. Weller, 1959; Jones and Leopold, 1967); they are the maladaptive results of several parasites laying in the same nest.

Precocial Birds

The Wood Duck Aix sponsa

Among precocial birds, intraspecific brood parasitism has been most thoroughly studied in the North American wood duck. It nests in tree holes or nest boxes and starts breeding when one year old, laying about a dozen eggs, usually one per day. Parasitism ('dump nesting') is common, with total clutch size often above 20 eggs, sometimes even exceeding 50 (Figure 7.5; Grice and Rogers, 1965; Jones and Leopold, 1967; Morse and Wight, 1969; Bellrose, 1976; Clawson *et al.*, 1979). As in most ducks, the hen incubates alone; the male leaves the family towards the end of the 30-day incubation period. Pair formation takes place during the autumn and winter, and the drake follows his mate to the breeding ground, often close to her birth site. The nesting season is extended, clutches being started over 140 days in the south of the range, and 60 days in the north. Some hens successfully breed twice per season, the only duck known to do so (Bellrose, 1976).

Many nests are destroyed by predators, and often only 40-50 per cent of all clutches hatch. Predation on nesting females can be high: 103 of 1,579 hens in Illinois were killed by racoons, and other predators took additional females, while males suffered negligible mortality. Clearly, parasitic hens may avoid serious hazards by not incubating or caring for the brood. The annual adult mortality is usually 50-60 per cent, higher than for any other North American duck. Due to the breeding hazards it is probably highest for females; the wood duck like most Anatinae shows an excess of adult males (Bellrose *et al*, 1964; Bellrose, 1976).

The hens accompany their broods for 5-6 weeks. Survivorship per chick may be similar in small and large broods, including parasitised

Figure 7.5: Frequency distribution of clutch sizes in 82 normal and 120 parasitised nests of wood ducks *Aix sponsa* in Oregon

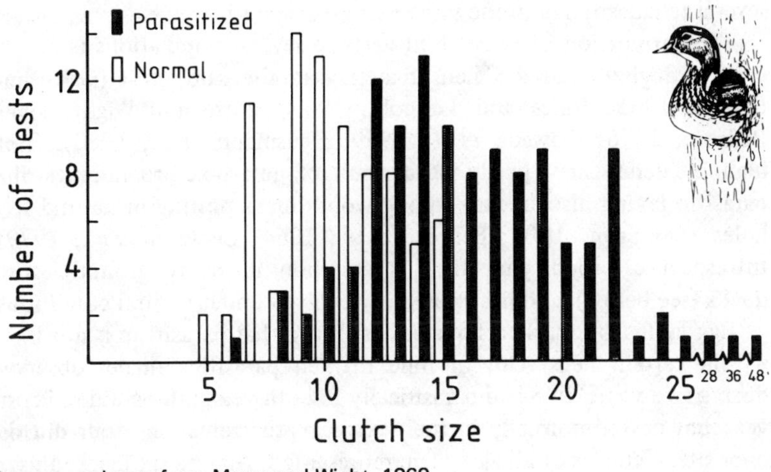

Source: redrawn from Morse and Wight, 1969

ones (Heusmann, 1972; Clawson *et al.*, 1979). However, there are three interfering temporal trends: with later hatching date, offspring survival declines, as does clutch size and proportion of parasitised nests (Grice and Rogers, 1965; Clawson *et al.*, 1979). Therefore, even if offspring mortality does increase with brood size and parasitism, such a trend may be masked by higher survival in early broods, which also tend to be larger and more frequently parasitised than late broods. Experimental manipulation of clutch size is desirable to test whether offspring survival for a given hatching date declines with increasing brood size also in precocial species (Safriel, 1975).

Hatching is often reduced under parasitism; more nests are then abandoned and a lower proportion of eggs hatch in successful nests, particularly if parasitic eggs are added after the host starts incubation (McLaughlin and Grice, 1952; Robinson, 1958; Grice and Rogers, 1965; Jones and Leopold, 1967; Morse and Wight, 1969; but see Clawson *et al.*, 1979). Below about 30 eggs, large clutch size in itself seems not to reduce hatching success (Morse and Wight, 1969).

The proportion of parasitised nests often increases with local population density, but competition for nest sites is not a sufficient explanation, because there are often many empty nest boxes when parasitism is common (Morse and Wight, 1969; Heusmann *et al.*, 1980; for other ducks see McCamant and Bolen, 1979; Andersson and Eriksson, 1982).

An alternative reason might be that improved breeding (due for example to nest box programmes) raises the production of young females which return to nest where they were born. This might cause several females to lay in the same nest (see below).

The proportion of parasitised nests in several populations is 50 per cent or higher, and 4-5 hens may lay in the same nest (Grice and Rogers, 1965; Jones and Leopold, 1967; Morse and Wight, 1969; Bellrose, 1976; Clawson *et al.*, 1979; Heusmann *et al.*, 1980). Such high incidence may partly be a result of nest box programmes, but parasitic laying also occurs among wood ducks nesting in natural tree holes (Sampson, 1901; Bellrose *et al.*, 1964; Clawson *et al.*, 1979). Intraspecific brood parasitism is common in many ground-nesting ducks (see below), and has apparently evolved under natural conditions.

Egg laying by marked females has shown that parasitism is not fixed within certain hens. Out of nine marked parasitic females observed during two years, two laid parasitically in both years; three did so in one year but nested normally in the next year; the remaining four did the opposite. Females of all ages were represented. Among six marked parasites which laid more than one egg in the study area, two apparently laid all eggs in one nest; the other four laid in two or more nests (Clawson *et al.*, 1979). Some hens laid parasitically in three or four nests (Heusmann *et al.*, 1980). Several ducks nested normally after parasitic laying earlier in the season (12 out of 16 marked parasitic hens, Heusmann *et al.*, 1980; and 6 out of 14 hens, Clawson *et al.*, 1979). Some females skipped several days of laying in their own nests, at least one of them for laying parasitically elsewhere. To summarise, parasitic laying in the wood duck is phenotypically flexible, and is not fixed within individuals or age classes.

Hosts often tolerate visits by other females: parasites laid when the host was present in the nest in 20 per cent of the cases. The resident female sometimes fought with the intruder; one hen laid an egg in the water outside the nest, apparently after being repulsed (Clawson *et al.*, 1979; Heusmann *et al.*, 1980). Whether the host tolerates the intruder or not possibly depends on their relatedness (see below). Normal females on average spent 90 minutes on their nests for laying an egg, whereas parasite sojourn time was only 19 minutes (Clawson *et al.*, 1979); this should reduce the risk of encounters with a possibly hostile host. Females were more likely to enter boxes that were empty or contained incomplete clutches than boxes where another female was incubating.

Does prior presence of eggs make it more likely that another female will lay in the nest? There was no significant difference in laying at control nests compared to nests 'pre-baited' with three eggs (Heusmann *et al.*, 1980). However, wood ducks seemed attracted by the presence of another pair, and when one moved, other pairs often followed to the new nest.

Is the wood duck representative of other Anatinae? Its breeding in nest boxes has probably influenced the results: predation may have been higher in some types of nest boxes than in natural tree holes, but in predator-proof boxes, reproductive success was higher than in natural nests (Bellrose *et al.*, 1964; Grice and Rogers, 1965; Clawson *et al.*, 1979; Heusmann *et al.*, 1980). This has sometimes led to situations with more pairs than nest boxes, and therefore frequent parasitism. But scarcity of suitable nests is not the only cause of parasitism, for it is common also when empty nest boxes are in excess. The most intense parasitism, affecting 65 per cent of all nests, occurred when only 58 per cent of all boxes were occupied (Morse and Wight, 1969).

Other Precocial Species

Among ducks, intraspecific brood parasitism is not restricted to hole-nesters. 'Almost all species harbor some individuals that lay one or more eggs in the nest of other hens of the same species' (Bellrose, 1976), and parasitic laying is common in several open-nesting ducks. The most thoroughly studied species is the redhead (*Aythya americana*) (Weller, 1959; Lokemoen, 1966; Joyner, 1976), with frequent intra- as well as interspecific parasitism. 'Dump' clutches occur with up to 87 eggs; 13 hens have been trapped on a nest (Weller, 1959). In one study, at least 36 per cent of the redhead nests contained eggs from several females (Lokemoen, 1966). Where other host species were common, only 5-10 per cent of the redhead hens nested normally (Weller, 1959). Except for the frequent laying in nests of other species, brood parasitism in the redhead resembles that in the wood duck.

Brood parasitism within species is common in several other open-nesting ducks, for example, the fulvous tree duck (*Dendrocygna bicolor*) (Dickey and van Rossem, 1923), red-crested pochard (*Netta rufina*) (Hellebrekers and Voous, 1964), pochard (*Aythya ferina*) (Bezzel, 1961; Rutschke and Lehmann, 1975), tufted duck (*A. fuligula*) (Hildén, 1964; Fredriksson, 1968; Newton and Campbell, 1975), and ruddy duck (*Oxyura jamaicensis*) (Joyner, 1976). This list of references is by no means complete. Some of the species, particularly *Netta* and *Oxyura*, often parasitise other anatids. In addition to the wood duck, several other hole-nesting ducks are also often parasitic within species, notably the black-bellied tree duck (*D. autumnalis*) (McCamant and Bolen, 1979).

Eggs have often been scattered around parasitised nests of open-nesting ducks. Weller (1959) concluded that the eggs had been lost accidentally; there is no evidence that ducks evict eggs laid by conspecific females, but the possibility remains open. Another possibility is that the eggs had been laid by parasites that were repulsed from the nest by the host. In certain communally nesting birds, a female may remove one or more eggs laid by others and hence increase the proportion of own offspring in the brood (see under 'Optimal parasite clutch', p. 212). In ostriches (*Struthio camelus*), the incubating hen recognises her own eggs and selectively pushes out many others (Bertram, 1979).

Intraspecific brood parasitism is also common in another group of precocial birds: Galliformes, especially pheasants and quail (Weller, 1959; Yom-Tov, 1980; Ehmann, 1981). Brood parasitism in this group parallels that in open-nesting ducks, but has received less attention under natural conditions. Intensive studies of brood parasitism in pheasants and quail in the wild could provide a check of the conclusions based mainly on ducks. In particular, it would be instructive to know if host and parasite are often genetically related, as may be the case among ducks (see below). This could be so, for example, if quail sisters tend to remain together until breeding.

Kin Selection and Intraspecific Brood Parasitism in Ducks

Ducks make up more than half the number of birds with frequent brood parasitism recorded within species (Yom-Tov, 1980). Why this predominance in Anatinae? In contrast with other birds, there are indications that host and parasite are often close kin among ducks. Population structure and breeding ecology in waterfowl may permit host as well as parasite to gain inclusive fitness through brood parasitism, which should facilitate its evolution. This is one possible reason why it is more common among ducks than other birds (Andersson and Eriksson, 1982; Andersson, in preparation).

The critical feature which makes kin selection among female ducks likely is their strong philopatry. In most birds males tend to breed closer to their natal area than do females (Baker, 1978; Greenwood, 1980), whereas the situation is reversed in waterfowl (Andersson, in preparation). If young females tend to return and nest at their birth site, for example a tree-hole or a protected islet, host and parasite may often be close kin. Since nest predation is high in ducks, returning to the birth site should be adaptive in ensuring that the young hen nests where breeding has proved successful at least once. This might lead to breeding attempts in the same nest by sisters, or by mother and daughter. In four cases of intraspecific brood parasitism documented in detail in canvasbacks (*Aythya valisineria*), two involved mother and daughter, and the other cases concerned sisters (Michael Anderson, personal communication).

Even if the parasite is a close relative, it may be deleterious for the host to become parasitised and hence probably produce fewer own offspring. But in ducks, a number of other benefits make it likely that two closely-related hens can gain in inclusive fitness by nesting together as 'host' and 'parasite': (a) Incubation and brood care raises female mortality considerably in ducks (see e.g. the wood duck above). Most breeding risks are avoided by a parasite; this benefits the spread not only of her own genes, but also those of the host, if closely related. (b) Egg laying draws heavily on the nutrient reserves of a duck (Owen and Reinecke, 1979; Drobney, 1980; Krapu, 1981). Host as well as parasite might therefore be more likely to survive if laying smaller clutches than normal females. (c) If host and parasite each produce fewer eggs, the clutch takes less time to lay. This reduces the period of exposure to nest predators, and the ducklings hatch earlier, which improves their survival (Perrins, 1970; Grice and Rogers, 1965; Dow and Fredga, in preparation).

These advantages might override the disadvantage of producing fewer own offspring, and favour the evolution of brood parasitism between close relatives. Consider the following one-locus diploid genetic model (from Andersson, in preparation), which demonstrates that the process is logically plausible; a more realistic model necessarily becomes more complicated. Assume the following: in a parasitic nesting, each female lays only b' eggs as compared to b eggs in a normal nesting. The survival S of each offspring until the next breeding season is independent of clutch size. The survival of the host female until the next breeding season is Pk, where k = probability of surviving incubation and brood care, and P = probability of surviving the rest of the year. For a parasite, which avoids incubation and brood care, mortality during the breeding season is negligible, so she survives with probability P until the next breeding season.

The allele A makes the host tolerate being parasitised by a close relative such as daughter or sister, which it recognises. For reasons suggested above, females tend to return and breed in the nest where they were born. If the nest is unoccupied, the female nests normally: if it is occupied by a relative which tolerates being parasitised, she lays parasitically; if it is occupied by a rejecting hen, she searches for another empty nest and breeds normally. For simplicity, the possibility that several relatives may parasitise the same nest is disregarded. The alternative allele *a* codes for resistance to parasitism. As the host has already familiarised herself with the site and started breeding, the present value of the nest is highest for the host; this asymmetry should make her likely to win a contest and chase away an undesired parasite (Maynard Smith, 1982).

Will parasitism spread in the population under these assumptions? That is, will a rare allele *A* increase in relative frequency from one season to the next, as compared to the alternative allele *a* which prevents parasitism? It can be shown (Andersson, in preparation) that the condition for spread of an initially rare allele *A* is

$$r > \frac{1}{\dfrac{P2(1-k)}{S\,(b-b')} - 1} \tag{4}$$

where r is the relatedness between host and parasite. If this inequality is satisfied, the rare allele *A* has a selective advantage and will increase in relative frequency. Equation (4) shows how closely related the two females must be for parasitism to increase. It is favoured by a high risk of mortality during incubation and brood care (small k) and by high adult survival P and low juvenile survival S. A tendency in host and parasite to lay fewer own eggs (b') in a parasitic than in a normal nesting (b) works against parasitism. Some of these points are illustrated in Figure 7.6. Equation (4) in several respects is unrealistic (for example, it assumes that the brood survives even if the host dies) but it demonstrates that evolution of parasitism through kin selection is logically possible, and it shows how different mortality factors influence the process. A more realistic model indicates that kin selection is likely to be important for the evolution of brood parasitism in ducks (Andersson, in preparation). This case qualifies as kin selection for 'altruistic' behaviour, as the host reduces her own reproduction and cares for the offspring of a related parasite.

There are different opinions as to whether close relatives such as mother–daughter should be embraced by definitions of 'altruism' and 'kin selection' (Bertram, 1982; Dawkins, 1982). If not, parasitism involving mother and daughter is not a case of kin selection, whereas parasitism between sisters is. However, with regard to the selective process and degree of relatedness there is no difference, so the issue here seems semantic rather than fundamental.

Figure 7.6: Conditions for spread of an initially rare allele *A* for parasitism between two related females. In the region above and to the right of each curve, inequality (4) is satisfied. The curves show the combinations of clutch size per female (b') and coefficient of relatedness r between them, necessary for the allele *A* to spread at different levels of (nesting) survivorship k for the host. The degree of relatedness required to favour the allele *A* increases with increasing survival k and decreasing clutch size b'. The curves were calculated with $S = 0.07$; $P = 0.95$, and $b = 10$; see main text for explanation of these parameters

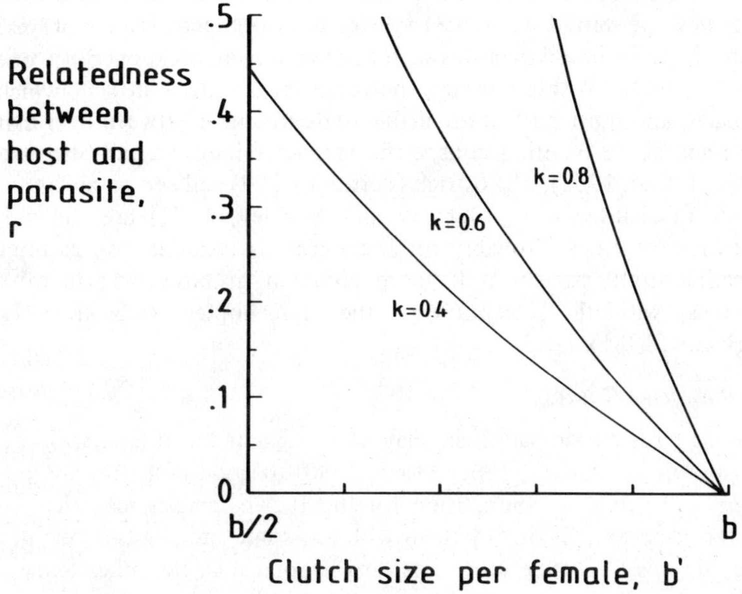

Evolution of Intraspecific Brood Parasitism

Taxonomic Occurrence

By definition, brood parasitism can occur only among animals with parental care. Differences in the possibility of placing offspring in the care of a host seem to explain much of the variation between species in the occurrence of parasitism. In some groups it may not be possible, because parental recognition of own progeny develops quickly after their birth, and foreign young are discriminated. Further, among animals in which offspring develop rapidly, a parasite may not be able to place her progeny in the care of the host before its young gain a a competitive advantage. Brood parasitism, therefore, is probably not feasible in most mammals, amphibia, reptiles or fish with parental care.

In birds, on the other hand, a parasite has a better chance of inserting its offspring among those of another parent. Most birds lay a clutch of several large, energy-rich eggs, usually only one per day. Moreover, incubation and embryonic development usually starts first when the clutch is almost complete. During these days or weeks, parasitic laying is potentially successful, a situation unique for birds. In addition, eggs are rather silent and birds have poor olfaction, so only visual (and possibly tactile) cues can be used for discrimination. Such cues have permitted many host species to evolve recognition and rejection of eggs of brood parasites such as cowbirds and cuckoos (Rothstein, 1975, 1982). Within species, however, eggs differ little between females, and most birds seem unable to discriminate between their own eggs and those of other conspecific females (Hoogland and Sherman, 1976; Lanier, 1982). The ostrich (Bertram, 1979) and certain alcids and larids (Tschantz, 1959; Buckley and Buckley, 1972) are the only known exceptions. Probably for these reasons intra- as well as interspecific brood parasitism is more common in birds than in other animals, with the exception of the Hymenoptera (Wilson, 1971; Eickwort, 1981).

Evolutionary Routes

Among birds, brood parasitism may have originated in at least two ways (Hamilton and Orians, 1965; Evans, 1980). If good nest sites are rare, there will often be competition for them; two females may then lay in the same nest. If one of them withdraws but later finds a new nest site, she may produce more offspring in total than the other females. Selection will then favour any genetic disposition for such behaviour, and parasitism may be used even when there are unoccupied nest sites. This hypothesis predicts that brood parasitism within species should occur particularly among hole-nesters and other birds with a shortage of breeding sites, as appears to be the case (see below).

Another possible route starts from nest failure, females often losing their nest before laying a full clutch. A female which then tends to lay in another nest of the same species should be favoured by selection (Hamilton and Orians, 1965). The high frequency of brood parasitism in ground-nesters which suffer high nest predation (such as many ducks and quail) is in line with this hypothesis. Additional support is that catastrophes such as flooding of nests often leads to increased parasitism (Yom-Tov, 1980).

Both processes may have been important in hole-nesting ducks, as they nest in sites of limited availability, yet with high risk of predation.

Ecological Background

What conditions favour a high frequency of parasitism, once the trait has arisen in a population?

Large Clutches. Species with large clutches are common among intraspecific brood parasites, probably because parasitic females then have many days available to find nests and lay eggs before the host begins to incubate (Yom-Tov, 1980). If females lay one egg per day and a parasite needs on average t days to find a nest after the host has started laying, the relative success of the parasite rises with the host's own final clutch size b. If the parasite lays $b-t$ eggs, its relative contribution $(b-t)/(2b-t)$ increases towards $1/2$ as $b \rightarrow \infty$; hence larger clutch sizes should make parasitism more favourable.

Precocial Young. Most of the birds with more than occasional brood parasitism have precocial young. The reason may be that precocial species with frequent parasitism also have large clutches, which facilitates parasitism (Yom-Tov, 1980). Another possibility is that being precocial, the young require little parental care; hence the disadvantage of being parasitised might be small (Michael Anderson, personal communication). Hosts could even gain from parasitism if the risk of predation per young is reduced in large broods (see p. 209). However, the dominance of precocial species is mainly due to one group, the waterfowl (Anatidae), which make up more than half of Yom-Tov's (1980) list. It is doubtful whether intraspecific brood parasitism is over-represented among precocial species excluding ducks. In addition, if hosts reduce their own clutch considerably, the disadvantage of being parasitised may be similar whether the young are precocial or altricial, and reduction of the host's own clutch seems to be the rule.

Long Egg-laying Season. A long laying season makes nests available for parasitism over an extended period, and permits parasites to produce more eggs than if all hosts lay in close synchrony. Yom-Tov (1980) therefore suggested that parasitism selects for synchronised breeding.

Why does parasitism not lead to synchronised nesting among ducks which have extended laying seasons (Bellrose, 1976)? Possibly because being parasitised may be less deleterious (or even advantageous) in ducks than in other birds, for reasons suggested in the section on kin selection.

Competition for Nest Sites. Many intraspecific parasites nest in sites of limited availability, such as tree holes (starling, bluebird, several ducks), islets inaccessible to mammalian predators or protected by colonies of larids (gulls), or patches of dense cover for ground-nesting species (ducks, pheasants and quail). Shortage of suitable sites may make it more likely that several females will lay in the same nest. Increases in parasitism with population density are in line with this hypothesis (Yom-Tov, 1980; Andersson and Eriksson, 1982), but it is not the only possible one. In hole-nesting ducks, nest-box programmes have led to improved reproduction and population increases; many young females might then return to lay in the nest where they were born (see p. 216), even if there are other empty sites. Competition is therefore not necessary to increase parasitism, which can be common although the majority of nest boxes are empty (Morse and Wight, 1969; McCamant and Bolen, 1979; Heusmann *et al*., 1980; Andersson and Eriksson, 1982).

Easily Discovered Nests. The nests of hole-breeding species are more easily found than those of many open-nesters; this may make parasitism more likely. Among open-nesting ducks, brood parasitism seems particularly common where good sites are rare and nests are therefore easy to discover, for example in island colonies of larids (Andersson and Eriksson, 1982). Among Phasianidae, parasitism has also been related to shortage of nesting cover (Stoddard, 1931), which probably makes nests easy to find.

Lack of Territorial Defence. Owls, woodpeckers and many passerines nest in tree holes, yet brood parasitism seems rare within these species compared to starlings and hole-nesting ducks. Why?

Differences in territoriality are probably important. Contrary to most other hole-nesters, ducks leave the nest undefended during the laying period, which they mainly spend foraging in aquatic habitats, accompanied by the drake. As he needs less food than the laying female, the drake could perhaps stay behind to defend the nest, but mate guarding may be necessary to ensure that he fertilises the eggs.

The frequent nest parasitism in starlings is in line with this explanation, since they usually do not guard the nest during the laying period (Johnny Karlsson, personal communication). This also holds for bobwhite quail (*Colinus virginianus*) (Rosene, 1969).

Relatedness Between Females. The hypothesis was raised above that kin selection between host and parasite can favour the evolution of brood parasitism within species of ducks. However, even if this turns out to be correct, ducks possess most or all of the previous features which seem to favour brood parasitism. Its prevalence among ducks, therefore, is probably due to a combination of several enhancing traits; their relative importance remains to be clarified in the field.

Brood Parasitism Within and Between Species. Several birds parasitise both their own and other species, and interspecific brood parasitism may have evolved via parasitism within species (Hamilton and Orians, 1965; Payne, 1977a). Many ducks are parasitic both within and between species (Weller, 1959; Hellebrekers and Voous, 1964; Joyner, 1976). One anatid, the black-headed duck (*Heteronetta atricapilla*), has become an obligate parasite of other birds (Weller, 1968). Certain *Coccyzus* cuckoos which usually nest in the normal way sometimes parasitise their own as well as other species (Nolan and Thompson, 1975). Brood parasitism between species of wasps (Hymenoptera) may also have been enhanced by parasitism within species (Wilson, 1971; Eickwort, 1975). Intraspecific parasitism in general might evolve more easily if suitable hosts of other species are also available, and vice versa.

Conclusions

Brood parasitism within non-co-operative species (a female having her offspring raised through parental care by another conspecific female) is known from a few insects and about a hundred species of birds. However, this elusive behaviour is probably more widespread; detailed studies of egg laying and parental care by individually marked animals are required to clarify the characteristics and tactics of parasites and hosts. A good start has been made, especially on the starling, wood duck, redhead and some other ducks and on certain Hymenoptera, but most hypotheses discussed remain to be tested, and several crucial topics to be clarified. The entire egg-laying sequence of individually marked females of known age and origin needs to be followed, preferably for a number of years. If possible, the origin of each egg in the local population should be determined by a combination of direct observations of egg laying and biochemical analysis of offspring genotypes. Other data required for an appraisal of the fitness of hosts and parasites are offspring survival in relation to clutch size, relative success of parasitic and normal eggs, and the effect on adult mortality of parental care for different numbers of offspring. To test the hypothesis that kin selection is important for intraspecific brood parasitism in ducks, populations need to be studied long enough for determination of the relatedness between hosts and parasites.

Through such analyses, it might be possible to demonstrate the identity, age, laying tactics, reproductive success and survival of parasitic and normal females, and some of the responses of parasitised hosts. Field experiments based on addition or removal of eggs or young should permit tests of several of the hypotheses discussed. Biochemical methods for determination of parentage should help to show if host and parasite are close kin, and if intraspecific parasitism is more common in animals with parental care than is presently known (as seems likely from the great potential advantages for the parasite).

Acknowledgements

I thank the following persons for kindly sending manuscripts, allowing me to quote unpublished work, or replying helpfully otherwise: Michael Anderson, Peter Blums, Hilary Dow, John Eadie, Christopher Edwards, Marianne Ehmann, Peter Evans, Christopher Feare, Patricia Gowaty, Johnny Karlsson, Kenneth McKaye, Ronald Mumme, Marion Petrie, John Rogers and Rodney Sayler. I also thank Stephen Emlen, Mats Eriksson and Robert Payne for helpful comments on the manuscript, Christopher Feare on the starling section, and Aino Wahlström for drawing the figures. The study was supported by the Swedish Natural Sciences Research Council.

References

Alcock, J. (1975). Social interactions in the solitary wasp *Ceceris simplex* (Hymenoptera sphecidae). *Behaviour 54*: 142-52

Andersson, M. (in preparation). Brood parasitism and kin selection in ducks Anatinae

—and Eriksson, M.O.G. (1982). Nest parasitism in goldeneyes *Bucephala clangula*: some evolutionary aspects. *Amer. Natur. 120*: 1-16

Baker, R.R. (1978). *The Evolutionary Ecology of Animal Migration*. London, Hodder and Stoughton

Ball, I.J., Gilmer, D.S., Cowardin, L.M. and Riechman, J.H. (1975). Survival of wood duck and mallard broods in North-Central Minnesota. *J. Wildl. Manage. 39*: 776-80

Bellrose, F.C. (1976). *Ducks, Geese and Swans of North America*. Harrisburg, Penn., Stackpole Books

— Johnson, K.L. and Meyers, T.U. (1964). Relative value of natural cavities and nesting houses for wood ducks. *J. Wildl. Manage. 28*: 661-76

—Scott, T.G., Hawkins, A.S. and Low, J.R. (1961). Sex ratios and age ratios in North American ducks. *Illinois Nat. Hist. Surv. Bull. 27*: 391-474

Bennet, F.D. (1966). Notes on the biology of *Stelis* (*Odontostelis*) *bilineolata* (Spinola), a parasite of *Euglossa cordata* (Linnaeus) (Hymenoptera: Apoidea:

Megachilidae). *J. New York Entomol. Soc. 74*: 72-9

Bertram, B.C.R. (1979). Ostriches recognise their own eggs and discard others. *Nature 279*: 233-4

— (1982). Problems with altruism. In King's College Sociobiology Group (ed.), *Current Problems in Sociobiology*: 251-67. Cambridge, Cambridge University Press

Bezzel, E. (1961). Über Mischgelege bei Enten. *Vogelwelt 82*: 97-101

Blanchard, F.B. (1934). The relation of the female four-toed salamander to her nest. *Copeia 1934*: 137-8

Bohart, G.E. (1955). Gradual nest supersedure within the genus *Osmia* (Hymenoptera, Apoidea). *Proc. Entomol. Soc. Washington 57*: 203-4

Brockmann, H.J. (1980). Diversity in the nesting behavior of mud-daubers (*Trypoxylon politum* Say; Sphecidae). *Fla. Entomol. 63*: 53-64

—, Grafen, A. and Dawkins, R. (1979). Evolutionarily stable nesting strategy in a digger wasp. *J. theor. Biol. 77*: 473-96

Buckley, P.A. and Buckley, F.G. (1972). Individual egg and chick recognition by adult royal terns (*Sterna maxima maxima*.). *Anim. Behav. 20*: 457-62

Clawson, E.L., Hartman, G.W. and Fredrickson, L.H. (1979). Dump nesting in a Missouri wood duck population. *J. Wildl. Manage. 43*: 347-55

Dawkins, R. (1982). *The Extended Phenotype*. Oxford, Freeman

Delnicki, D.E., Bolen, E.G. and Cottam, C. (1976). An unusual clutch size of the black-bellied whistling duck. *Wilson Bull. 88*: 347-8

Dickey, D.R. and van Rossem, A.J. (1923). The fulvous tree-ducks of Buena Vista lake. *Condor 25*: 39-50

Dow, H. and Fredga, S. (in preparation). Factors influencing reproductive output in the goldeneye *Bucephala clangula*

Drobney, R.D. (1980). Reproductive bioenergetics of wood ducks. *Auk 97*: 480-90

Dunnet, G.M. (1955). The breeding of the starling *Sturnus vulgaris* in relation to its food supply. *Ibis 97*: 619-62

Eadie, J. McA. and Lumsden, H.G. (in preparation). Is nest parasitism always deleterious to goldeneye?

Ehmann, M. (1981). Organisation sociale et stratégies reproductrices du faisan commun (*Phasianus colchicus*). Ph. D. Thesis. France, University of Rennes

Eickwort, G.C. (1975). Gregarious nesting of the mason bee *Hoplitis anthocopoides* and the evolution of parasitism and sociality among Megachilid bees. *Evolution 29*: 142-50

— (1981). Presocial insects. In H.R. Hermann (ed.), *Social Insects*. vol. II: 199-280. New York, Academic Press

Eriksson, M.O.G. (1979). Aspects of the breeding biology of the goldeneye *Bucephala clangula. Holarct. Ecol. 2*: 186-94

Evans, P.G.H. (1980). Population genetics of the European starling *Sturnus vulgaris*. Unpubl. DPhil. thesis, Univ. Oxford

Fabre, J.H. (1914). *The Mason-bees*. London, Hodder and Stoughton

Feare, C.J. (1983). *The Starling*. Oxford, Oxford University Press

— Spencer, P.L. and Constantine, D.A.T. (1982). Time of egglaying of starlings *Sturnus vulgaris. Ibis 124*: 174-8

Fredriksson, K.A. (1968). Observations on parasitic nesting in the tufted duck (*Aythya fuligula*). *Ornis Fenn. 45*: 127-30 (in Swedish; English summary)

Gillespie, J.H. (1977). Natural selection for variances in offspring numbers: a new evolutionary principle. *Amer. Natur. 111*: 1010-14

Gould, S.J. and Lewontin, R.C. (1979). The spandrels of San Marco and the Panglossian paradigm: a critique of the adaptationist programme. *Proc. Roy. Soc. Lond. B 205*: 581-98

Gowaty, P.A. and Karlin, A.A. (in press). Multiple maternity and paternity in

apparently monogamous eastern bluebirds (*Sialia sialis*): electrophoretic evidence. *Behav. Ecol. Sociobiol.*
Greenwood, P.J. (1980). Mating systems, philopatry and dispersal in birds and mammals. *Anim. Behav. 28*: 1140-62
Grenquist, P. (1963). Hatching losses of common goldeneyes in the Finnish Archipelago. *Proc. Int. Ornithol. Congr. 13*: 685-9
Grice, D. and Rogers, J.P. (1965). The wood duck in Massachusetts. Final Report, Federal Aid in Wildlife Restoration, Project no. W-19-R, Mass. Div. of Fisheries and Game, Westboro, Mass.
Hamilton, W.D. (1964). The genetical evolution of social behaviour. I and II. *J. theor. Biol. 7*: 1-52
— (1971). Selection of selfish and altruistic behavior in some extreme models. In J.F. Eisenberg and W.S. Dillon (eds.), *Man and Beast: comparative social behavior*: 57-91. Washington, Smithsonian Inst. Press
Hamilton, W.J. III and Orians, G.H. (1965). Evolution of brood parasitism in altricial birds. *Condor 67*: 361-82
Hellebrekers, W.Ph.J. and Voous, K.H. (1964). Nestparasitisme van de krooneend. *Limosa 37*: 5-11
Heusmann, H.W. (1972). Survival of wood duck broods from dump nests. *J. Wildl. Manage. 36*: 620-4
— Bellville, R. and Burrell, R.G. (1980). Further observations on dump nesting by wood ducks. *J. Wildl. Manage. 44*: 908-15
Hildén, O. (1964). Ecology of duck populations in the island group of Valassaaret, Gulf of Bothnia. *Ann. Zool. Fenn. 1*: 153-279
Holmes, W.G. and Sherman, P.W. (1983). Kin recognition in animals. *Amer. Sci. 71*: 46-55
Hoogland, J.L. and Sherman, P.W. (1976). Advantages and disadvantages of bank swallow (*Riparia riparia*) coloniality. *Ecol. Monogr. 46*: 33-58
Jones, R.E. and Leopold, A.S. (1967). Nesting interference in a dense population of wood ducks. *J. Wildl. Manage. 31*: 221-8
Joyner, D.E. (1976). Effects of interspecific nest parasitism by redheads and ruddy ducks. *J. Wildl. Manage. 40*: 33-8
Karlsson, J. (1983). Breeding of the starling (*Sturnus vulgaris*). Ph. D. Thesis. Sweden, University of Lund
Krapu, G.L. (1981). The role of nutrient reserves in mallard reproduction. *Auk 98*: 29-38
Lack, D. (1948). Natural selection and family size in the starling. *Evolution 2*: 95-110
Lanier, G.A., Jr. (1982). A test for conspecific egg discrimination in three species of colonial passerine birds. *Auk 99*: 519-25
Lokemoen, J.T. (1966). Breeding ecology of the redhead duck in western Montana. *J. Wildl. Manage. 30*: 668-81
Manwell, C.L.C. and Baker, M.A. (1975). Molecular genetics of avian proteins. XIII. Protein polymorphism in three species of Australian passerines. *Austr. J. Biol. Sci. 28*: 545-57
Maynard Smith, J. (1978). Optimization theory in evolution. *Ann. Rev. Ecol. System. 9*: 31-56
—(1982). *Evolution and the Theory of Games*. Cambridge, Cambridge University Press
McCamant, R.E. and Bolen, E.G. (1979). A 12-year study of nest box utilization by black-bellied whistling ducks. *J. Wildl. Manage. 43*: 939-43
McKaye, K.R. (1981). Natural selection and the evolution of interspecific brood care in fishes. In R.D. Alexander and D.W. Tinkle (eds.). *Natural selection and Social Behavior: recent research and new theory*: 173-83. New York, Chiron Press

— and McKaye, N.M. (1977). Communal care and kidnapping of young by parental cichlids, *Evolution 31*: 674-81

McLaughlin, C.L. and Grice, D. (1952). The effectiveness of large-scale erection of wood duck boxes as a management procedure. *Trans. N. Amer. Wildl. Conf. 17*: 242-59

Michenor, C.D. (1979). *The Social Behavior of the Bees*. Cambridge, Mass., Harvard University Press

Morse, T.E. and Wight, H.M. (1969). Dump nesting and its effects on production in wood ducks. *J. Wildl. Manage. 33*: 284-93

Mumme, R.L., Koenig, W.D. and Pitelka, F.A. (in press). Reproductive competition and communal nesting in the acorn woodpecker: sisters destroy each other's eggs. *Nature*

Munro, J. and Bedard, J. (1977). Gull predation and crèching behaviour in the common eider. *J. Anim. Ecol. 46*: 799-810

Newton, I. and Campbell, C.R.G. (1975). Breeding of ducks at Loch Leven, Kinross. *Wildfowl 26*: 83-103

Nolan, V., Jr. and Thompson, C.F. (1975). The occurrence and significance of anomalous reproductive activities in two North American non-parasitic cuckoos, *Coccyzus* spp. *Ibis 117*: 496-503

Nudds, T.D. (1980). Canvasback tolerance of redhead parasitism: an observation and a hypothesis. *Wilson Bull. 92*: 414

Owen, R.B. and Reinecke, K.J. (1979). Bioenergetics of breeding dabbling ducks. in T.A. Bookhout (ed.), *Waterfowl and Wetlands – an Integrated Review. Proc. 1977 Symp., Madison, WI, NC Sect, The Wildl. Soc*: 71-93

Payne, R.B. (1977a). The ecology of brood parasitism in birds. *Annu. Rev. Ecol. Syst. 8*: 1-28

— (1977b). Clutch size, egg size, and the consequences of single vs. multiple parasitism in parasitic finches. *Ecology 58*: 500-13

Perrins, C.M. (1970). The timing of birds' breeding seasons. *Ibis 112*: 242-55

Power, H.W., Litovich, E. and Lombardo, M.P. (1981). Male starlings delay incubation to avoid being cuckolded. *Auk 98*: 386-9

Ribbink, A.J., Marsh, A.C. and Marsh, B.A. (1981). Nest-building and communal care of young by *Tilapia rendalli* Dumeril (Pisces, Cichlidae) in Lake Malawi. *Env. Biol. Fish. 6*: 219-22

Robinson, R.H. (1958). Use of nest boxes by wood ducks in the San Joaquin Valley, California. *Condor 60*: 256-7

Rosene, W. (1969). *The Bobwhite Quail*. New Brunswick, Rutgers University Press

Rothstein, S.I. (1975). An experimental and teleonomic investigation of avian brood parasitism. *Condor 77*: 250-71

—(1982). Successes and failures in avian egg and nestling recognition with comments on the utility of optimality reasoning. *Amer. Zool. 22*: 547-60

Rubenstein, D.I. (1982). Risk, uncertainty and evolutionary strategies. In King's College Sociobiology Group (ed.), *Current Problems in Sociobiology*: 91-111. Cambridge, Cambridge University Press

Rutschke, E. and Lehmann, R. (1975). Zür Fortpflanzungsbiologie der Tafelente (*Aythya ferina*) bei optimalen Ernährungsbedingungen. *Beitr. Vogelkd. 21*: 439-46

Safriel, U.N. (1975). On the significance of clutch size in nidifugous birds. *Ecology 56*: 703-8

Sampson, W.B. (1901). An exceptional set of eggs of the wood duck. *Condor 3*: 95

Sowls, L.K. (1950). *Prairie Ducks*. Harrisburg, Penn., Stackpole

Stoddard, H.L. (1931). *The Bobwhite Quail*. New York, Scribner's

Tinbergen, J.M. (1981). Foraging decisions in starlings (*Sturnus vulgaris* L.). *Ardea 69*: 1-67

Tschantz, B. (1959). Zür Brutbiologie der Trottellumme (*Uria aalge* Pont.). *Behaviour 14*: 1-100
Vehrencamp, S.L. (1977). Relative fecundity and parental effort in communally nesting anis, *Crotophaga sulcirostris. Science 197*: 403-5
Verheyen, R.F. (1980). Breeding strategies of the starling. In E.N. Wright, I.R. Inglis and C.J. Feare (eds.), *Bird Problems in Agriculture*: 69-82. London, BCPC Publications
Weller, M.W. (1959). Parasitic egg laying in the redhead (*Aythya americana*) and other North American Anatidae. *Ecol. Monogr. 29*: 333-65
—(1968). The breeding biology of the parasitic black-headed duck. *Living Bird 7*: 169-208
Westerterp, K., Gortmarker, W. and Wijngaarden, H. (1982). An energetic optimum in brood-raising in the starling *Sturnus vulgaris*: an experimental study. *Ardea 70*: 153-62
Wilson, E.O. (1971). *The Insect Societies*. Cambridge, Mass., Belknap Press of Harvard University Press
Yom-Tov, Y. (1980). Intraspecific nest parasitism in birds. *Biol. Rev. 55*: 93-108
—Dunnet, G.M. and Anderson, A. (1974). Intraspecific nest parasitism in the starling *Sturnus vulgaris. Ibis 116*: 87-90

8 EXPLOITATION IN CO-OPERATIVE SOCIETIES: MODELS OF FITNESS BIASING IN CO-OPERATIVE BREEDERS

Sandra L. Vehrencamp

In most human and non-human societies in which groups of individuals undertake collective and co-operative activities, elements of competition and conflict are usually also present (Alexander, 1974). Where groups of individuals are more successful at foraging, brood care or predator defence than solitary individuals, an opportunity exists for dominant individuals to exploit the mutual benefits of the group to their own advantage. This will result in a biased division of resources, mates, offspring production or risk-taking that favours the dominants. Subordinates would be expected to resist such exploitation, setting the stage for conflict. The relative leverage of dominants and subordinates will ultimately determine the degree of biasing and exploitation. Where subordinates have strong leverage, the level of exploitation and biasing will be low, and a relatively egalitarian society will result; benefits will be divided roughly equally or in proportion to the risk or effort taken. Where dominants have strong leverage, exploitation and biasing will be high, and a despotic society will result; benefits will accrue disproportionately to a few individuals in the group at the expense of others. The question arises, 'What determines the relative leverage of dominants and subordinates?'

It seems likely that the opposing forces of competition and co-operation within most groups reach a stable equilibrium that determines both the optimal group size and the degree of exploitation. Selection acts simultaneously on the stronger, dominant members of the group to secure more resources for themselves at the expense of subordinates, and on subordinates to leave the group when excessively manipulated if they can do better elsewhere. When it is to the advantage of the dominant to maintain the group, the dominant will ultimately be limited in the degree of bias it can impose by the options available to subordinates outside the group. By assuming that the conflict between dominants and subordinates is resolved in favour of the dominant within the limits of the subordinate's ability to leave, one can solve for the optimal degree of bias within social groups using optimality theory (Maynard Smith, 1978). External, ecological factors affect the net benefit of grouping and the options for leaving, and thus establish the extent of exploitation by dominants.

Co-operatively breeding societies are an excellent model system for investigating the question of exploitation for several reasons. Firstly, a direct measure of fitness biasing can be obtained where offspring production is the co-operative endeavour. For co-operative foraging systems, on the other hand, one would need to translate resource acquisition into a more direct fitness measure. Secondly, it may be easier to meet the critical condition of this model in the context of co-operative brood care, namely, that dominants can bias fitness to their advantage at little extra cost. In co-operative foraging or mate defence, dominants may not be able to control resource allocation as completely, or may have to expend considerable energy to do so. Thirdly, there is a wide range of easily measured exploitation levels among co-operative breeders in a variety of taxa. In birds, there are several low-bias co-operative breeders such as ostriches, Tasmanian native hen, and anis, in which females, males, or both sexes, respectively, co-operate in brood care in a relatively egalitarian manner (Bertram, 1979; Ridpath, 1972; Davis, 1940). These stand in contrast to the highly biased helper-at-the-nest species such as scrub jays in which one pair breeds and all other group members provide only brood care (Woolfenden, 1975). Similar contrasts also exist in the mammals. For example, lions, hyenas and elephants exhibit communal suckling of young among co-operatively breeding females (Bertram, 1975; Mills, 1982; Laws *et al.*, 1975), whereas most canids (jackals, wolves, wild dogs), a few monogamous primates, and naked mole rats possess a highly despotic helper-at-the-den system equivalent to the birds (Moehlman, 1979, Harrington and Paquet, 1981; Malcolm and Marten, 1982; Kleinman, 1973; Jarvis, 1981). Even within the mongoose family, there are contrasting examples of communal and helper breeding systems (Rood, 1974, 1975, 1980). In the social insects, a similar distinction has been made between the communal or semi-social route to eusociality (e.g. bees: Lin and Michener, 1972) and the familial route (e.g. ants and wasps: Wilson, 1971).

In this chapter I expand on models developed in earlier publications for predicting the maximum amount of fitness biasing possible in a group (Vehrencamp, 1979, 1980, 1983). Three specific cases of the model involving different degrees of relatedness for group members are outlined and several examples are discussed. The degree of relatedness has a large impact on the level of biasing. To determine whether groups should be composed of relatives or non-relatives, I develop a second type of optimisation model that seeks the best biasing *and* group composition strategy from the point of view of a dominant individual and a set of his relatives. The options available to the dominant are: (1) to retain the relatives and impose no bias; (2) to retain the relatives and impose the maximum possible bias; (3) to disperse the relatives into other groups, recruit non-relatives, and impose no bias on the recruits; (4) to disperse the relatives, recruit non-relatives, and impose the maximum possible bias; and (5) the dominant and all relatives breed solitarily. The specific conditions that favour the evolution of each situation are outlined and examples provided. These models demonstrate the ways in which dominance and exploitative behaviour can affect aspects of population structure such as group size, dispersal pattern, group composition, and fitness biasing within groups.

The key idea presented here, that the subordinates' options outside the group limit the dominants' power, is not a new one. It has been argued verbally by Alexander (1974), and is implicit in other treatments (e.g. Baker, 1978; Harris, 1978). I have quantified the argument, provided a simple graphical model, and explored some of the logical consequences of the idea. This process has both identified some of the variables involved in grouping and biasing, and also allowed the idea to be quantitatively tested with existing field data.

The Model

A simultaneous solution to dominance interactions, resource allocation within groups, group size and group composition would require a many-individual repeated game, and the inclusion of kin considerations would render the analysis even more difficult. I have, therefore, made several simplifying assumptions in order to examine the key variable of interest, the degree of fitness biasing within groups. The next section clarifies these assumptions and their implications. The following section then describes three cases of the model: Case I. Non-related group members: Case II. Related group members; and Case III. Groups composed of parents and offspring. Subsequently I consider the effect of relaxing many of the assumptions.

Assumptions

(1) It is assumed that in each group there is a single manipulator who can appropriate resources from other individuals in the group. Since

dominance interactions are generally felt to cause such distinctions between group members, the manipulator shall be called the dominant and its fitness indicated by W_a, and the usurped group members shall be called subordinates and their fitness indicated by W_ω. It should be noted that a highly structured dominance system is not required, but only that some individuals in a group can further their own fitness at the expense of other individuals. The initial model assumes that all subordinates in a group have the same fitness. Variations of this and all subsequent assumptions are treated in a later section.

(2) It is assumed that the mean *per capita* fitness of the group depends only on the group size k ($k > 1$) and is denoted by $\bar{W}(k)$. The total output of the group is therefore equal to $k\ \bar{W}(k)$. Implicit in this assumption is the restriction that there are no behaviours available to either dominants or subordinates for changing the functional relationship between $\bar{W}$ and k.

(3) It is assumed that the dominant's only strategy is to redistribute the total fitness in the group to its own benefit. For each unit of fitness (e.g. offspring and/or reduced risk) that the dominant can usurp from each subordinate, I assume that one unit of fitness accrues to the dominant. In other words, the ratio of the benefit to the dominant and the cost to the subordinate is one, and the game is zero-sum. This implies that (a) there is no cost to the dominant of dominating, and (b) the total number of offspring gained by the group is independent of who are the breeders.

(4) It is assumed that a subordinate's only strategy to avoid being manipulated is to leave the group. In principle, a departing subordinate could either operate on its own or join another group. Initially, I assume that a subordinate joining another group retains its same status and is subjected to the same manipulation process. (The effect of relaxing this assumption will be discussed in a later section.) The only alternatives available to a subordinate are therefore to stay in a biased group or to leave and breed alone. Subordinates will not tolerate a biased fitness in a group of size k, $W_\omega(k)$, that goes below the fitness it could expect in the solitary situation, $\bar{W}(1)$. If the solitary strategy involves dispersing and establishing a new territory or locating an unoccupied habitat patch, then the fitness value of solitary living must be devalued by the probability of dispersing successfully, d. d is a function of both the cost of travelling to a new site and the availability of unoccupied 'slots' relative to the number of individuals competing for them. For most of the social organisms discussed here, travel costs are assumed to be negligible, and habitat availability is presumed to be the

primary factor affecting dispersal costs. This assumption allows d to be operationally measured as the number of 'slots' available in the population, divided by the number of potential dispersers.

The coefficient of variation (CV) of individual fitnesses (W's) within a group shall be used here as an index of the degree of bias. The lower the value of $W_{\omega}(k)$ relative to $\overline{W}(k)$, the greater the difference between dominant and subordinates, and the greater the CV of fitnesses. For the models presented here, a CV between zero and 100 per cent indicates no to moderate bias, while a CV above 300 per cent will be considered a high degree of bias.

The operational measure of fitness will depend on the particular species and the type of co-operative behaviour being investigated. Where offspring production is the biased fitness component, reproductive success per breeding attempt or per breeding season is the simplest measure of fitness. If survivorship varies as a function of group size or status in the group, a lifetime summation of fitness must be used. If individuals regularly change status during their lifetime then a summation of two or more breeding seasons will also be necessary. In this case, there may be a greater degree of bias exhibited by group members in the short term compared to the long term (e.g. baboons: Hausfater, 1975; wasps: West-Eberhard, 1978).

The objective of the model is to determine the allocation of resources or fitness to dominants and subordinates that maximises the inclusive fitness of the dominant. The model establishes the *maximum* degree of biasing that a dominant should favour, given the ecologically determined $\overline{W}(k)$ function and d. The model can be falsified only by finding a society that is *more* biased than would be predicted on the basis of the model. A slightly less biased society may still be operating under a set of modified assumptions as described in a subsequent section. The discovery of a strongly egalitarian society for which the model predicted a significant bias would certainly undermine the basic notion that ecological factors and options for subordinates outside the group are the main factors limiting the power of potential manipulators in social groups.

Case I: Unrelated Group Members

For groups of unrelated individuals, the maximum degree of bias favoured by a dominant is illustrated in Figure 8.1. Given that the *per capita* fitness $\overline{W}(k)$ rises to a peak at some intermediate group size and then declines, and that $d = 1.0$, groups can form wherever $\overline{W}(k)$ is greater than $\overline{W}(1)$. The lowest a dominant can push a subordinate's fitness is $\bar{W}(1)$, thus the $W_\omega(k)$ curve is depicted by a horizontal line intersecting $\bar{W}(1)$ and,

$$W_\omega(k) = d\,\bar{W}(1) \tag{1}$$

If the cost of dispersing is sufficiently high, i.e. $d < 1.0$, grouping could conceivably occur when $\bar{W}(k)$ is less than $\bar{W}(1)$. However, it will be shown later that groups composed of non-relatives are likely to form only when d is close to 1.0, and thus d has little impact on the non-relative case.

The increment in the dominant's fitness due to biasing is the difference between $\bar{W}(k)$ and $W_\omega(k)$ times the number of subordinates in the group, so that the dominant's personal fitness is given by:

$$W_a(k) = k\,\bar{W}(k) - (k-1)\,W_\omega(k) \tag{2}$$

The greater the difference in fitness for groups and solitary individuals, the greater the bias can be. If $\bar{W}(k)$ is equal to $\bar{W}(1)$ groups without bias might form. Note that the optimal group size preferred by the dominant is generally larger than the group size that permits the highest *per capita* fitness (Figure 8.1). Whether the dominant can increase group size by encouraging recruitment depends on the species and contexts. It can always reduce an over-large group to its optimum by lowering the fitness of excess individuals below $\bar{W}(1)$.

Female anis (*Crotophaga* spp) are an excellent example of co-operative breeding among unrelated individuals. Communal groups consist of 2 to 4 females monogamously mated to an equal number of males, but solitary breeding pairs also occur. Group-living females lay eggs in a communal clutch, and nest construction, incubation and nestling care are shared by both males and females. Despite the appearance of co-operation and non-aggression, there are competitive interactions within the group resulting from the limited number of eggs and nestlings a group can raise. Females toss out eggs that appear in the nest before they themselves have begun to lay, creating a bias in egg ownership that consistently favours the last layer (Vehrencamp, 1977). The laying order within a group is stable from nest to nest and appears to be determined by dominance interactions among the females. To ascertain whether the fitness bias follows the predictions of the model, lifetime reproductive success has been calculated for first-laying and last-laying females as a function of group size (Figure 8.2). Annual reproductive success declines with increasing

Figure 8.1: Illustration of the maximum degree of bias in fitness favoured by dominants as a function of group size for Case I: unrelated group members. $\bar{W}(k)$ is the *per capita* of average fitness of group members, $W_\alpha(k)$ is the fitness of the single dominant, and $W_\omega(k)$ is the fitness of all subordinates

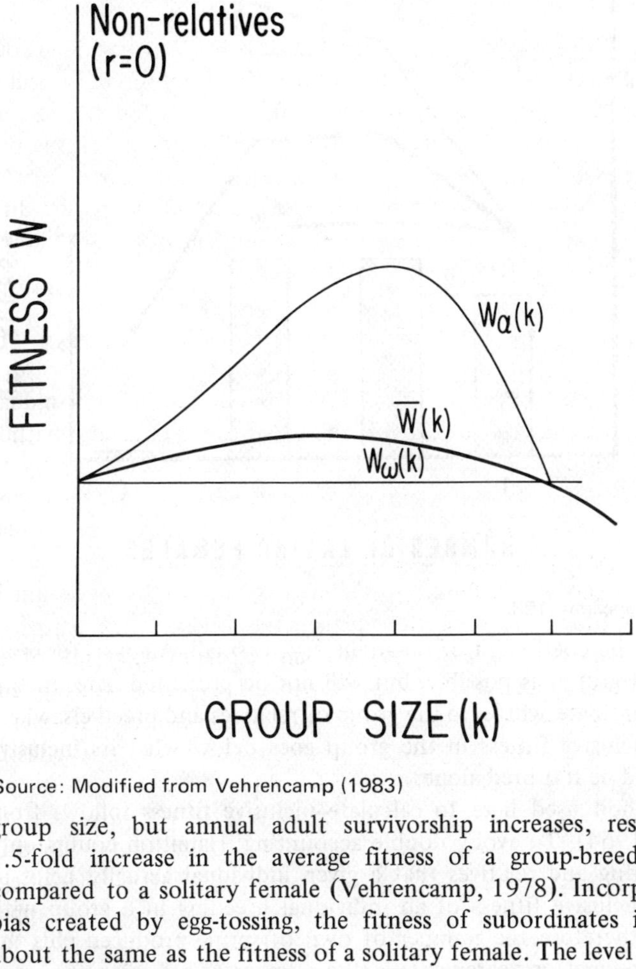

Source: Modified from Vehrencamp (1983)

group size, but annual adult survivorship increases, resulting in a 1.5-fold increase in the average fitness of a group-breeding female, compared to a solitary female (Vehrencamp, 1978). Incorporating the bias created by egg-tossing, the fitness of subordinates in groups is about the same as the fitness of a solitary female. The level of exploitation among female anis therefore follows the predictions of the model very closely.

Case II: Related Group Members

The above model can be easily modified to examine the maximum bias attainable in homogeneous groups of same-generation relatives such as siblings or cousins and in groups of mixed relatives if the within-group

Figure 8.2: Degree of bias in lifetime fitness for female groove-billed anis. The frequency distribution of group sizes is shown in the histogram below

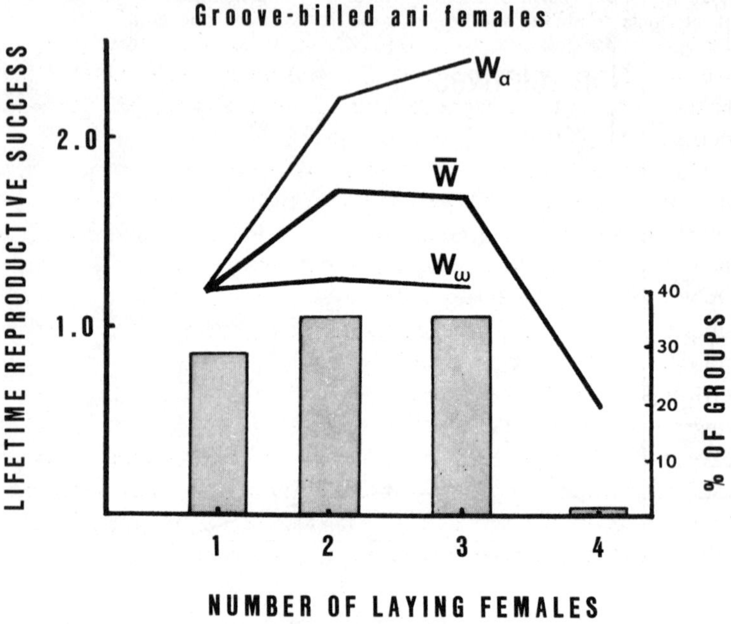

Source: Vehrencamp (1983)

variance of the coefficient of relatedness, *r*, is small. Analysis for more complicated groups is possible, but will not be presented here. In this case, a subordinate will leave the group of relatives and breed elsewhere when its inclusive fitness in the group goes below what its inclusive fitness would be if it bred alone.

The method used here to calculate inclusive fitness follows from Hamilton (1964). To avoid 'double accounting' Hamilton counts only those offspring and relatives that a given individual actually helps to raise. The inclusive fitness of an individual breeding in a group with relatives is therefore the number of own offspring produced plus the change (increment or decrement) in the number of offspring produced by relatives as a result of the individual's presence in the group, devalued by the degree of relatedness. This value would then be compared to the number of offspring raised by the individual if it bred alone.

The coefficient of relatedness, *r*, used here is the fraction of genes 'identical by descent' between the individual in question and the group members with which it is co-operating. This avoids the need to specify two r's, that between the individual and its own offspring and that between the individual and its relatives' offspring (West-Eberhard, 1975; Vehrencamp, 1979). In some cases it may be preferable to use the two-r method, but in most instances the resulting measure of inclusive fitness is one-half the value obtained by the one-r method. Since I make only relative comparisons of inclusive fitness here, the simpler method is employed.

One can determine the maximum bias in fitness attainable in a group of relatives related by r by using the inclusive fitness formula to calculate the lowest personal fitness a subordinate will tolerate before leaving the group. This problem can be approached in two ways. The first and simplest (Case II A) assumes that when conditions favour leaving, all subordinates in the group emigrate and the dominant as well as the subordinates breed alone. If the group of relatives forms on the natal territory, either by retention of offspring into the kin-group or by sib-group inheritance of the parental territory, then individuals that leave must undergo the cost of dispersal whereas group members do not. The fitness benefit of breeding solitarily must therefore be devalued by d, the probability of dispersing successfully. If subordinates leave the group, their inclusive fitness is:

$$IF(1) = d\,\bar{W}(1)$$

but if the subordinates stay their inclusive fitness is:

$$IF_\omega(k) = W_\omega(k) + r\,[k\,\bar{W}(k) - W_\omega(k) - (k-1)\,d\,\bar{W}(1)]$$

Setting $IF_\omega(k)$ equal to $IF(1)$ and solving for $W_\omega(k)$ gives:

$$W_\omega(k) = \frac{(1+kr-r)\,d\,\bar{W}(1) - kr\,\bar{W}(k)}{(1-r)} \tag{3a}$$

and by definition, $0 \leqq W_\omega(k) \leqq \bar{W}(k)$. Other patterns of dispersal give a slightly different result. If groups of relatives form by the joint dispersal of sib-groups, and these groups have a higher probability of dispersing successfully, $d(k)$, than solitary dispersers, $d(1)$, then $W_\omega(k)$ becomes:

$$W_\omega(k) = \frac{(1+kr-r)\,\bar{W}(1)\,d(1) - kr\,\bar{W}(k)\,d(k)}{(1-r)} \tag{3b}$$

The personal fitness of the dominant is calculated from (2), and its inclusive fitness is given by:

$$IF_{\alpha}(k) = k\,\bar{W}(k) + (k-1)\,[(r-1)\,(W_{\omega}(k) - rd\bar{W}(1)] \quad (4)$$

Using the same $\bar{W}(k)$ curve as in Case I, the maximum bias in personal fitness attainable in a group of relatives related by 0.25 has been plotted in Figure 8.3. Note that $W_{\omega}(k)$ can go well below $\bar{W}(1)$, and that the bias in fitness is correspondingly higher than for the

Figure 8.3: Illustrations of the maximum degree of bias favoured by dominants as a function of group size for Case II of the model: related group members. $\bar{W}(k)$, $W_{\alpha}(k)$ and $W_{\omega}(k)$ are the personal fitnesses of an average group member, the dominant, and subordinates, respectively, as in Figure 8.1: (a) maximum bias for relatives related by $r = 0.25$ and a moderate benefit to grouping, where all subordinates leave the group simultaneously (Case II A); (b) as in (a), but with a large benefit to grouping; (c) as in (a), but with a low degree of relatedness $r = 0.1$; (d) as in (a), but subordinates leave singly (Case II B)

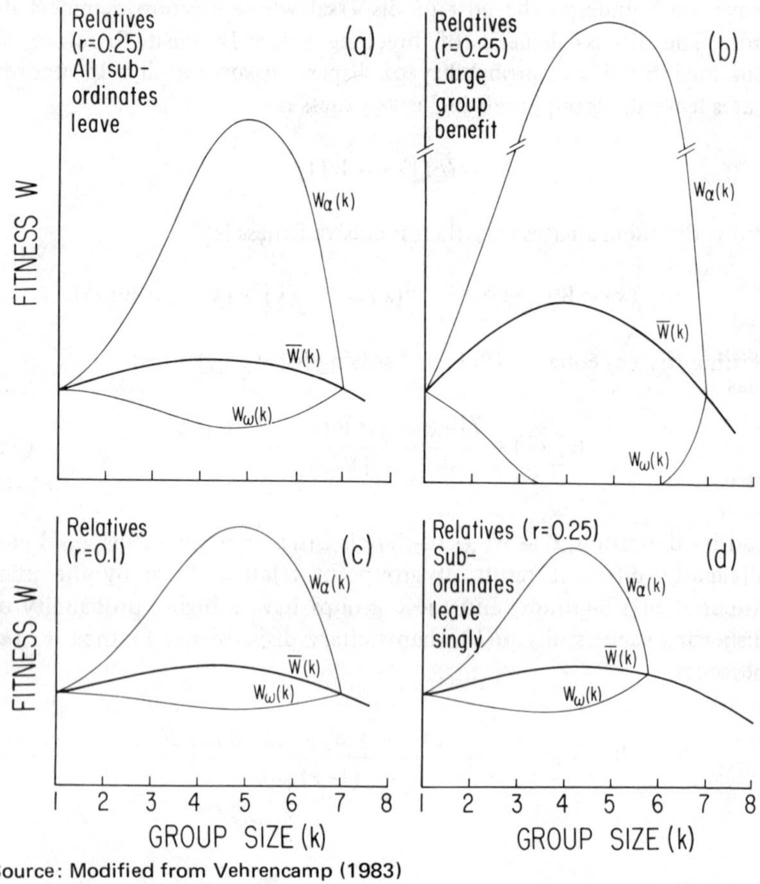

Source: Modified from Vehrencamp (1983)

equivalent group of non-relatives. As the difference between $\bar{W}(k)$ and $\bar{W}(1)$ increases, the bias can be much greater (Figure 8.3b). In this example, group breeding is twice as productive as solitary breeding and the bias is extreme, with subordinates producing no offspring themselves. The lower the degree of relatedness, the less the bias (Figure 8.3c). If there is a substantial cost of dispersing from the natal territory (d is small), the degree of bias can increase. The relationship between the degree of bias and d, the cost of leaving, is shown in Figure 8.4 for different coefficients of relatedness. Groups of relatives can even form when $\bar{W}(k) < \bar{W}(1)$ if d is small. As in Case I, the dominant prefers a group size slightly larger than the optimum *per capita* group size.

The second approach to this problem (Case II B) assumes that only one subordinate at a time leaves the group to breed solitarily, leaving a group of k–1 relatives. The inclusive fitness of a subordinate therefore becomes:

$$IF_\omega(k) = W_\omega(k) + r[k\,\bar{W}(k) - W_\omega(k) - (k-1)\,\bar{W}(k-1)]$$

The lowest fitness a subordinate will tolerate is thus:

$$W_\omega(k) = \frac{d\,\bar{W}(1) - r\,W_i(k)}{(1-r)} \qquad (5)$$

where $W_i(k)$ is the increment in total group production due to the kth member, or $k\,\bar{W}(k) - (k-1)\,\bar{W}(k-1)$. Once $W_\omega(k)$ is known, $W_a(k)$ and $IF_a(k)$ are calculated in the same manner as above (by (2) and (4), respectively). This approach gives a qualitatively similar result, but the bias is slightly less and all of the relatives prefer the group size with the greatest *per capita* fitness (compare Figure 8.3d to Figure 8.3a).

Species in which siblings breed co-operatively are the best examples of Case II. Data from Metcalf and Whitt (1977a, b) on the wasp *Polistes metricus* provide the most complete information from the point of view of the model. *P. metricus* is primarily a solitary breeder, but occasionally two-foundress nests occur. Since females generally return to the site of the nest in which they were born, two females attempting to nest at the same location are generally sibs. As a result of the haplodiploid genetic system these sisters are related by an r of 0.63. Metcalf and Whitt used electrophoretic techniques, extensive nest observations and brood counts to measure precisely the bias of egg ownership, adult survivorship and reproductive success of single-foundress and dual-

Figure 8.4: Degree of bias (coefficient of variation in personal fitnesses of group members) as a function of the coefficient of relatedness, *r*, and the probability of dispersing successfully, *d*, with a ratio of group benefit to solitary breeding of 1.3, the leaving subordinates are assumed to breed either solitarily or to remain subordinate in a new group

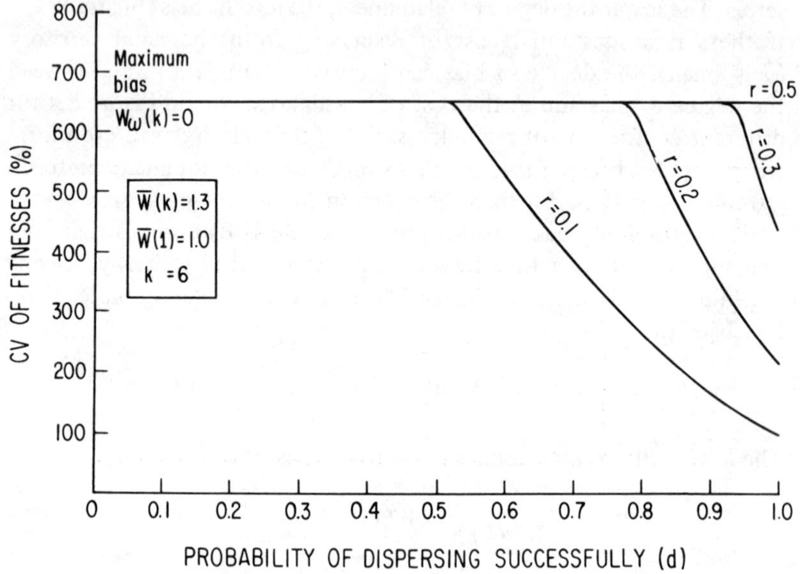

Source: Vehrencamp (1983)

foundress nests. The first female to arrive at the site is usually alpha. I assume that a late-arriving (subordinate) female never has the opportunity to become a dominant at another nest, and that her only alternative is to breed solitarily. The alpha female lays 87 per cent of the eggs in the joint clutch. The beta female performs most of the foraging and therefore has lower survivorship than the alpha female. The average lifetime fitness of dual-female nesters is greater than that of solitary nesters (Figure 8.5), but the lifetime fitness of the beta female in a duo is less than that of a solitary female. However, the inclusive fitness of a beta in a duo is still higher than her inclusive fitness would be if the two females bred solitarily. These results fit very well with the predictions of Case II of the model. Similar conclusions have been drawn by Noonan (1981) for *Polistes fuscatus*, but she suggests that beta females are smaller and inferior compared to alpha females and have a reduced potential for reproducing successfully as solitary foundresses.

The acorn woodpecker is another example of sibling co-operation, but sufficient information is not yet available to test the model

Figure 8.5: Degree of skew in lifetime reproductive success for single- and dual-foundress nests of the wasp *Polistes metricus*. Relative inclusive fitnesses for the three possible strategies are: $IF_{\alpha} = 1.83$, $IF_{\beta} = 1.39$, $IF(1) = 1.0$

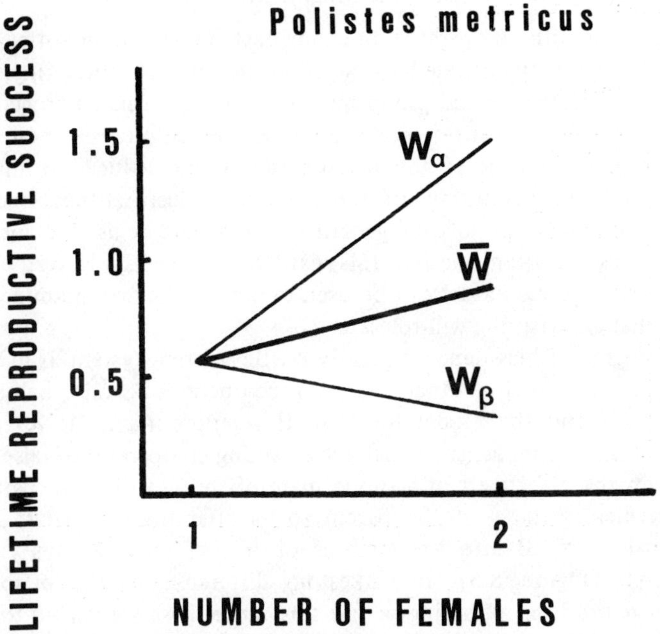

Source: Modified from Metcalfe and Whitt (1977a)

quantitatively. Co-operative groups consist of 1 to 2 females and 1 to 4 males (Koenig, 1981a). Both sexes disperse in sibling groups of like sex. Breeding vacancies are scarce and highly contested, and it is very likely that a group of birds has a greater chance of winning a slot than a solitary bird (Koenig, 1981b). Another advantage of group living is improved survivorship. *Per capita* reproductive success declines with increasing group size for both sexes, more steeply for females than for males (Koenig *et al.*, 1983). However, there does appear to be a slight net advantage to group breeding for both sexes (Koenig, personal communication). Both male and female siblings co-operate and contribute gametically to the breeding effort, but there are potential biases among them. Co-operating females toss each other's eggs out of the nest, much as in the anis (Koenig *et al.*, 1983). All of the sibling males in a group may copulate with the breeding female(s), but there is a dominance order among them which creates biased access to mates (Stacey, 1979b; Joste *et al.*, 1982). A quantitative test of the model will have to await the publication of the relevant demographic and fitness values.

Case III: Groups Composed of Parent and Offspring

The situation with co-operative breeding between parent and offspring, where the parent is assumed to be dominant over the offspring, is a special case of the related group member model. Because two generations are involved, a two-r model is preferable. Inferences about the total lifetime strategy of the parents cannot be deduced from the model, as 'double-counting' of the parental inclusive fitness occurs. Strictly speaking, a population genetic model should be used to analyse such 'games' between relatives (Maynard Smith, 1982). Nevertheless, the inclusive fitness model can be used to predict the maximum degree of bias that an offspring will tolerate.

The degree of bias depends greatly on the mating system. If mating is promiscuous, such that an offspring co-operates to raise half-sibs, then $r = 0.5$ and the model for Case II is appropriate. However, if mating is monogamous, such that the offspring co-operates to raise full sibs, then the effective r of a subordinate offspring to its two parents is 1.0, whereas the r of the parent to its offspring is equal to 0.5. Substituting $r = 1.0$ into the analysis of $W_{\omega}(k)$ above always yields $W_{\omega}(k) = 0$ (Figure 8.5). It makes no difference to the offspring whether it produces offspring or full sibs, since these are related to the same degree. However, an offspring will prefer to breed solitarily when the *per capita* fitness of the group $\overline{W}(k)$, is less than $d\,\overline{W}(1)$ (assumptions as in Case II A) or the benefit it provides to the parents, $W_i(k)$, is less than $d\,\overline{W}(1)$ (assumptions as in Case II B). The predicted result of such interaction between monogamous parents and their offspring is therefore either helping behaviour (maximum bias) or solitary breeding. A parent would of course not want to bias all of its offspring to $W_{\omega} = 0$ all of the time, otherwise its inclusive fitness on a lifetime basis would be zero. The net outcome of parent/offspring interaction will therefore depend on whether the parents can maximise their production of reproducing offspring by exploiting some proportion of their total offspring or by biasing all offspring for a limited proportion of their lives.

The vertebrate examples of this type of interaction come from monogamous species with male parental care and strict territoriality. Male parental care is significant not only because it leads to monogamy but also because the adaptations that allow males to care also allow other non-breeders to contribute to parental care as helpers. Territoriality is significant because it limits the number of pairs that can breed in the available habitat. This restricts the entry of newly-matured individuals into the breeding population by increasing the cost of dispersal, and selects for temporary retention of the young on the natal territory. The retained young aid their parents in the production of additional young. The effort of the helpers is usually not great enough to result in a *per capita* increase in production, but because the bias imposed by the parents is complete ($W_\omega = 0$), parental fitness is higher with helpers than without helpers. Examples include marmosets and tamarins, canids such as jackals, coyotes and wolves, naked mole rats, and many species of birds. The best quantitative data are provided by Rowley's (1981) study of the splendid wren, *Malurus splendens*. In this species it is primarily male offspring that remain to help. Fledging success is plotted in Figure 8.6 as a function of the number of helpers in the group. *Per capita* reproductive success declines with increasing group size, but total group output, or parental fitness, increases. Yearling males have about a 10 per cent chance of obtaining a breeding territory ($d = 0.1$), older helpers have a 41 per cent chance ($d = 0.41$), and females have close to a 100 per cent chance due to very high adult female mortality. Females do not meet the conditions of the model required to evolve helping, but younger males do. As males become older, their chances of obtaining a territory increase and eventually they disperse and breed on their own.

The invertebrate examples of co-operation between parent and offspring are also characterised by monogamy, but territoriality is not apparent in these species (e.g., termites and Hymenoptera: Wilson, 1971). The number of helpers is often immense and there is a division of labour among colony members that improves reproductive and foraging efficiency. The breeding female in many species designates certain offspring as sterile helpers and others as reproductives. It is more likely that the conditions for group formation are met by a direct advantage to grouping ($\overline{W}(k) > \overline{W}(1)$), rather than a high cost to dispersing (low d).

Relaxing the Assumptions

In this section I discuss the effects of relaxing or modifying some of the assumptions of the model.

(1) Relaxing the assumption that a single manipulator usurps all other group members to the same degree: This assumption was included to simplify the mathematics of the model. It is in fact the true situation in many strongly biased societies (e.g. helpers-at-the-nest in birds and canids, most eusocial Hymenoptera), but in other systems there is a

Figure 8.6: Degree of bias in annual reproductive success for the splendid wren. The frequency distribution of group sizes is shown in the histogram below

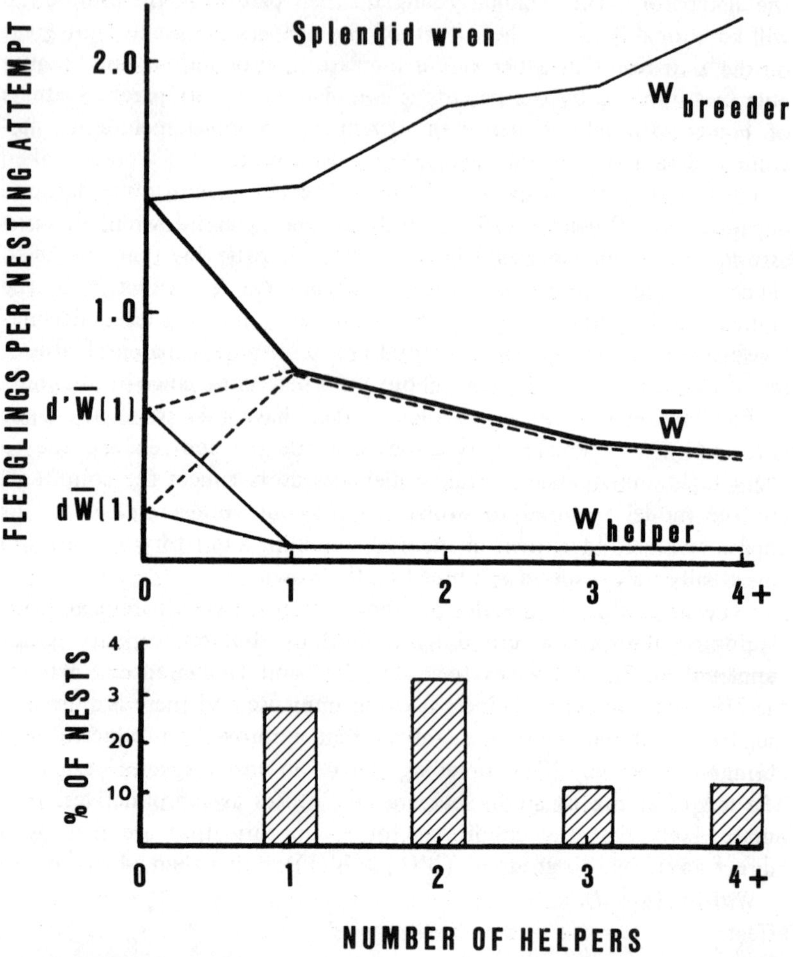

Source: Modified from Rowley (1981)

linear hierarchy with a range of dominants and subordinates. The model only establishes the maximum level of exploitation that can be imposed on any one subordinate. If some subordinates in the group exhibit fitnesses between $\overline{W}(1)$ and $\overline{W}(k)$, and there are several dominants in the group with fitnesses above $\overline{W}(k)$, then the degree of bias (*CV* of fitnesses) within the group will be less than the bias established by the model. Such deviations from the maximum predicted bias must ultimately be due to significant costs of dominating, such that a single dominant cannot reduce the fitness of all other group members to the minimum level. The factors that determine whether dominance will be complete or partial will vary from species to species depending on the contexts, mechanisms of dominance, and nature of the resource being partitioned. A game-theoretic treatment of this problem would be of great interest, (e.g. Barnard and Sibly, 1981; Sibly, Ch. 9 this volume), but is beyond the scope of this analysis.

(2) Relaxing the assumption that *per capita* fitness of the group depends only on group size, regardless of the degree of bias: This assumption specifies that a group of a given size in a particular ecological context has a fixed total potential for offspring production, foraging and/or defence which is redistributed among group members by biasing. It may be the case that biasing not only redistributes this total, but it also *alters* the effective productivity or efficiency so as to change $\overline{W}(k)$. There are three ways to incorporate changes in efficiency. One is to impose extrinsic changes in efficiency on the system and look for changes in optimal bias. A second possibility is to solve for the optimal level of efficiency in systems where the degree of bias if fixed. This approach could be applied to helper-at-the-nest systems, for example where high dispersal costs force subordinates to remain in groups and accept the maximum degree of bias, but where subordinates provide only the minimum amount of aid necessary to justify group membership. The third approach is to let both the degree of bias and the efficiency level be strategic variables controlled by individual group members. The first approach (outlined below) is easily added to the current model. The second will be treated in a subsequent publication. The third approach may be the most realistic for many animal systems, but is beset by the analytical complexities mentioned in (1) above, and has not been attempted here.

Within the constraints of the current model, one can examine the effects of an increase or decrease in the efficiency of biased groups relative to unbiased groups on the optimal degree of bias. An efficiency factor is incorporated into the existing model by multiplying all $\overline{W}(k)$ values by e, the total fitness of a biased group divided by the total fitness of an unbiased group, as follows:

$$W_{\omega}(k) = \frac{(1+kr-r)\, d\, \overline{W}(1) - ekr\, \overline{W}(k)}{(1-r)} \tag{6}$$

$$W_\alpha(k) = e\,k\,\overline{W}(k) - (k-1)\,W_\omega(k) \tag{7}$$

If $e < 1$, then a biased group is less efficient than a non-biased group. The effect of this reduced efficiency on the model is to lower the value of $\overline{W}(k)$ relative to $\overline{W}(1)$, and this in turn greatly reduces the degree of bias. If e is small enough, unbiased groups may form, particularly among relatives as discussed later. Possible situations leading to reduced efficiency in biased groups are: (a) a high cost of dominating, such that the dominant 'dissipates' fitness trying to extract resources from unwilling subordinates; (b) a high cost of resisting, such that subordinates 'dissipate' fitness trying to defend resources; (c) coalition formation by subordinates leading to sabotage or the threat of mutiny; and (d) physiological, morphological or time limitations on achieving the optimal degree of specialisation into workers versus reproducers. On the other hand, if $e > 1$, then biased groups are more productive than unbiased groups. This could occur if the specialisation of group members into foragers, defenders and reproducers resulted in greater reproductive efficiency than non-specialisation. The outcome is clearly an increase in the degree of bias.

(3) Relaxing the assumption that subordinates that leave breed solitarily: The model can be easily altered to examine the degree of bias possible when subordinates that leave join a new group with some probability of becoming dominant. Although this probability may take a variety of intermediate values, an examination of the two end-points of the continuum shows the trends. If the leaving subordinate has *no* chance of becoming dominant (once a subordinate, always a subordinate), then the calculation of $W_\omega(k)$ is the same as for the case of solitary breeding (see (5) and Figure 8.4). If the leaving subordinate has a *random* chance of becoming dominant in a new group, and all groups are approximately the same size, then this equation becomes:

$$W_\omega(k) = \frac{\overline{W}(k)\,[d+dkr-dr-kr]}{(1-r)} \tag{8}$$

In this situation, the degree of bias attainable is much less than that which occurs when subordinates can only breed solitarily, and depends primarily on the probability of dispersing successfully, d. In Figure 8.7, the maximum degree of bias is plotted for different values of d and r for the random dominance situation with the same $\overline{W}(k)$ as in Figure 8.4. As before, the higher the r, the greater the bias. For the case of non-relatives, effectively no bias is possible because d is close to 1.0 ($W_\omega(k) \cong \bar{W}(k)$). Interestingly, if $d = 1.0$ (i.e. there is no cost to leaving) *and* dominance is random, then no bias is possible ($W_\omega(k) = \bar{W}(k)$), regardless of the degree of relatedness.

Figure 8.7: Degree of bias as a function of r and d with a benefit ratio of 1:3, but where the leaving subordinates are assumed to join a new group with a random chance of becoming dominant

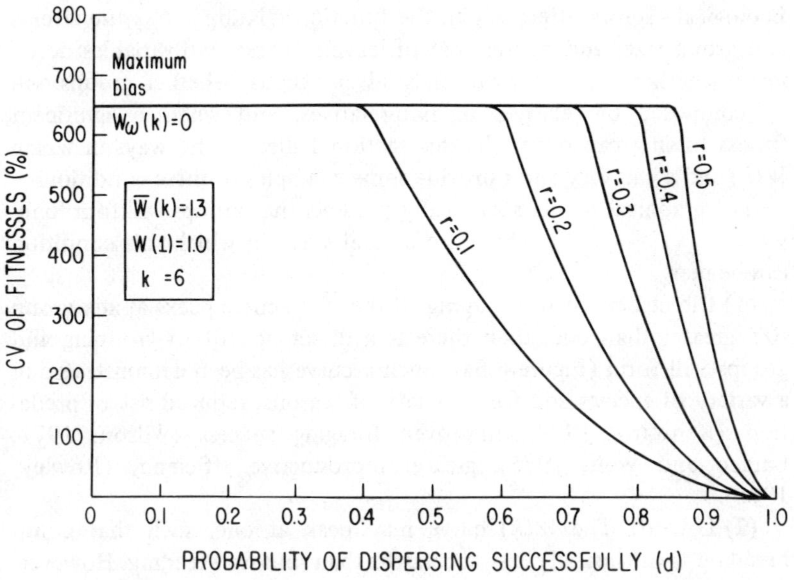

Source: Vehrencamp (1983)

In many group-living species, the probability of a leaving subordinate becoming dominant in a new group should lie between these two points and will depend on the way in which groups are formed. If the disperser must join an established group, it will have the disadvantage of being a newcomer, and may only be accepted into the group if it remains subordinate. If emigrants band together to form new groups, on the other hand, then each may have a random chance of becoming dominant. The probability of a subordinate being dominant may also depend on whether the reference group is related or unrelated. A subordinate in a group of non-relatives has by definition already dispersed once and its status in a new group is likely to be the same as its status in the current group. However, an individual in a group of relatives has not yet dispersed, and while it may be naturally subordinate to an older relative or parents, it may have a random chance of becoming dominant in a non-related group. Finally, if new immigrants *must* be dominant in order to enter a group (e.g. harem male takeover in polygynous mammalian societies), then the cost of dispersing into a new group is implicitly high, and solitary existence or subordinance are the most readily available alternatives for most individuals.

Effects of Ecological Factors

Ecological factors affect $\overline{W}(k)$, the function relating *per capita* fitness and group size, and d, the cost of leaving. These two variables determine whether group formation is advantageous, whether groups will be composed of relatives or non-relatives, and whether significant fitness biasing can occur. In this section I discuss the ways in which $\overline{W}(k)$ and d can vary and I provide some examples of these conditions.

The graphical model specifically predicts that groups will form only when $\overline{W}(k) > d\,\overline{W}(1)$. There are several ways in which this condition can be met:

(1) Direct benefit to grouping: If the $\overline{W}(k)$ curve peaks at any group size greater than one, then there is a direct benefit to grouping and groups will form (Figure 8.8a). Such a curve has been demonstrated in a variety of species and for a variety of reasons: reduced risk of predation (Kenward, 1978), improved foraging success (Wilson, 1971; Caraco and Wolf, 1975), greater reproductive efficiency (Rowley, 1978).

(2) Low d: The $\overline{W}(k)$ curve may peak at one, such that group breeding is intrinsically less successful than solitary breeding. However, if d is sufficiently low, the solitary breeding option becomes difficult to attain, and group formation on the natal territory is favoured (Figure 8.8b, dashed line). This situation can only generate groups of relatives (sibs and/or parents). A low d can result from several ecological situations: (a) in territorial species, a limited availability of breeding habitat, nest sites, or food storage sites plus a high adult survivorship relative to the number of offspring produced; (b) a skewed sex ratio, leading to a low probability of obtaining mates for the more numerous sex (Tasmanian hen: Maynard Smith and Ridpath, 1972; pied kingfisher: Reyer, 1980); (c) other types of costs associated with leaving, such as having to give up a good breeding site or abandon a breeding attempt already in progress, especially if the season is short (Bekoff and Wells, 1982). A decreasing $\overline{W}(k)$ curve occurs when groups are less productive on a *per capita* basis than solitary breeders. This may result when groups are more conspicuous than solitary breeders and

Figure 8.8: Four ways to generate a benefit to grouping: (a) direct group benefit; (b) low probability of dispersing to the optimal solitary situation; (c) limited primary habitat with no disadvantage to grouping; and (d) a direct benefit to grouping occurring unpredictably in poor years only

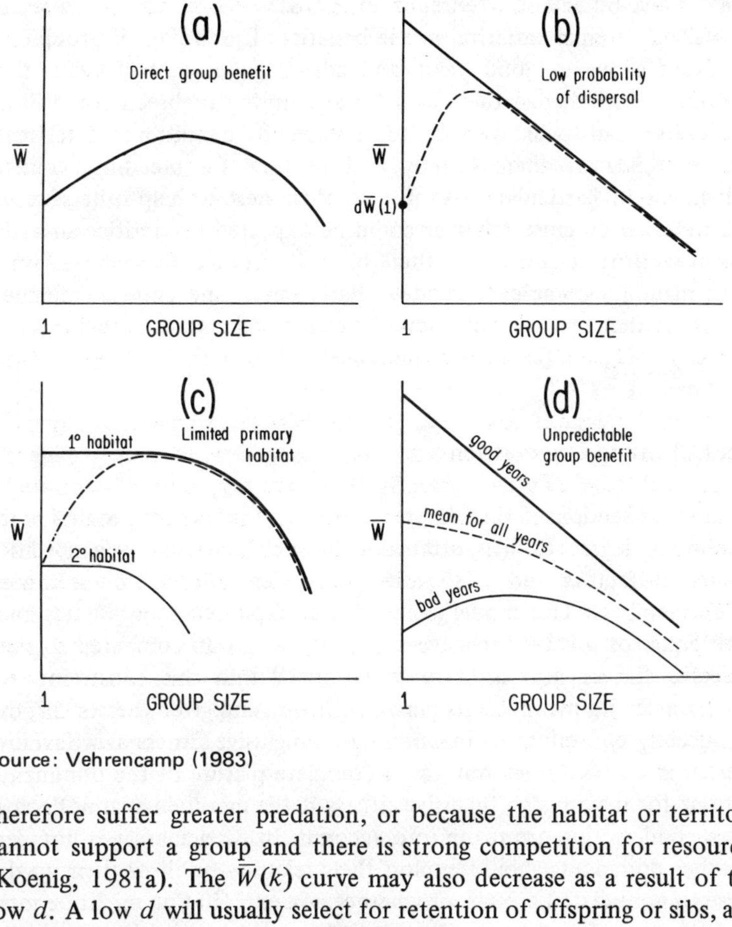

Source: Vehrencamp (1983)

therefore suffer greater predation, or because the habitat or territory cannot support a group and there is strong competition for resources (Koenig, 1981a). The $\overline{W}(k)$ curve may also decrease as a result of the low d. A low d will usually select for retention of offspring or sibs, and the ensuing high degree of relatedness will very likely lead to a maximally-biased breeding system. Helpers should cut back their effort to the minimum required to remain in the group. Since the dominant breeder is producing all of the offspring, it can still reap a benefit even though the *per capita* group output declines.

(3) Limited primary habitat: In the situation illustrated in Figure 8.8c, there is a primary habitat that is limited and not greatly detrimental to grouping over some range of group sizes, and a secondary

habitat that still supports breeders and is not limited. The two types of habitats must be close by or arranged in a mosaic so that dispersers have a choice between them. Large groups will occur on good territories, small groups or solitary individuals on the poorer territories (e.g. Groove-billed anis: Vehrencamp, 1978).

(4) High annual variation in the benefits of grouping: If grouping is disadvantageous in good years and advantageous in bad years, then individuals will hedge their bets by attempting to breed solitarily in all years, but will switch to groups if conditions deteriorate (Figure 8.8d). If there is only a short time for breeding, younger individuals in particular may give up their nest to help others during bad times. Only close relatives could be expected to sacrifice an entire season's effort to aid each other. In white-fronted bee-eaters, family clans maintain complex extended relationships, and younger members will abort their nest to aid dominant clan members, often their parents (Emlen, 1981, and personal communication).

Optimal Group Composition

In the first section of this chapter I outlined a model for predicting the maximum degree of bias attainable in social groups given specified ecological factors and a specified coefficient of relatedness among group members. This model assumed that exploited subordinates must have equal or greater inclusive fitness in the group compared to their inclusive fitness in a solitary situation. Within this constraint, the dominant could maximise its personal fitness, and given these rules, this was usually equivalent to maximising its inclusive fitness as well. However, this analysis does not give a complete picture of the dominant's options for maximising inclusive fitness if the modified assumptions of the second section are taken into account. In particular, it is not clear whether dominants should exploit their relatives or allow them to disperse to non-related groups, or whether non-exploitation might be more advantageous than exploitation in some circumstances. In this section I examine a wider range of options from the point of view of the dominant individual. Following the logic of the model, I assume that the dominant individual (manipulator) also has control over group composition, and regulates both the degree of bias and the composition of groups to maximise its inclusive fitness. This approach suggests that a set of behavioural rules in the context of a few specified ecological conditions can determine the structure and dispersal pattern of a population.

For this analysis, it is assumed that groups are initially composed of relatives as a result of a common natal area, e.g. sib groups, parents and offspring. A dominance order is established, and one manipulator emerges who can control the fitness bias within the group. The dominant may then either retain these relatives and impose any level of bias up to the maximum allowable, or it can encourage the relatives to disperse and breed elsewhere, recruit new unrelated members, and impose any allowable level of bias on them. The dominant need not force its relatives out of the natal group, it can merely increase the degree of bias such that dispersal is preferred. Five specific grouping and biasing strategies are established; these are listed in Table 8.1 along with the equation for the dominant's inclusive fitness in each situation.

Table 8.1: Alternative strategies from the dominant's point of view

		Relative IF_{α} =
A	The dominant and all relatives live solitarily	$\overline{W}(1)$
B	The dominant retains the relatives and imposes no bias	$\overline{W}(k) + r(k-1)\ (1-d)\ \overline{W}(k)$
C	The dominant retains the relatives and imposes the maximum allowable bias	$ek\overline{W}(k) - (1-r)\ (k-1)\ W_{\omega}(k) - r\,d\ (1-k)\ \overline{W}(k)$
D	The dominant disperses the relatives, recruits an equal number of non-relatives, and imposes no bias	$\overline{W}(k)$
E	The dominant disperses the relatives, recruits an equal number of non-relatives, and imposes the maximum allowable bias	$e\overline{W}(k) + (k-1)\ [e\,\overline{W}(k) - \overline{W}(1)]$

The assumptions of the basic model have been modified as follows. For strategies B, C, D and E, I assume that subordinates join groups and have a random chance of becoming dominant when $\overline{W}(k) > \overline{W}(1)$. The efficiency factor, e, is incorporated into the inclusive fitness equations for the two biasing strategies C and E. Finally, group size is fixed (here k=6) and is assumed to be the same for biased and unbiased groups of both relatives and non-relatives. The key variables I systematically vary are: r, the coefficient of relatedness between the initial set of relatives; the ratio $\overline{W}(k)/\overline{W}(1)$, an ecological function that determines the relative benefit of grouping compared to solitary living, and d, the probability of dispersing successfully. For each set of variables, the alternative strategy with the highest $IF_{\alpha}(k)$ is denoted the optimal strategy. The results of this analysis are summarised below, and some of the more interesting relationships are illustrated.

When d, the probability of dispersing successfully, is very low (0.3), the best strategy for the dominant is to retain the relatives and bias their fitness to the maximum allowable (Figure 8.9). This is true even when $\bar{W}(1) > \bar{W}(k)$ and $e < 1.0$. For intermediate values of d (0.6), it is still optimal to retain and bias relatives, but only if the efficiency is not too low and $\bar{W}(k)$ is not much lower than $\bar{W}(1)$. When d is high (0.9), strong biasing of relatives is not possible (Figure 8.4), and other strategies become optimal.

When d is high (0.9) the optimal strategy changes as a function of the relative values of $\bar{W}(k)$ and $\bar{W}(1)$. Solitary breeding (strategy A) is optimal if $\bar{W}(1) > \bar{W}(k)$. Biased breeding with relatives (strategy C) is optimal when $\bar{W}(k)$ is slightly greater than $\bar{W}(1)$ but the degree of bias is low $(0 < W_{\omega}(k) \leq W(1)$. Biasing of non-relatives (strategy E) is optimal for $\bar{W}(k) >> \bar{W}(1)$ (Figures 8.10 and 8.11).

When d is high and e, the efficiency of biased groups, is less than 0.7, then unbiased breeding with very close relatives (strategy B) is optimal

Figure 8.9: Optimal social strategy from the point of view of the dominant when r is high, d is low and biasing efficiency (e) is 1.0

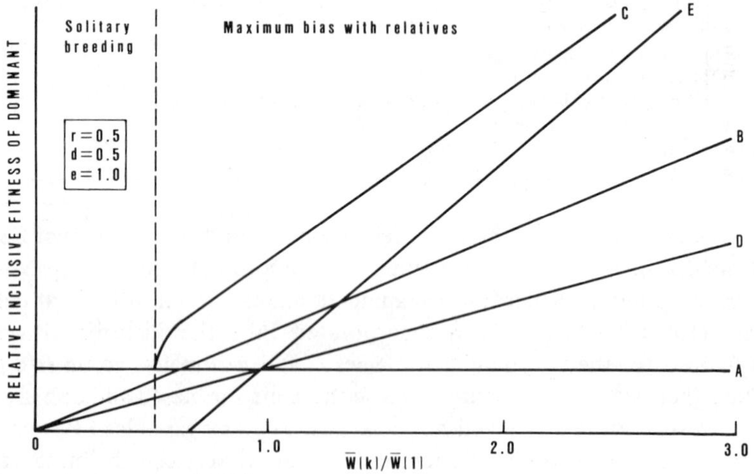

Source: Vehrencamp (1983)

Figure 8.10: Optimal social strategy from the point of view of the dominant when r is high, d is high and biasing efficiency (e) is 1.0

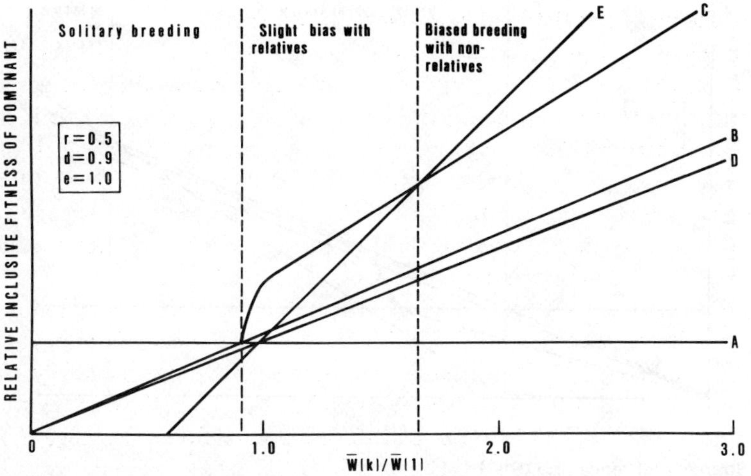

Source: Vehrencamp (1983)

Figure 8.11: Optimal social strategy from the point of view of the dominant when r is low, d is high and biasing efficiency is 1.0

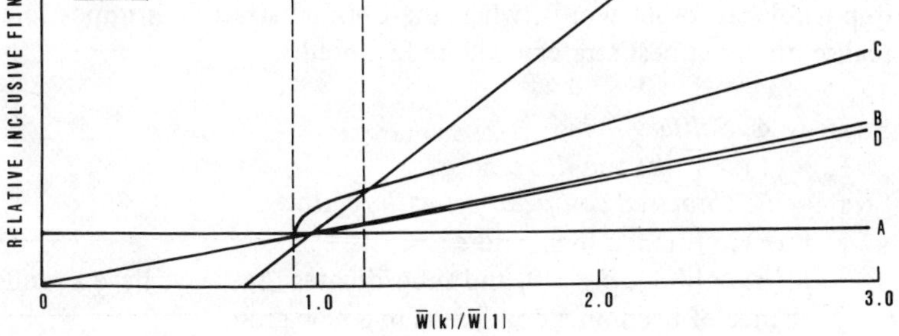

Source: Vehrencamp (1983)

Figure 8.12: Optimal social strategy from the point of view of the dominant when r is high, d is high, and biasing efficiency is significantly less than 1.0

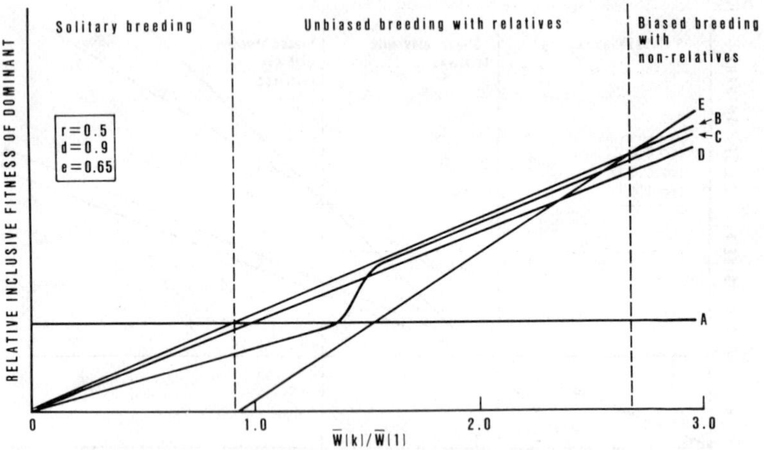

Source: Vehrencamp (1983)

for most $\overline{W}(k) > \overline{W}(1)$. If relatedness is low, biasing of non-relatives becomes optimal when $\overline{W}(k) >> \overline{W}(1)$ (Figure 8.12).

The degree of relatedness, r, has little effect on the optimal strategy, but does affect the degree of bias when groups of relatives are favoured. A $W_{\omega}(k) = 0$ is usually only possible when r is high (0.5-1.0)..

As a means of summarising the predictions of the bias model, the effects of relaxing some of its assumptions, and the results of the above group composition analysis, I shall list the sets of conditions for the five types of societies. Note that in some cases groups of non-relatives occur only because there is no possibility for interaction among relatives, and, similarly, unbiased groups occur only because biasing is impossible. In other words, when the optimal strategy is impossible to realise, the next best strategy will be favoured.

Strategy *A*: *Solitary living*

1. $\overline{W}(1) > \overline{W}(k)$ and d is high

Strategy *B*: *Unbiased co-operation with relatives*

1. $\overline{W}(k) > \overline{W}(1)$, d high, and $e << 1.0$
2. $\overline{W}(k) > \overline{W}(1)$, $d = 1.0$, and subordinates that leave have a random chance of becoming dominants in a new group.

Strategy *C*: *Biased co-operation with relatives*

1. d very low, $e < 1.0$, and $\overline{W}(k) >$ or $< \overline{W}(1)$

2. d intermediate, $e \cong 1.0$, and $\overline{W}(k) >$ or $\cong \overline{W}(1)$
3. $\overline{W}(k) \geq \overline{W}(1)$, d high, and $e \cong 1.0$ (only low to moderate degree of bias possible)

Strategy D: *Unbiased co-operation with non-relatives*

1. No relatives are available because brood size is one, adult mortality is high, there is no postnatal parental care, or the maturation period is long, $\overline{W}(k) > \overline{W}(1)$, and $e << 1.0$.
2. No relatives available, $\overline{W}(k) > \overline{W}$ (1), $d = 1.0$, and subordinates that leave have a random chance of becoming dominants in a new group.
3. Optimal group size much larger than number of available relatives, and $e << 1.0$.
4. Optimal group size much larger than number of available relatives, and domination not possible because of large or scattered group.

Strategy E: *Biased co-operation with non-relatives*

1. $\overline{W}(k) >> \overline{W}(1)$, d high, r high, and $e \cong 1.0$
2. $\overline{W}(k) >> \overline{W}(1)$, d high, r low, and $e \cong 1.0$
3. No relatives are available, $\overline{W}(k) > \overline{W}(1)$, and $e \cong 1.0$
4. Optimal group size much larger than the number of available relatives, and $e \cong 1.0$.

There are very few field studies for which all of the critical parameter values listed here have been measured. Those studies for which the parameters are available are presented below as a test of the model, and other suggestive studies are also discussed. It is hoped that this discussion will stimulate field investigators to measure the relevant parameters. The most revealing types of field studies will be comparisons of closely related species that exhibit different group composition and biasing strategies.

Communal nesting groove-billed anis provide a good test of the model because males and females exhibit slightly different biasing and dispersal strategies, and because they are a good contrast to the helper-at-the-nest species. Parameter values have been estimated from an earlier study (Vehrencamp, 1978); subsequent studies of this same population will hopefully provide a more accurate and refined test of the model. A rough estimate of the average coefficient of relatedness between a pre-dispersal yearling and adults in a typical natal group is $r = 0.4$. This results from the fact that only half of the adults in the group are its true parents on average, and that one of the parents may have disappeared in the interim. The efficiency of a biased group, e, is assumed to be 1.0, since the average number of incubated eggs per female is the same for all group sizes. The cost of laying additional eggs that are tossed out does not appear to cause a decline in fitness for biased versus non-biased groups. The probability of a yearling dispersing successfully, d, is high, varying from 0.7 to 1.0 and averaging 0.8 for the two years analysed here. This is a result of the relatively high adult mortality for the whole population (30-40 per cent per year), the occasional turnover of groups that creates vacated territories, and the low population-wide production of young (0.6 to 0.9 yearlings per pair per year). Annual reproductive success decreases with increasing group size. For females, this is offset by an increase in adult survivorship, leading to a 1.5-fold increase in lifetime reproductive success for group members compared to pairs. For males, there is no increase in adult survivorship, so group breeding is 0.6 to 0.8 times as successful as solitary breeding. The theoretical curves from the model that come closest to the conditions for anis are depicted in Figure 8.10. The female ratio of $\overline{W}(k)/\overline{W}(1)$ falls near the intersection of the biased-with-relatives and biased-with-non-relatives curves. In fact, females always disperse from their natal groups, and exhibit the maximum bias predicted for such non-relatives. Even though retention would be equally viable, females may disperse to avoid inbreeding, since males are frequently retained. For males, the ratio of $\overline{W}(k)/\overline{W}(1)$ falls into the disperse-and-breed-solitarily region, but unbiased breeding with either relatives or non-relatives gives nearly the same inclusive fitness. In the field, males commonly do disperse and establish single-pair territories. However, males also immigrate into new groups, and some are retained on their natal territories as breeders. Regardless of the degree of relatedness, the fitness of males in communal groups is unbiased. In fact, the behaviour of the males strategically undoes the slight bias in male egg ownership created by their females' egg-tossing. The male mated to the last-laying female usually performs all of the nocturnal incubation and as a result suffers slightly higher mortality than the other males in the group. Lifetime reproductive success for group males consequently appears to be equal.

It is the reasonable availability of vacated breeding slots in the anis that favours communal nesting over helping, not any significant increase in reproductive output for larger groups. This can be seen by comparing parameter values of anis to those of helper species. Anis do not possess adaptations that make them intrinsically more effective at group breeding than the helper species. In most helper species, *per capita* reproductive success declines with increasing group size, as it does in the anis (Koenig, 1981a). Where survivorship has been measured, it tends to increase for groups with more helpers, as in the anis (Woolfenden, 1981). The key differentiating characteristic of helper species is the low probability of dispersal for yearlings ($d = 0.1$-0.5) (Rowley, 1981; Woolfenden, 1975, 1978). This is due to habitat saturation, i.e. a large overproduction of young relative to the number of vacated slots. This is argued to result from some combination of the following factors: high adult survivorship, stable population size, lack of long-distance migration, restricted or patchy breeding habitat, and/or lack of a marginal habitat that can support non-breeders (Fry, 1972; Ricklefs, 1975; Brown, 1978; Stacey, 1979a; Koenig and Pitelka, 1981; Emlen, 1982). Habitat saturation in a monogamous, territorial species leads to retention of offspring, high parental leverage, and biased co-operative breeding.

This is not to say that intrinsic *per capita* benefits to group breeding do not play any role in the evolution of egalitarian versus despotic societies. Even in birds there are situations in which group co-operation is advantageous. Sib-group dispersal in the Tasmanian hen (Maynard Smith and Ridpath, 1972) and the acorn woodpecker (Koenig, 1981b) may be more successful than solitary dispersal, and leads to moderately biased co-operation as predicted by the model. Communal laying in ratites may be advantageous because the optimal brood size is larger than the number of eggs a single female can lay, and since there is no territoriality or dispersal costs, slightly biased co-operation among non-relatives is favoured (Bruning, 1974; Bertram, 1980). However, group advantages may be a more conspicuous and important factor in mammalian and insect co-operative breeding. Among the social mammalian carnivores and social Hymenoptera, survival and reproduction are often very poor for solitary individuals. Within these taxa there are examples of biased and non-biased societies. Dispersal costs and territory availability do not seem to be important in explaining these differences. This leaves coefficients of relatedness and efficiency of biasing as the most likely factors.

In the mammalian examples, the egalitarian societies are characterised by communal suckling among (usually) related females, e.g. lions, feral house cats, hyenas, banded mongooses, coati and mice, whereas the despotic societies employ provisioning and guarding of the breeding female's young by non-breeding helpers, e.g. wild dogs, wolves, dwarf mongooses and naked mole rats (Macdonald and Moehlman, 1982; Sayler and Salmon, 1971; Jarvis, 1981). If co-operative suckling is required to achieve a *per capita* reproductive benefit, then biasing may prove to reduce efficiency, since non-breeding females cannot lactate.

Provisioning, on the other hand, can easily be performed by non-breeders. In addition, there may be mating system differences among these species that affect coefficients of relatedness and dispersal strategies. It should also be noted that there is a potential relationship between group size and the coefficient of relatedness: as group size increases, average relatedness tends to decrease (Bertram, 1976). Group size could therefore affect dispersal, mating system and biasing strategies. Precise field data on co-operative mammals are rapidly accumulating and should soon provide a quantitative test of the model.

In the social Hymenoptera, biased breeding among close relatives is the rule, and there are very few instances of unbiased or communal breeding among relatives (Wilson, 1971; West-Eberhard, 1975). Relatively unbiased co-operation is primarily found in some bees where unrelated females nest jointly and there is no overlap of generations (Michener, 1969). There are usually dominance interactions within these groups, resulting in slight biases as might be expected from the model. For insects, dominants appear to be able to impose a fitness bias at little or no cost and no loss of group breeding efficiency. Where grouping is advantageous, the degree of bias is determined by the coefficient of relatedness. Relatedness is determined by the presence or absence of an overlap of generations, so that mildly biased societies of non-relatives occur when relatives are not available.

Scope and Applicability of the Model

One goal of this fitness biasing model for co-operative breeders was to integrate the concepts of kin selection for worker altruism (Hamilton, 1964), mutualism (Lin and Michener, 1972), and parental manipulation (Alexander, 1974) in the evolution of eusociality in animal species (see also West-Eberhard, 1981). Recent population genetic models have tended to favour the manipulation hypothesis (Charnov, 1978; Charlesworth, 1978; Craig, 1979). They show that the conditions required to select for a gene for manipulation by dominants are much easier to meet than those required to select for altruism by subordinates. However, the conditions for subordinate altruism must at least be minimally met before the stronger force of manipulation can act. Field data from primitively eusocial Hymenoptera that retain breeder/worker flexibility also support the notion of manipulation (Michener and Brothers, 1974). The current model combines these ideas to predict the exact degree of fitness biasing dominants can impose, identifies some of the ecological conditions under which biasing can occur, and specifies conditions for unbiased co-operation and for co-operation between non-relatives. This model rests on the strong assumption that whenever bias is the best economic strategy for a potential dominant, it *will* occur and do so at the level predicted only by the mean group output curve, the payoffs to a subordinte of leaving, and the relatedness of the individuals involved. The question is then how many of the known variations in group size, group composition (relatives versus non-relatives), and within-group bias can be explained with only these few parameters and assumptions.

The model was originally developed with co-operatively nesting birds in mind and it appears to fit those systems for which adequate data are available relatively well. The two best-documented systems are at opposite ends of the biasing continuum. Anis possess a low-bias, joint-nesting breeding system because dispersal options for subordinates are good and group nesting is advantageous (at least for females). These two factors lead to groups of non-relatives in which dominants have little leverage to enforce strong biasing. In contrast, helpers-at-the-nest species, such as the splendid wren, possess a maximally-biased breeding system because dispersal options for youngsters are poor. Youngsters benefit by remaining on the parental territory until dispersal options improve, but the high coefficient of relatedness among group members coupled with the lack of options outside the group provide parents with sufficient leverage to force the youngsters to help if they do stay. Systems with intermediate degrees of bias are most likely to occur where siblings disperse and breed co-operatively, such as are found in the Tasmanian native hen (Ridpath, 1972; Maynard Smith and Ridpath, 1972) and the acorn woodpecker (Stacey, 1979a, b; Koenig, 1981; Koenig *et al.*, 1982). Whether these intermediate systems are also accounted for by the model will have to await publication of the relevant parameter values.

The model also appears appropriate for the one strongly biased insect system for which all parameter values are known. For wasps such as *Polistes metricus* and other hymenopterans where sisters ($r \cong 0.66$) breed co-operatively, available parameter values predict a system of strong to maximum bias. In *P. metricus*, this outcome follows largely from the fact that females prefer to nest at the site where they were raised. It is not clear whether social breeding in such insects has arisen because good nesting sites are limited or because of direct advantages to group breeding. Contrasts between *P. metricus* and species with unrelated co-operating females following the format outlined here ought to be quite informative in this respect. Mammals also show a wealth of co-operative breeding systems ranging from high bias to no bias and from related to unrelated group members (MacDonald and Moehlman, 1982). Strongly biased systems analagous to avian helpers-at-the-nest appear to exhibit the same kinds of habitat saturation constraints as the birds, and therefore limited high-quality breeding habitat and low probability of dispersal are the conditions favouring temporary retention of young on the natal territory and strong parental leverage. Other species exhibit low- or non-biased co-operation, sometimes among related females and sometimes among unrelated females. In many of these species, there are significant direct advantages to group living and breeding via co-operative hunting, co-operative suckling, group defence, etc. Inefficiency problems associated with biased communal suckling may be responsible for the lack of fitness biasing. It remains to be seen whether these species will meet the exact predictions of the model.

A critical issue in the current model is the cost of domination. The model is best applied to situations where the cost of domination is low or negligible. This assumption is apparently met in many cases. Hymenoptera queens, for example, employ behavioural means to subordinate workers where group size is small, but employ inexpensive hormonal suppression where group size is large and behavioural methods would be much less effective (Wilson, 1971). There is no evidence for hormonal control of subordinates in birds, but groups are small and behavioural domination and direct protection of female mates and the nest prevent sabotage by subordinates (Mumme *et al.*, 1983). Olfactory suppression of ovarian cycling has been documented in several mammalian species (Hearn, 1978; Katz and Epple, 1980). For communal-suckling mammals, it is tempting to speculate that one of the benefits of the complex hormonal regulation of lactation is the prevention of manipulation by dominant females. However, in two group-breeding mammals, dingos and dwarf mongooses, several females may produce pups, but the dominant female kills the pups of subordinates, subordinates then suckle the pups of the dominant female (Corbett, in Macdonald and Moehlman, 1982; Rood, 1980). From these examples one must conclude that where prevention of breeding by subordinates is advantageous to dominants, animals will find a low-cost method of doing it. However there are many examples where dominance is costly and dominants are regularly deposed or turned over, to be replaced by another. This is especially true for males defending mates in multi-male, multi-female groups (Hausfater, 1975), and for defence of food resources against many individuals. An interesting reversal of the model occurs in co-operative winter territories of the pied wagtail (Davies and Houston, 1981). A territorial bird (dominant) only allows a co-resident (subordinate) into its territory when food is sufficiently abundant. The food is shared 50-50, but the subordinate contributes less than half of the territory defence costs. The subordinate therefore derives a greater net benefit from the territory than the dominant (as long as it is allowed to stay), and the dominant roughly breaks even in the group compared to when it is alone. The cost of dominating is therefore a question worthy of further investigation.

A related issue is the cost of subordination and the kinds of work a dominant can demand from a subordinate. In eusocial insects, where queens relegate a certain fraction of their offspring to worker status for their entire lives, there is seemingly no limit; workers are sterile and provide very energetically expensive and risky tasks such as foraging and colony defence. In birds, helpers provision nestlings and stand guard, but may be reluctant to attack predators and never incubate. In mammals, young non-breeding helpers bring food to pups and in some cases stand guard, but are not the individuals who bring down the large prey in co-operative hunts and do not feed pups by lactation. These differences between invertebrates can probably be explained on the basis of life-history patterns. In low-fecundity vertebrates, where individuals help only while young and temporarily prevented from obtaining a territory, subordinates are 'hopeful reproductives' and should attempt to maximise their survival. They would be expected to avoid any risky forms of helping, and instead attempt to maximise their inclusive fitness with low-risk aid. Naked mole rats appear to be a major vertebrate exception (Jarvis, 1981). Insects, on the other hand, are high-fecundity organisms that can afford to relegate a certain fraction of their offspring to lifelong, costly helping.

The final questions are how the model can be falsified and what alternatives are available. I have found it difficult to envisage any alternative cause of strongly biased societies which is not incorporated, at least qualitatively, in the model. For example, among vertebrates workers have suggested that maximum bias occurs in helpers-at-the-nest as a result of high dispersal costs (leading to retention of offspring) coupled with high inbreeding costs (Koenig and Pitelka, 1979; Stacey, 1979b). Dispersal costs are already incorporated into the model. Inbreeding costs can be quantified and incorporated as an increase in efficiency for biased groups compared to unbiased (inbred) groups, i.e. $e > 1$.

Viewed this way, the model predicts the strong bias observed. The only tests of the model involving strong bias would thus seem to be quantitative comparisons between predicted and observed values of group size, amount of bias and group composition.

On the other hand, the frequency of occurrence of egalitarian societies can be used as a test of the model as defined. There are two ways one might obtain egalitarian societies. Egalitarian societies of the first kind are those in which economic conditions do not favour bias when viewed from the dominant's position. This can occur whenever (a) $\bar{W}(k) \simeq d\bar{W}(1)$, or (b) the cost and disruption due to enforcing bias is strong enough to reduce the effective $\bar{W}(k)$ to a value commensurate with $dW(1)$ (i.e. $e \ll 1$). Egalitarian societies of the second kind are those in which bias is economically preferable to the dominant *but* is not enforced. Reasons for non-enforcement include: (a) individuals within the group are too homogeneous in fighting ability for the establishment of stable dominance; (b) enforcement entails physiological controls that are impossible (e.g. production of workers versus queens, or lactation by non-reproductives); (c) enforcement entails conditions of proximity that are impossible (e.g. the major hen of communally-nesting ostriches performs the diurnal incubation and removes many eggs of secondary females from the nest (Bertram, 1979), but in rheas the male performs all incubation and effectively prevents egg-tossing by communally-laying females (Bruning, 1974); (d) subordinates form coalitions to counter-bias enforcement by dominants (an act that potentially can be countered by coalitions of dominants); and (e) subordinates and dominants 'agree' to avoid bias for some common good (e.g. to achieve the *Pareto* optimum of maximum *per capita* output or other group-selected type optima).

The current model only allows for egalitarian societies of the first kind. It would be possible to relax the model assumption that bias is usually enforceable so as to incorporate the five conditions leading to egalitarian societies of the second kind. In fact, I have examined some of the consequences of doing so for model predictions in an earlier section. But while the model would become more general, it would also become less interesting. This model is only useful *if* bias is frequently favoured *and if* it is usually enforceable. If egalitarian societies of the second kind occur very frequently in nature, then the kind of model outlined here does not provide useful predictions. But if, as I suspect, bias and its enforcement are the rule, then the model will have widespread utility in explaining group size, levels of bias, and group composition, and will do so with a single synthetic formulation.

Acknowledgements

This chapter extends a model first developed in earlier publications and I thank Baillière Tindall Press for permission to use material from these publications. I also wish to thank J. Maynard Smith, W. Koenig, R. Mumme, B. Bowen, R. Koford, J. Bradbury and anonymous reviewers for helpful discussion of the model.

References

Alexander, R.D. (1974). The evolution of social behaviour. *Ann. Rev. Ecol. Syst. 5*: 325-83

Baker, M.C. (1978). Flocking and feeding in the great tit *Parus major* – an important consideration. *Amer. Nat. 112*: 779-81

Barnard, C.J. and Sibly, R.M. (1981). Producers and scroungers: a general model and its application to captive flocks of house sparrows. *Anim. Behav. 29*: 543-50

Bekoff, M. and Wells, M.C. (1982). Behavioural ecology of coyotes: Social organization, rearing patterns, space use, and resource defense. *Z. Tierpsychol. 60*: 281-305

Bertram, B.C.R. (1975). Social factors influencing reproduction in wild lions. *J. Zool. Lond. 177*: 463-82

—(1976). Kin selection in lions and in evolution. In P.P.G. Bateson and R.A. Hinde (eds.), *Growing Points in Ethology*: 281-301. Cambridge, Cambridge University Press

—(1979). Ostriches recognize their own eggs and discard others. *Nature, Lond. 279*: 233-4

—(1980). Breeding system and strategies of ostriches. *Proc. 17th Intl. Ornithol. Congr. 17*: 890-4

Brown, J.L. (1978). Avian communal breeding systems. *Ann. Rev. Ecol. Syst. 9*: 123-55

Bruning, D. (1974). Social structure and reproductive behavior in the greater rhea. *Living Bird 13*: 251-94

Caraco, T. and Wolf, L.L. (1975). Ecological determinants of group sizes of foraging lions. *Amer. Nat. 109*: 343-51

Charlesworth, B. (1978). Some models of the evolution of altruistic behavior between siblings. *J. theor. Biol. 72*: 297-319

Charnov, E.L. (1978). Evolution of eusocial behavior: offspring choice or parental parasitism? *J. theor. Biol. 75*: 451-65

Craig, R. (1979). Parental manipulation, kin selection, and the evolution of altruism. *Evolution 33*: 319-34

Davies, N.B. and Houston, A.I. (1981). Owners and satellites: the economics of territory defense in the pied wagtail, *Motacilla alba. J. An. Ecol. 50*: 157-80

Davis, D.E. (1940). Social nesting habits of Smooth-billed Anis. *Auk 57*: 179-218

Emlen, S.T. (1981). Altruism, kinship, and reciprocity in the white-fronted bee-eater. In R.D. Alexander and D.W. Tinkle (eds.), *Natural Selection and Social Behavior: Recent Research and New Theory*: 217-230. New York, Chiron Press

—(1982). The evolution of helping. I. An ecological constraints model. *Amer. Nat. 119*: 29-39

Fox, M.W. (1975). *The Wild Canids*. New York, Van Nostrand Reinhold

Fry, C.H. (1972). The social organization of bee-eaters (*Meropidae*) and cooperative breeding in hot-climate birds. *Ibis 114*: 1-14

Hamilton, W.D. (1964). The genetical evolution of social behavior, I and II. *J. theor. Biol. 7*: 1-52

Harrington, F.H. and Paquet, P.C. (eds.) (1981). *Wolves of the World: Perspectives in Behavior, Ecology and Conservation*. Park Ridge, New Jersey, Noyes

Harris, M. (1978). *Cannibals and Kings: the Origins of Cultures*. New York, Vintage Press

Hausfater, G. (1975). Dominance and reproduction in baboons (*Papio cynocephalus*). *Contrib. to Primatol. 7*: 1-150

Hearn, J.P. (1978). The endocrinology of reproduction in the common marmoset, *Calithrix jacchus*. In D.G. Kleinman (ed.), *The Biology and Conservation of the Callithricidae*: 163. Washington, D.C., Smithsonian Institution Press

Jarvis, J.U.M. (1981). Eusociality in a mammal: cooperative breeding in naked mole-rat colonies. *Science 212*: 571-3

Joste, N., Koenig, W.D., Mumme, R.L. and Pitelka, F.A. (1982). Intragroup dynamics of a cooperative breeder: an analysis of reproductive roles in the acorn woodpecker. *Behav. Ecol. Sociobiol. 11*: 195-201

Katz, Y. and Epple, G. (1980). Social influence on urinary estradiol cyclicity of female *Saquinus fuscicollis* (*Callitrichidae*). *Proc. 8th Intl. Congr. Primatol*: 219. Florence

Kenward, R.E. (1978). Hawks and doves: factors affecting success and selection in goshawk attacks on wood-pigeons. *J. Anim. Ecol. 47*: 449-60

Kleinman, D.G. (ed.) (1978). *The Biology and Conservation of the Callitrichidae*. Washington, D.C., Smithsonian Institution Press

Koenig, W.D. (1981a). Reproductive success, group size, and evolution of co-operative breeding in the acorn woodpecker. *Amer. Nat. 117*: 421-33

—(1981b). Space competition in the acorn woodpecker: power struggle in a cooperative breeder. *Anim. Behav. 29*: 396-409

—and Pitelka, F.A. (1979). Relatedness and inbreeding avoidance: counterploys in the communally nesting acorn woodpecker. *Science, N.Y. 206*: 1103-5

—and Pitelka, F.A. (1981). Ecological factors and kin selection in the evolution of cooperative breeding in birds. In R.D. Alexander and D.W. Tinkle (eds.), *Natural Selection and Social Behavior: Recent Research and New Theory*: 261-80. New York, Chiron Press

—Mumme, R.L. and Pitelka, F.A. (1983). Female roles in cooperatively breeding acorn woodpeckers. In S.K. Wasser (ed.), *Social Behavior of Female Vertebrates*: 235-61. New York, Academic Press

Laws, R.M., Parker, I.S.C. and Johnstone, R.C.B. (1975). *Elephants and Their Ecology: The Ecology of Elephants in N. Bunyoro, Uganda*. Oxford, Clarendon Press

Lin, N. and Michener, C.D. (1972). Evolution of sociality in insects. *Q. Rev. Biol. 47*: 131-59

Macdonald, D.W. and Moehlman, P.D. (1982). Cooperation, altruism, and restraint in the reproduction of carnivores. In P.P.G. Bateson and P.H. Klopfer (eds.), *Perspectives in Ethology*, Vol. 5: 433-67. New York, Plenum Press

Malcolm, J.R. and Marten, K. (1982). Natural selection and the communal rearing of pups in African wild dogs (*Lycaon pictus*). *Behav. Ecol. Sociobiol. 10*: 1-14

Maynard Smith, J. (1978). Optimization theory in evolution. *Ann. Rev. Ecol. Syst. 9*: 31-56

—(1982). *Evolution and the Theory of Games*. Cambridge, Cambridge University Press

—and Ridpath, M.G. (1972). Wife sharing in the Tasmanian native hen, *Tribonyx mortierii*: a case of kin selection? *Amer. Nat. 106*: 447-52

Metcalf, R.A. and Whitt, G.S. (1977a). Intra-nest relatedness in the social wasp *Polistes metricus. Behav. Ecol. Sociobiol. 2*: 339-52
—and Whitt, G.S. (1977b). Relative inclusive fitness in the social wasp *Polistes metricus. Behav. Ecol. Sociobiol. 2*: 353-60
Michener, C.D. (1969). Comparative social behavior of bees. *Ann. Rev. Entom. 14*: 299-342
——and Brothers, D.H. (1974). Were workers of eusocial Hymenoptera initially altruistic or oppressed? *Proc. Natl. Acad. Scie. USA 71*: 671-4
Mills, M.G.L. (1982). The mating system of the brown hyena, *Hyaena brunnea* in the southern Kalahari. *Behav. Ecol. Sociobiol. 10*: 131-6
Moehlman, P.D. (1979). Jackal helpers and pup survival. *Nature Lond. 277*: 382-3
Mumme, R.L., Koenig, W.D. and Pitelka, F.A. (in press 1983). Mate-guarding in the acorn woodpecker: within-group reproductive competition in a cooperative breeder. *Anim. Behav.*
Noonan, K.M. (1981). Individual strategies of inclusive-fitness-maximizing *Polistes fuscatus* foundresses. In R.D. Alexander and D.W. Tinkle (eds.), *Natural Selection and Social Behavior: Recent Research and Theory*: 18-44. New York, Chiron Press
Reyer, H.-U. (1980). Flexible helper structure as an ecological adaptation in the Pied Kingfisher (*Ceryle rudis*). *Behav. Ecol. Sociobiol. 6*: 219-28
Ricklefs, R.E. (1975). The evolution of cooperative breeding in birds. *Ibis 117*: 531-4
Ridpath, M.G. (1972). The Tasmanian native hen *Tribonyx mortierii*. I.-III. *CSIRO Wildl. Res. 17*: 1-118
Rood, J.P. (1974). Banded mongoose males guard young. *Nature, Lond. 248*: 176
—(1975). Population dynamics and food habits of the banded mongoose. *E. Afr. Wildl. J. 13*: 89-111
—(1980). Mating relationships and breeding suppression in the dwarf mongoose. *Anim. Behav. 28*: 143-50
Rowley, I. (1978). Communal activities among White-winged Choughs *Corcorax melanorhamphus. Ibis 120*: 178-97
—(1981). The communal way of life in the splendid wren *Malurus splendens. Z. Tierpsychol. 55*: 228-67
Sayler, A. and Salmon, S. (1971). An ethological analysis of communal nursing by the house mouse. *Behavior 40*: 62-85
Stacey, P.B. (1979a). Habitat saturation and communal breeding in the acorn woodpecker. *Anim. Behav. 27*: 1153-66
—(1979b). Kinship, promiscuity, and communal breeding in the acorn woodpecker. *Behav. Ecol. Sociobiol. 6*: 53-66
Vehrencamp, S.L. (1977). Relative fecundity and parental effort in communally nesting anis, *Crotophaga sulcirostris. Science, N.Y. 197*: 403-5
—(1978). The adaptive significance of communal nesting in groove-billed anis (*Crotophaga sulcrostris*). *Beh. Ecol. Sociobiol. 4*: 1-33
—(1979). The roles of individual, kin, and group selection in the evolution of sociality. In P. Marler and J.G. Vandenbergh (eds.), *Handbook of Behavior and Communication*, Vol. 3: 351-94. New York, Plenum Press
—(1980). To skew or not to skew? *Proc. 17th Congress Internat. Ornithol. 17*: 869-74
—(1983). A model for the evolution of despotic versus egalitarian societies. *Anim. Behav. 31*: 667-82
West-Eberhard, M.H. (1975). The evolution of social behavior by kin selection. *Q. Rev. Biol. 50*: 1-34
—(1978). Temporary queens in *Metapolybia* wasps: nonreproductive helpers without altruism? *Science N.Y. 200*: 441-3
—(1981). Intragroup selection and the evolution of insect societies. In

R.D. Alexander and D.W. Tinkle (eds.), *Natural Selection and Social Behavior: Recent Research and New Theory*: 3-17. New York, Chiron Press
Wilson, E.O. (1971). *The Insect Societies*. Cambridge, Belknap Press
Woolfenden, G.E. (1975). Florida scrub jay helpers at the nest. *Auk 92*: 1-15
—(1981). Selfish behavior in Florida scrub jays. In R.D. Alexander and D.W. Tinkle (eds.), *Natural Selection and Social Behavior: Recent Research and New Theory*: 257-260. New York, Chiron Press
—and Fitzpatrick, J.W. (1978). The inheritance of territory in group-breeding birds. *Bioscience, 28*: 104-8

9 MODELS OF PRODUCER/SCROUNGER RELATIONSHIPS BETWEEN AND WITHIN SPECIES

Richard Sibly

This chapter is about models and modelling more than it is about animals, and the reason for this is that most of the fields covered in this book are moving towards explicit modelling of producer/scrounger relationships (indeed infectious disease has a noteworthy history in this area; Anderson, 1982). Scientific progress is guided and channelled by the types of model that are used and investigated, so initial choice of model *may* have far-reaching consequences. There are several different types of model that could be used, so it is worth spending a little time getting to know them, and considering the relationships between them.

The simplest way to model producer/scrounger relationships would be to suppose scroungers obtained a payoff dependent solely on the producer:scrounger ratio. Consider the gull-lapwing association as an example. As described in Barnard (Chapter 4) gulls steal food unearthed by lapwings, and one can imagine that the greater the ratio of gulls to lapwings the less the payoff to individual gulls. Hence the gull payoff curve must decline with the proportion of gulls in the flock, as in Figure 9.1a. However, at low frequencies of gulls the gulls may perhaps be rate-limited in the extent of their scrounging (plateau at left-hand side of Figure 9.1a). If individual lapwings were able to retain smaller earthworms but lost all larger worms to gulls then the payoff to gulls would be proportional to

$$\frac{\text{no. of lapwings}}{\text{no. of gulls}} = \frac{1-p}{p}$$

where p is the proportion of gulls in the gull-lapwing feeding flocks. This curve forms the right-hand side of Figure 9.1a. If lapwings only ate the smaller worms they unearthed (see Figure 4.7) then the lapwing payoff curve would be a horizontal line. However, Barnard suggests that in fact lapwings are able to ingest more in the absence of gulls (Figure 9.1b). As it stands, the model does not allow the possibility of flock-size effects, yet these are likely to be important (Barnard, personal communication.) Extension of the model to incorporate flock-size effects would render it a two-dimensional model, as discussed in the next section.

Superimposing the two payoff curves in Figures 9.1a and 9.1b as in 9.1c, it appears that at high gull:lapwing ratios lapwings may do better than gulls, but at low gull:lapwing ratios gulls may do better than lapwings. Other things being equal (notably reproductive success in the breeding season), if payoff were measured in units of Darwinian fitness, flock composition would adjust over several generations to an evolutionarily stable point as shown in Figure 9.1d.

Obviously this is an overly simple model which raises more questions than it answers. For example the breeding biology of the two species may differ, even to the extent that one species regularly outperforms the other in the breeding season but fails to perform so well over winter. Another complication arises from variation between feeding flocks, since flocks may feed in areas differing in productivity. However, if flocks do feed on areas of similar productivity, then individual gulls may well distribute themselves between feeding flocks according to an ideal free distribution (Fretwell, 1972) just as herring gulls do between refuse tips (Sibly and McCleery, 1983), which would result in the same proportion of gulls in every gull-lapwing feeding flock – an easily tested proposition. Overall the simple model depicted in Figure 9.1 may help us to investigate gull-lapwing associations, but there are too many uncontrolled variables for it to be useful in studying the population dynamics of gulls and lapwings.

In other feeding associations, however, producer and scrounger tactics are adopted by conspecifics (as in house sparrows: Barnard and Sibly, 1981; Barnard, Ch. 4 this volume), so in principle it is possible that individuals change tactics in the short term according to which tactic profits them more. Barnard and Sibly (1981) call this an 'opportunistic' strategy. Thus producers might switch to scrounging when scroungers do better than producers and vice versa. In this case the evolutionarily stable flock composition would be reached in the short term.

If individuals do not change tactics opportunistically as described above, then we may ask why not, since they would apparently obtain a greater payoff if they did. One possibility is that animals adopting inferior tactics would change tactics if they could, but they are prevented from doing so by force. Thus would-be harem-holders are only successful in obtaining a harem if they can defeat other contenders

Figure 9.1: Simple model of the gull-lapwing feeding association. Lapwings can be seen as 'producers' of food which may be stolen by 'scrounger' gulls. (a) Payoff to gulls. The greater the ratio of gulls to lapwings the less the payoff to individual gulls; however at low proportions of gulls the gulls may perhaps be rate-limited in the extent of their scrounging (plateau at left-hand side). (b) Payoff to lapwings. Lapwings change their foraging method in the presence of gulls and then only eat the smaller worms they unearth (Barnard, Ch. 4 this volume). Although larger worms may be taken by gulls the smaller worms are not. Hence payoff to lapwings is a step function of the ratio of gulls to lapwings. It must be emphasised that the curves shown are speculative: discovery of their actual shapes must await further field research (for a summary of what is known now see Barnard, Ch. 4 this volume). (c) (a) and (b) superimposed. The equal payoff point can now be identified. (d) Adjustment of flock composition to an evolutionarily stable point

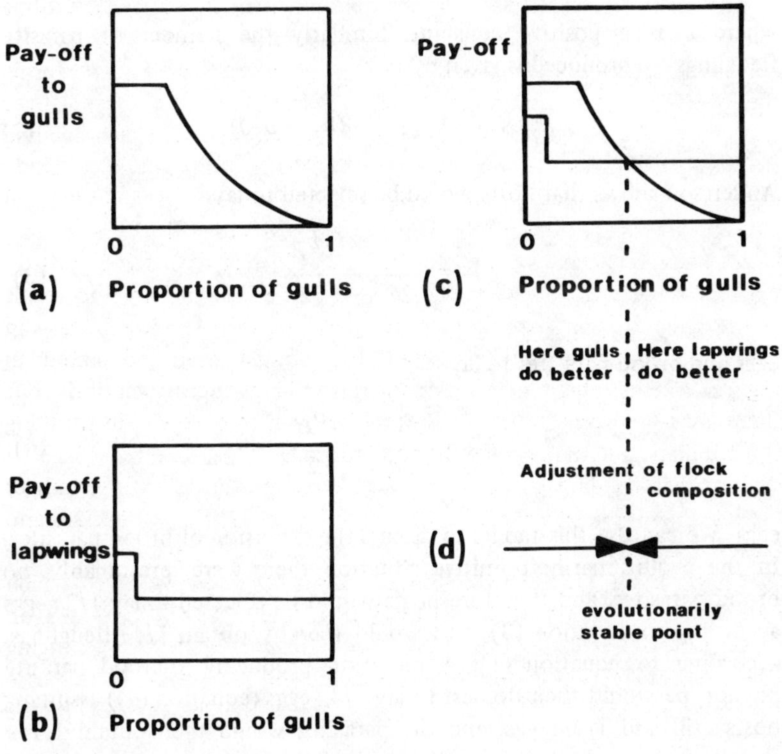

in fights. In this case the existence of two strategies is dependent on phenotypic variation in fighting ability, such as might result from phenotypic variation in size. Phenotypic variation in size might depend on age (if size increases with age) or on essentially random factors acting during development. Size can affect strategy in other ways too. Strategies which depend on an animal's phenotypic characteristics are called 'phenotype-limited strategies' (Parker, 1982) and are discussed at greater length by Parker (Ch. 5 this volume) and Arak (Ch. 6 this volume).

Simple models can also be used to examine the evolution of brood parasitism, as described by Andersson (Ch. 7 this volume). Andersson supposes that the number of host fledglings F_H produced depends on the number of eggs b_H laid by the host and on the number of eggs b_P laid by the brood parasite. In his simplest model (Andersson's equation (1)) the relationship is

$$F_H(b_H) = b_H(1 - a\,(b_H + b_P)) \quad (1)$$

where a is a positive constant. Similarly the number of parasite fledglings F_P produced is given by

$$F_P(b_P) = b_P(1 - a\,(b_H + b_P)) \quad (2)$$

Andersson shows that hosts would be selected to lay

$$b_H^* = \frac{1}{2a} - \frac{b_P}{2} \quad (3)$$

eggs and brood parasites to lay

$$b_P^* = \frac{1}{2a} - \frac{b_H}{2} \quad (4)$$

eggs. We can use this model to discuss the evolution of brood parasites. In the evolutionarily primitive situation there were, presumably, no brood parasites, and therefore hosts would be selected to lay $1/2a$ eggs according to equation (3), and would thereby obtain $1/4a$ fledglings, according to equation (1). A mutation producing a brood parasite phenotype would then do best to lay $1/4a$ eggs (equation (4)) assuming hosts still laid $1/2a$ eggs, and the parasites would then obtain $1/16a$ fledglings, i.e. only a quarter the number obtained by an unparasitised host. Parasitised hosts would obtain $1/8a$ fledglings (from equation (1)). i.e. twice as many as the brood parasites. Thus other things being equal the brood parasite mutation would initially be selected against, though, as discussed by Andersson, there are a number of other factors to be considered.

Games Theory and Producer/Scrounger Relationships

What is the relationship of these models to evolutionary games theory as described by Maynard Smith (1982)? Evolutionary games theory, at least in its simple form, gives rise to payoff curves which are straight lines, i.e. linear functions of the proportion of individuals playing the scrounger strategy. This arises as follows. In the simplest case, the hawk-dove game (Maynard Smith, 1982, Ch. 2) when hawk meets dove, the hawk achieves a payoff $E(H, D)$ and the dove obtains $E(D, H)$. When hawk meets hawk each achieves $E(H, H)$, when dove meets dove each achieves $E(D, D)$. It is supposed that individuals from an infinite population meet one other individual at random. If the proportion of hawks in the population is p, then the average payoff to hawks will depend on the proportion of hawks which encounter doves $(1-p)$, and on the proportion that encounter hawks (p). In hawk-dove encounters hawks obtain payoff $E(H, D)$ whereas in hawk-hawk encounters hawks obtain payoff $E(H, H)$. Hence average payoff obtained by hawks is

$$p\,E(H, H) + (1-p)\,E(H, D)$$
$$\text{i.e. } E(H, D) + p\,(E(H, H) - E(H, D))$$

Written in this form it is clear that the payoff to individual hawks is a linear function of p, the proportion of hawks in the population. A similar argument applies to doves. Hence the 'payoff curves' of this evolutionary game are straight lines. The population dynamic implications of models of this type have been examined by Taylor and Jonker (1978), Zeeman (1979) and Maynard Smith (1982, Appendix D).

Two-dimensional Models

The models considered so far have supposed that payoff depends only on the proportion of individuals playing each strategy, not on the absolute numbers. However, it would be necessary to consider the absolute numbers in the group if, for example, the animals were subject to predation and there were group-size effects on predation risk to individual group members, or if there were group-size effects on individual foraging success (see Bertram, 1978, for a review). A simple model of this type is shown in Figure 9.2a. Here it is supposed that the payoff to scroungers increases with the number of producers and increased opportunity to parasitise. The payoff to producers, however, decreases with increasing numbers of scroungers, because more and more of the fruits of their investment are usurped. An important difference between the payoff curves is that the scrounger surface starts at zero because, when there are no producers, no resource is available, but the curve for producers may only reach zero when very large numbers of scroungers are present. The payoff to producers of having more producers around may also increase slightly if there are advantages, such as protection from predators (Hamilton, 1971; Lazarus, 1972) or thermal insulation (e.g. Brenner, 1965), which are independent of the resource being exploited. The payoff may decline again with very high numbers if there are density-dependent effects such as competition and interference (e.g. Goss-Custard, 1980).

Where the two payoff curves intersect, both types fare equally well: to one side of the intersection producers do better, to the other, scroungers do better. We can call this the *equal payoff point*. The equal payoff point represents the stable mixture of producers and scroungers in selective terms to which groups which contain the two types should converge (Figure 9.2b).

The model so far (Figure 9.2a) is essentially the same as that in Figure 9.1, the only difference being that instead of labelling the x-axis 'proportion of scroungers' we now label it 'number of scroungers', but since the total number of scroungers plus producers is fixed at six in Figure 9.2a, the two variables are equivalent.

Now consider the effect of allowing group size to vary. This means adding another dimension to the payoff diagram shown in Figure 9.2a, so that payoff is now to be plotted as a function of two variables. These two variables could be 'proportion of scroungers' and 'group size', but it will be more convenient in what follows if the two variables represent simply 'number of producers' and 'number of scroungers'. Then a two-dimensional model represents the payoff as dependent on the number of producers and the number of scroungers in the group, giving rise to a payoff surface (Figure 9.2c). The same principles apply to payoff surfaces as to the payoff curves in Figure 9.2a, except now the intersection between the surfaces for producers and scroungers produces a line rather than a single point. The line of intersection can be mapped as an *equal payoff line* (EPOL) onto the two-dimensional plane between the producer/scrounger axes (Figure 9.2d) and groups should now 'track' the line if groups change in size (although groups of fixed size should converge to a single point as before). A new and important implication arising from the idea of an EPOL is that the ratio of producers to scroungers at equilibrium is likely to depend on group size. Depending on the shape of the two intersecting surfaces, the EPOL in the horizontal plane can describe a wide variety of curves all

Figure 9.2: (a) Payoff to individual producers and scroungers as a function of the producer:scrounger ratio in the group (here arbitrarily set at six individuals). It is supposed that the payoff to scroungers increases with the number of producers and increased opportunity to parasitise. The payoff to producers, however, decreases with increasing numbers of scroungers, because more and more of the fruits of their investment are usurped. The intersection of the two curves is a point representing equal payoffs to producers and scroungers; when strategies are conditional it is the point at which it would not pay any individual to change strategy. (b) The equal payoff point corresponding to the payoffs shown in (a). To the left of the equal payoff point producers do better than scroungers — so the number of producers should increase, adjusting group composition towards the equal payoff point. Conversely, to the right of the equal payoff point scroungers do better than producers, and again group composition should move towards the equal payoff point. (c) The payoff to individual producers and scroungers as a function of the numbers of producers and scroungers at a site yields two surfaces. The intersection of the surfaces is the equal payoff line (EPOL). (d) The projection of the EPOL onto the horizontal plane, giving the EPOL as a function of the number of producers and the number of scroungers. To the left of the EPOL producers do better than scroungers, to the right of the EPOL scroungers do better than producers, so in either case group composition should move towards the EPOL. Note: For simplicity the EPOL has been drawn as if non-integer numbers of producers and scroungers were possible. Restriction to integers gives a line to the right of that shown, usually as close as possible. The integer EPOL for a given flock size gives a ratio of scroungers to producers such that if any one changed strategy he would do worse

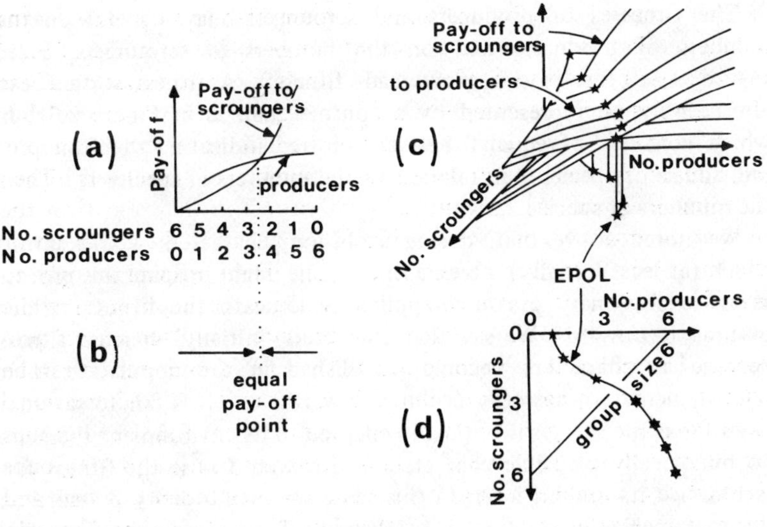

of which, except for straight lines through the origin, show a group size effect. The precise shapes of the surfaces may vary depending on the nature of the producer/scrounger relationship. In guarder/'sneak' relationships during mating, for example, the payoff to guarders (producers) might decrease monotonically as sneak frequency increases, especially if females are grouped. However, such variation does not affect the basic predictions which arise from the model.

Population Dynamics of Two Species

The evolution of interacting species is the sum of an almost infinite series of population interactions, and interactions between populations of different species forms the subject of population dynamics. There is a simple link between evolution and population dynamics, as follows. In evolution any genotype spreads which has a higher rate of increase than its competitors (Charlesworth, 1980), so evolution tends to maximise the rate of increase of the genotypes in a population. In a population without age structure rate of increase of a population of size n is $1/n\, dn/dt$, and we shall refer to this quantity as *fitness* (i.e. individual Darwinian fitness). The important point is that any genetically-coded character which increases fitness would spread in the population.

The fitnesses of producers and scroungers may depend on the numbers of producers and on the numbers of scroungers, as in Figure 9.2c (for 'payoff' now read 'fitness'). A fitness surface can alternatively be represented by a contour map as in Figure 9.3a, in which contours of equal fitness are plotted indicating the fitness of individuals of species 1 in relation to the numbers of species 1, n_1, and the numbers of species 2, n_2.

We noted above that, during evolution, characters would spread which (at least initially) increase fitness; one might imagine this process as an 'evolutionary pressure' applied underneath the fitness surface pushing it upwards. We say that characters 'initially' increase fitness because as characters become established in the population their rate of increase necessarily declines. Nevertheless, it is fair to say that in so far as the species is optimally adapted to its environment there are no biologically plausible characters which would raise the fitness surface above its present level. In this sense the evolutionary pressure on the two species interaction is to push the fitness surfaces of the two species upwards.

Alternative forms of the fitness surfaces of the two species are given in Figures 9.4a and b and Figures 9.5a and b. In all these cases the fitness of each species is affected by the numbers of the other species. Population interactions are generally classified as 'competition' if the fitness of each species is lowered by the presence of the other, and 'mutualism' if the fitness of each species is increased by the presence of the other (for further discussion see Roughgarden, 1979, p. 452). Producer-scrounger interactions would perhaps be those in which the fitness of species 1 is decreased by the presence of species 2, but the fitness of species 2 is increased by the presence of species 1. Within this definition would come predator-prey and plant-herbivore interactions.

The population dynamics of the two species interaction is usually studied by considering the rate of change of the population of one species $\frac{dn}{dt}$ in relation to numbers of each species, i.e. fitness $(\frac{1}{n}\frac{dn}{dt})$ multiplied by n. $\frac{dn}{dt}$ surfaces can be calculated for each species; the $\frac{dn}{dt}$ surfaces corresponding to the fitness surfaces in Figures 9.3a, b are shown in Figures 9.3c, d, respectively. The population equal payoff line (*population EPOL*) can be defined as the line on which $\frac{dn_1}{dt} = \frac{dn_2}{dt}$ so that $\frac{dn_1}{dt_2} = 1$. On the line on which $\frac{dn_1}{dn_2} = 1$ the direction of population change is 1.1, one member of species 1 being added (subtracted) for each member of species 2 added (or subtracted). The direction of motion on a plot of n_2 v. n_1 is thus 45° (either ascending or descending according as species numbers are increased or decreased). The population EPOL are shown as dotted lines in Figures 9.3c, d. The *equilibrium point* is defined as the point for which $\frac{dn_1}{dt} = 0$ and $\frac{dn_2}{dt} = 0$. Since at the equilibrium point $\frac{dn_1}{dt} = \frac{dn_2}{dt}$, the equilibrium point lies on the population EPOL. A natural question to ask of the dynamics of the population interaction is whether or not the equilibrium point is stable, in the sense that trajectories starting close to the equilibrium point lead to the equilibrium point, and this has been the subject of a considerable research effort in population dynamics (see, e.g. Begon and Mortimer, 1981 and May, 1981). The results of this analysis can be seen intuitively in Figure 9.6. Remembering that, on the population EPOL, trajectory direction is 45° to the horizontal it can be seen that, if the population EPOL has slope 45°, stability may result, as in Figure 9.6a or instability, as in Figure 9.6b. If the population EPOL has slope −45° then oscillations are likely as in Figures 9.6c, d.

It has proved useful in the analysis of these problems to consider the shapes of the zero-isoclines, defined as the line on which $\frac{dn_1}{dt} = 0$ and the line on which $\frac{dn_2}{dt} = 0$. Suppose these lines have slopes S_1 and S_2 respectively, in the neighbourhood of the equilibrium point, and suppose the slope of the population EPOL is S_3. Then (Sibly, unpublished),

Figure 9.3: The case of stable competition between two species. n_1 = number of species 1, n_2 = number of species 2. (a) The fitness of individuals of species 1 decreases as the numbers of species 1 increase and/or as the numbers of species 2 increase. (b) Similarly the fitness of individuals of species 2 decreases as n_1 increases and/or as n_2 increases. Thus the fitness of either species is lowered by the presence of the other — when this is true, species are said to be competing with one another (see text). In fact both species suffer intraspecific competition as well as interspecific competition. A possible example would be flour beetles. The interactions between the two species are governed by the equations

$$\frac{dn_1}{dt} = n_1\left(1 - \frac{n_1 + n_2}{10}\right), \quad \frac{dn_2}{dt} = n_2\left(1 - \frac{n_2}{5} - \frac{n_1}{20}\right)$$

In (c) and (d) the dashed line is the population EPOL. In (e) directions of motion are indicated. Motion is vertical on the zero-isocline $\frac{dn_1}{dt} = 0$; motion is horizontal on the zero-isocline $\frac{dn_2}{dt} = 0$; and motion is at 45° to the horizontal on the population EPOL

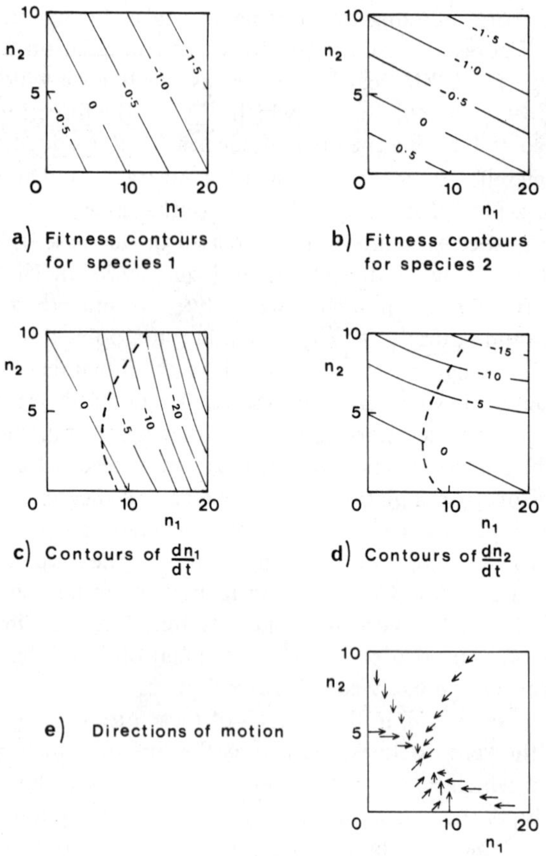

Figure 9.4: The case of stable oscillation involving two species. Species 1 suffers from both the presence of conspecifics and species 2. Species 2 gains from 1, perhaps by eating it, but shows no competition with conspecifics. A possible example would be insects (species 2) eating plants (species 1). The interaction is governed by the equations

$$\frac{dn_1}{dt} = n_1\left(1 - \frac{n_2}{5} - \frac{n_1}{50}\right), \quad \frac{dn_2}{dt} = n_2\left(-1 + \frac{n_1}{10}\right)$$

Notation as in Figure 9.3

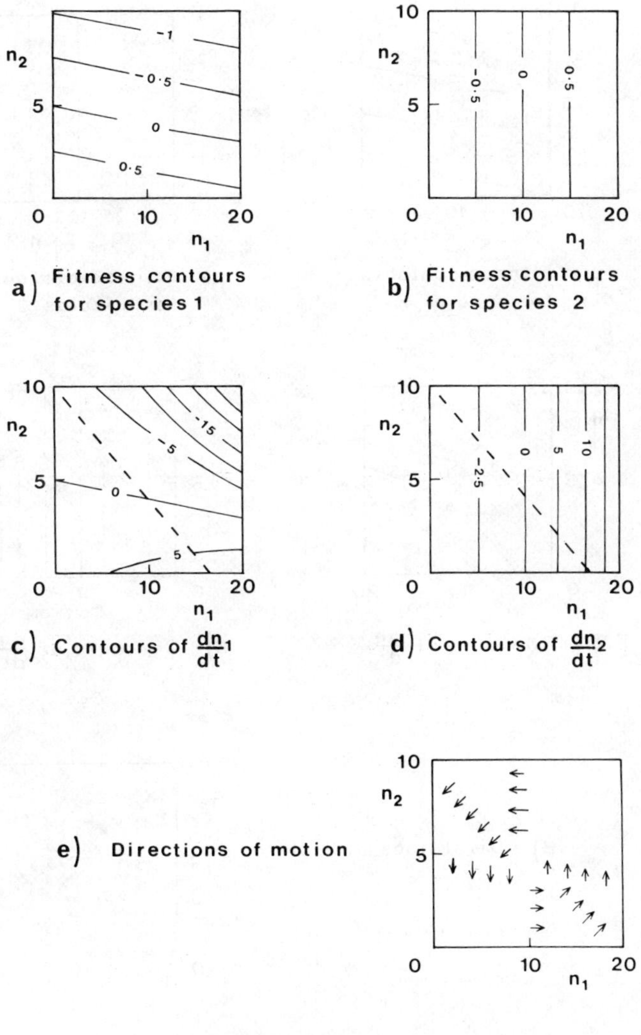

Figure 9.5: The case of unstable oscillation involving two species. Species 1 gains from the presence of conspecifics but suffers in the presence of species 2. Species 2 gains from 1 but show no competition with conspecifics. A possible example would be people (species 1) and mosquitoes in a tent. The interaction is governed by the equations

$$\frac{dn_1}{dt} = n_1\left(1 - \frac{n_2}{5} + \frac{n_1}{50}\right), \quad \frac{dn_2}{dt} = \frac{n_2(-1 + n_1)}{10}$$

Notation as in Figure 9.3

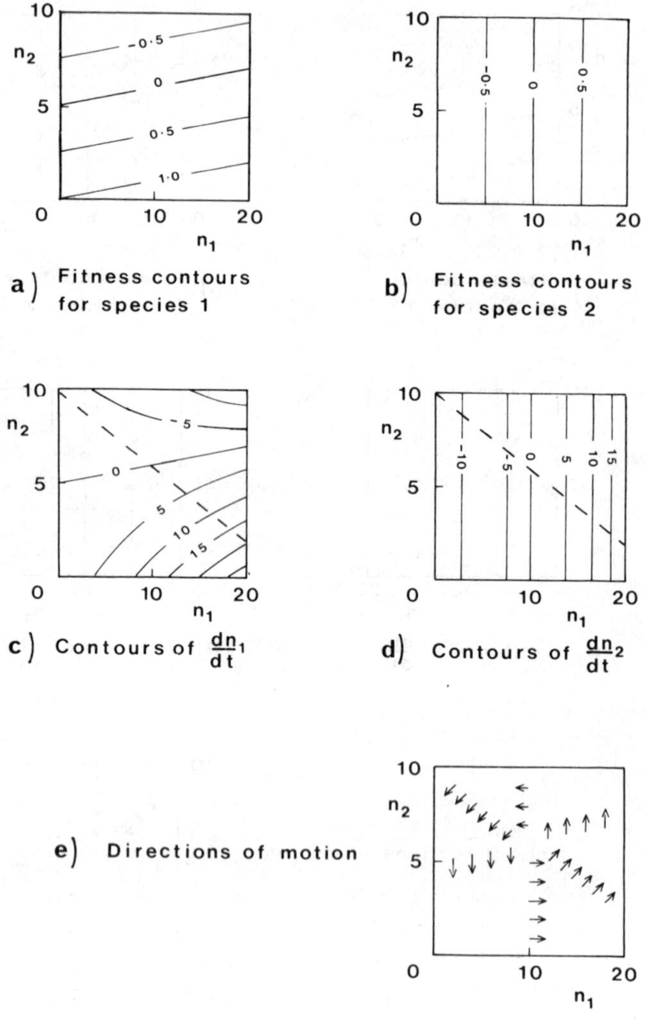

Figure 9.6: The four possible shapes of the population EPOL. See text for explanation

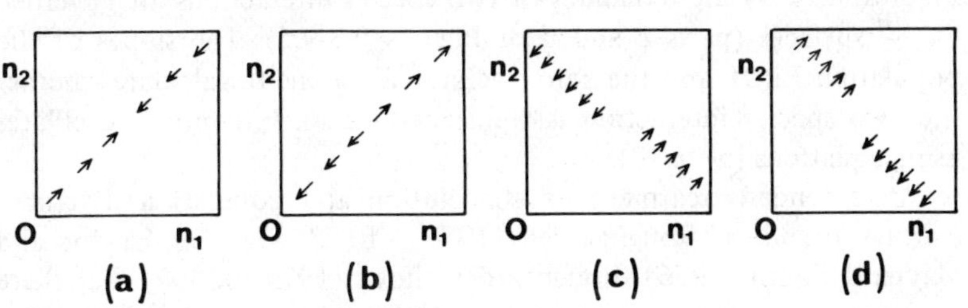

provided the fitness of species 1 is affected by the numbers of species 2 in the neighbourhood of the equilibrium point, the condition for oscillatory motion around the equilibrium point is that

$$\left(S_1 + \frac{S_3 - S_1}{S_3 - S_2}\right)^2 - 4S_2\frac{S_3 - S_1}{S_3 - S_2} < 0 \tag{5}$$

The stability condition for oscillatory motion is

$$\left(S_1 - \frac{S_3 - S_1}{S_3 - S_2}\right)\delta > 0 \tag{6}$$

Where δ is +1 if $\frac{dn_1}{dt}$ increases as n_2 increases, $\delta = -1$ otherwise. The stability conditions for non-oscillatory motion are (6) and

$$S_2 \frac{S_3 - S_1}{S_3 - S_2} > S_1 \frac{S_3 - S_1}{S_3 - S_2} \tag{7}$$

Using this method we can assess the stability and tendency to oscillate of each of the cases represented in Figures 9.3-9.5. Directions of motion are depicted in part e of each figure. Small arrows represent the directions of motion on the zero-isoclines ($\frac{dn_1}{dt} = 0$ and $\frac{dn_2}{dt} = 0$) and on the population EPOL. Checking conditions (5)-(7) shows that Figure 9.3e represents stable non-oscillatory motion (as is clear intuitively), Figure 9.4e represents stable oscillations, and Figure 9.5e represents unstable oscillations.

In this section we have explored the relationship between the payoff surfaces of Figure 9.2c and population dynamics. Whereas selection will act to increase the height of the fitness surface $\frac{1}{n}\frac{dn}{dt}$ (parts a and b of Figures 9.3-9.5) the dynamics of two species interactions are governed by $\frac{dn}{dt}$ surfaces (parts c and d of Figures 9.3-9.5). The slopes of the population EPOL and the zero-isoclines allow one to calculate whether the two species interaction is stable and whether or not it oscillates using equations (5) to (7).

An extended treatment of coevolution and population dynamics can be found in Roughgarden (1979, Ch. 23; see also Lawlor and Maynard Smith, 1976). Roughgarden shows (1979, p. 497) that there are no grounds for expecting the evolution of 'prudent predators' (or prudent parasites) which so adjust the exploitation of their prey that they maximise the abundance of the predator species. However Roughgarden accepts that parasites may evolve to reduce the harm they inflict upon their hosts, as in the case of the myxoma virus, carried by mosquitoes, which was introduced into the Australian rabbit population in 1950. Over the next 20 years both the rabbits and the virus evolved; the rabbits evolving more resistance to the virus and the virus evolving lower virulence (Roughgarden, 1979, p. 452). By analogy it might happen during coevolution that the impact of scroungers on producers was reduced; producers evolving more resistance to scroungers, and scroungers evolving to take less from individual producers.

Infectious Disease

Graphical methods can also be used in the analysis of the population dynamics of infectious diseases. The example to be discussed here is malaria, and the treatment is based on that of Aron and May (1982). The malarial parasite (a parasitic protozoon of the family Plasmodiidae) is transferred from mosquito to man, and at another stage of its life-cycle from man to mosquito, when a female anopheline mosquito bites a human host. The population dynamics of the malarial parasite are modelled in terms of the proportion of humans that are infected, and the proportion of female mosquitoes that are infected, using the following notation:

x is the proportion of the human population infected;
y is the proportion of the female mosquito population infected;
N is the size of the human population, assumed constant;
M is the size of the mosquito population, assumed constant;
$m = M/N$ is the number of female mosquitoes per human host;

a is the rate of biting on man by a single mosquito (number of bites per unit time);
b is the proportion of infected bites on man that produce an infection;
r is the *per capita* rate of recovery for humans;
μ is the *per capita* mortality rate for mosquitoes.

How will changes occur in the proportion of humans infected? It is supposed that infections are acquired at a net rate $\frac{dx}{dt}$ = (number of mosquito bites per person per unit time, $a\,M/N$) (probability that the biting mosquito is infected, y) × (probability that a bitten human is unaffected $(1-x)$) × (probability that an unaffected person when bitten will actually become infected, b) – (rate of recovery of humans from infection, rx). Hence

$$\frac{dx}{dt} = (a\,b\,\frac{M}{N})\,y\,(1-x) - rx \qquad (8)$$

How will changes occur in the proportion of mosquitoes infected? It is supposed that infections are acquired at a net rate $\frac{dy}{dt}$ = (number of bites per mosquito per unit time, a) × (probability that the biting mosquito is unaffected, $(1-y)$) × (probability that the bitten human is infected, x) – (death rate of infected mosquitoes, μy). Hence

$$\frac{dy}{dt} = a\,x\,(1-y) - \mu\,y \qquad (9)$$

Although this model is of course highly simplified (a discussion of the simplifications is provided by Aron and May, 1982), it nevertheless contains the essential elements of the transmission process. The rate of spread of infection in the human population can be calculated from equation (8), and is illustrated in Figure 9.7a. The rate of spread of infection in the mosquito population can be calculated from equation (9), and is illustrated in Figure 9.7b. Figures 9.7a and b are similar to Figures 9.3c and d, and the dynamic course of infection in the human and malarial populations can be calculated as in Figure 9.7c. It seems by inspection that in the case shown the disease is stable and non-oscillatory, and this can be confirmed by substitution in equations (5)-(7).

Estimation of Payoff Surfaces Using de Wit's Method

An elegant experimental method of analysing two species interactions

Figure 9.7: Simple model of the population dynamics of the malarial parasite in terms of the proportion of humans infected, x, and the proportion of female mosquitoes infected, y. The model is specified by the equations

$$\frac{dx}{dt} = 50y\,(1-x)-4x, \quad \frac{dy}{dt} = 40x\,(1-y)-20y$$

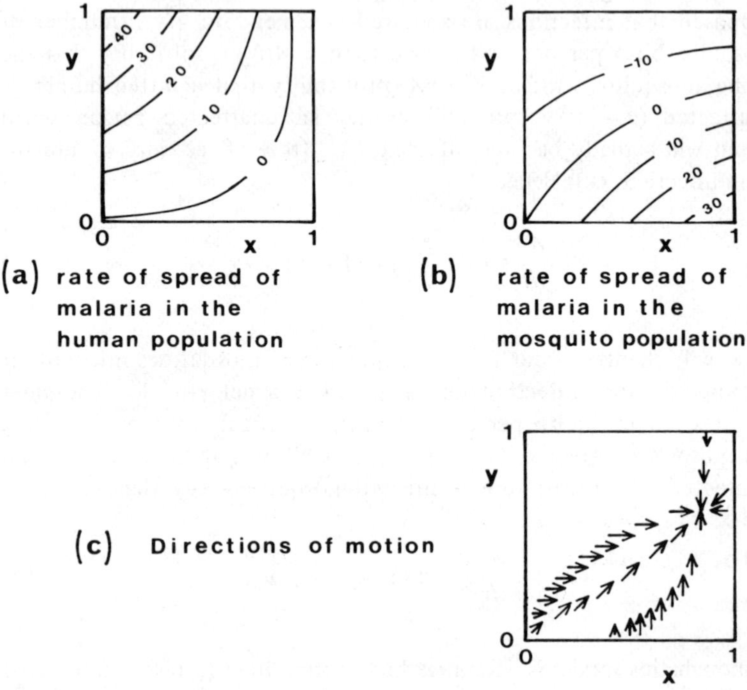

has been devised by de Wit (1960; for a simple exposition see Begon and Mortimer, 1981). de Wit's method involves placing together individuals of the two species in varying proportions, the total number of individuals being fixed, and observing the yield, ideally measured as the rate of increase (fitness) of each species. The fitness of each species can then be plotted against the initial proportion of each species, as in Figure 9.8. It should be noted that Figure 9.8 represents a vertical transect through two payoff surfaces, i.e. Figure 9.8 is exactly equivalent to Figure 9.2a. A series of de Wit diagrams obtained at different total densities would therefore provide a means of experimentally estimating the fitness surfaces of two interacting species. This

Figure 9.8: Fitness of two species of weevils, *Callosobruchus maculatus* (F.) and *Callosobruchus rhodesianus* (Pic.) as a function of the *rhodesianus*: *maculatus* ratio in the group (after Giga, 1982, Figure 5.7). Weevils of the two species were placed together in an enclosure in numbers detailed on the x-axis, and the progeny produced after one generation were recorded and used to calculate individual fitness. Note that total initial group size was 200 individuals in each experiment. Three replicates were obtained at each of five initial combinations of the two species

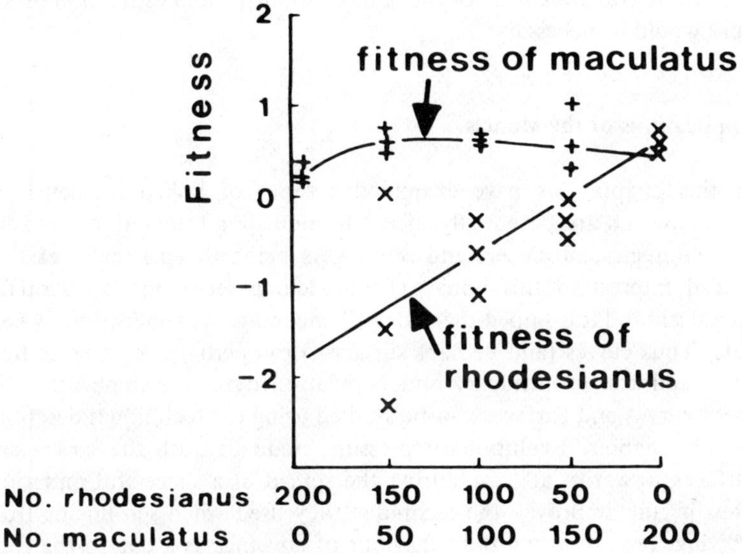

Source: Modified from Giga (1982)

is by analogy with the representation of a landscape by a series of parallel transects through it. Alternatively de Wit-type experiments could be carried out at a number of points in the n_1-n_2 plane (e.g. on a rectilinear grid) and then the fitness surface can be estimated by two-variable polynomial regression, i.e. by fitting equations of type

$$(\text{fitness of species 1}) = a_0 + a_1n_1 + a_2n_1^2 + a_3n_1^3 + \ldots + b_1n_2 + b_2n_2^2 + b_3n_2^3 + \ldots \quad (10)$$

$$(\text{fitness of species 2}) = c_0 + c_1n_1 + c_2n_1^2 + c_3n_1^3 + \ldots + d_1n_2 + d_2n_2^2 + d_3n_2^3 + \ldots \quad (11)$$

This method could also be used in behavioural ecology. In behavioural interactions the numbers of producers and the numbers of scroungers may change rapidly, as in feeding flocks of birds. Rapid changes in numbers may facilitate estimation of the payoff surfaces since a new point on the surface can be located every time the combination of producers and scroungers changes. For this method to work it is necessary that the 'true' payoff surfaces do not change in shape while the estimates are being obtained. This assumption would be reasonable if payoff were solely in terms of food intake, and if the food supply did not change. On the other hand, if the avoidance of predation were an important component of payoff a more sophisticated estimation procedure would be necessary.

Applications of the Models

In this chapter we have examined a series of linked mathematical models which are potentially of use in modelling interactions between two strategies, producers and scroungers being the particular case of special interest in this book. The modelling technique is essentially graphical, and it is hoped that this will encourage its application to field data. Thus curves (and perhaps surfaces!) can perhaps be fitted to field data, and the evolutionary and population dynamics implications of these curves and surfaces can be studied using the techniques described in this chapter. Evolutionary pressures tend to push the curves and surfaces upwards, at least during the spread of a successful mutation. Thus if gulls improved the technique they used when scrounging from lapwings (e.g. by using the behaviour of lapwings as a cue to the food the lapwings had obtained, see Barnard, Ch. 4 this volume) then the payoff to gulls using this technique would increase (at least at higher gull-lapwing ratios) and the payoff curve to gulls (Figure 9.1) would move upwards. Similar arguments apply to the surfaces (Figure 9.2c) considered in the two-dimensional case.

The population dynamics of two species interactions can also be studied using Figure 9.2c as a starting point. Thus the payoff surfaces of Figure 9.2c can be represented as contour maps showing the fitness of each species in relation to the number of individuals of each species as in Figures 9.3a, b (Figures 9.4a, b, Figures 9.5a, b). The zero-isoclines on these surfaces together with the population equal payoff line (EPOL) provide sufficient information to analyse the stability and oscillatory properties of the two species interaction, shown graphically in Figure 9.2e, using equations (5)-(7). A case of stable competition between two species is shown in Figure 9.3, stable oscillation in Figure 9.4, and unstable oscillation in Figure 9.5. Experimental estimation of the fitness surface is possible using an extension of the method of de Wit, which takes a transect through the fitness surface (Figure 9.8) exactly equivalent to the transect through the payoff surface shown in Figures 9.2a and 9.1c.

In summary, we have seen that it is possible in a two-dimensional model like Figure 9.2c to represent the payoffs to two strategies, producer and scrounger, as dependent on the number of producers and the number of scroungers in the group. The line of intersection can be mapped as an equal payoff line (EPOL) onto the two-dimensional plane between the producer-scrounger axes, and groups should 'track' the EPOL after any change in group size. We can envisage tracking of the EPOL occurring at different levels, as follows:

(i) *Evolutionary history*. If an evolutionarily primitive environment were populated only by producers, the 'scrounger' niche would be unfilled. Scroungers introduced to such an environment by mutation or immigration would spread rapidly to reach the EPOL. During subsequent coevolution the shapes of the payoff surfaces would change, and therefore the shape of the EPOL would change too.
(ii) *Population dynamics*. Two-species interactions on a much shorter time-scale form the subject of population dynamics. The interaction may be stable, about an 'equilibrium point' that lies on the EPOL, and it may or may not oscillate, depending on the slope of the population EPOL. Examples in this book relate to infectious disease including malaria (Kennedy, Ch.2 this volume, and this chapter), and to some brood parasites (Andersson, Ch. 7 this volume).
(iii) *Developmental history*. Individuals may be producers or scroungers for a limited part of their life-cycle. Male red deer, for example, are usually only scroungers on harem-guarding males when they are young bachelors (Clutton-Brock *et al.*, 1982) and may switch to harem-guarding and hence being producers later. Age also has an important effect on the tendency to initiate feeding interactions in a wide range of species (e.g. Watson, 1970; Ewer, 1971; Rowell, 1974). The theory of such cases is treated in depth by Parker (1982 and Ch. 5 this volume).
(iv) *Individual capability*. Individuals may choose to be scroungers if they are larger than average for the group and hence unlikely to be challenged. This could apply to gulls in gull-lapwing feeding associations. On the other hand in male-male mating associations the larger individuals are usually 'producers', presumably because they thereby obtain more matings (Arak, Ch. 6 this volume). The reason brood

parasites adopt different strategies from hosts may be that they differ in individual capability.

(v) *Learning*. Individuals may learn which is the more flexible strategy. See Harley (1981, plus exposition in Maynard Smith, 1982) for a theoretical treatment.

(vi) *Opportunity*. Whether or not an individual chooses to be a producer or a scrounger may depend on the opportunities for scrounging which are presented to it by other individuals as a result of instantaneous group composition (Brockmann and Barnard, 1979). I imagine that gulls in the gull-lapwing feeding association fall into this category. Surprisingly, house sparrows adopting 'searcher' and 'copier' strategies (Barnard, Ch. 4 this volume) do not change strategy opportunistically.

(vii) *Internal state*. Individuals which are, for example, more hungry than others may be prepared to risk injury to obtain food rather than to use up more energy searching.

In cases (iii), (iv), (v) and (vii) the individual's best strategy depends on its phenotype, so these would be called phenotype-limited strategies by Parker (Ch. 5 this volume).

Acknowledgements

I have been much helped in developing the ideas in this chapter by Dr Robert Smith, who also suggested to me the examples represented by Figures 9.3-9.5 and Figure 9.8. I am very grateful for this help. For constructive criticism of the manuscript and for many helpful suggestions through two revisions of the manuscript I am very grateful to Dr Chris Barnard and Dr Bill Sutherland.

References

Anderson, R.M. (ed.) (1982). *The Population Dynamics of Infectious Diseases: theory and applications*. London, Chapman and Hall

Aron, J.L. and May, R.M. (1982). The population dynamics of malaria. In R.M. Anderson (ed.), *The Population Dynamics of Infectious Diseases: theory and applications*. London, Chapman and Hall

Barnard, C.J. and Sibly, R.M. (1981). Producers and scroungers: a general model and its application to captive flocks of house sparrows. *Anim. Behav. 29*: 543-50

Begon, M. and Mortimer, M. (1981). *Population Ecology*. Oxford, Blackwell Scientific Publications

Bertram, B.C.R. (1978). Living in groups: predators and prey. In J.R. Krebs and N.B. Davies (eds.), *Behavioural Ecology*. Oxford, Blackwell Scientific Publications

Brenner, F.J. (1965). Metabolism and survival time of grouped starlings at various temperatures. *Wilson Bulletin 77*: 388-95

Brockmann, H.J. and Barnard, C.J. (1979). Kleptoparasitism in birds. *Anim. Behav. 27*: 487-514

Charlesworth, B. (1980). *Evolution in Age-structured Populations*. Cambridge, Cambridge University Press

Clutton-Brock, T.H., Guinness, F.E. and Albon, S.D. (1982). *Red Deer. Behaviour and ecology of two sexes*. Edinburgh, Edinburgh University Press

Ewer, R.F. (1971). The biology and behaviour of a free-living population of black rats (*Rattus rattus*). *Anim. Behav. Monogr. 4*: 127-74

Fretwell, S.D. (1972). *Populations in a Seasonal Environment*. Princeton, Princeton University Press

Giga, D.P. (1982). The comparative biology of four *Callosobruchus* species with particular reference to competition in *C. rhodesianus* and *C. maculatus*. Unpublished Ph.D. thesis, University of Reading

Goss-Custard, J.D. (1980). Competition for food and interference among waders. *Ardea 68*: 31-52

Hamilton, W.D. (1971). Geometry for the selfish herd. *J. theor. Biol. 31*: 295-311

Harley, C.B. (1981). Learning the evolutionarily stable strategy. *J. theor. Biol. 89*: 611-13

Lawlor, L.R. and Maynard Smith, J. (1976). The coevolution and stability of competing species. *Am. Nat. 110*: 79-99

Lazarus, J. (1972). Natural selection and the function of flocking in birds: a reply to Murton. *Ibis 114*: 556-8

May, R.M. (1981). Models for two interacting populations. In R.M. May (ed.), *Theoretical Ecology* (2nd ed.). Oxford, Blackwell Scientific Publications

Maynard Smith, J. (1982). *Evolution and the Theory of Games*. Cambridge, Cambridge University Press

Parker, G.A. (1982). Phenotype limited evolutionarily stable strategies. In King's College Sociobiology Group (ed.), *Current Problems in Sociobiology*. Cambridge, Cambridge University Press

Roughgarden, J. (1979). *Theory of Population Genetics and Evolutionary Ecology. an introduction*. New York, Macmillan

Rowell, T.W. (1974). The concept of social dominance. *Behav. Biol. 11*: 131-54

Sibly, R.M. and McCleery, R.H. (1983). The distribution between feeding sites of herring gulls breeding at Walney Island, U.K. *J. Anim. Ecol. 52*: 51-68

Taylor, P.D. and Jonker, L.B. (1978). Evolutionarily stable strategies and game dynamics. *Math. Biosc. 40*: 145-56

Watson, J.R. (1970). Dominance-subordination in cage groups of house sparrows. *Wilson Bull. 82*: 268-78

de Wit, C.T. (1960). On competition. *Verslagen van Landbouwkundige Ondersoekingen, 66*: 1-82

Zeeman, E.C. (1979). Population dynamics from game theory. In Z. Nitecki and C. Robinson (eds.), *Global Theory of Dynamical Systems*. Berlin, Springer-Verlag

INDEX

Author Index

Afchain, D. 58
Aho, J.M. 58
Alatalo, R. 154, 163, 164, 179, 181, 189
Albon, S.D. 31, 190, 287
Alcock, J. 137, 156, 158, 183, 189, 196, 224
Alexander, R.D. 166, 169, 189, 229, 231, 258, 263
Anderson, R.M. 34, 36, 37, 39, 43, 57, 58, 267
Andersson, M. 118, 123, 124, 180, 189, 204, 206, 207, 208, 213, 216, 217, 218, 222, 224, 270
Arak, A. 156, 157, 158, 167, 179, 180, 183, 189
Arme, C. 11, 31
Arnason, E. 109, 117, 118, 124
Arnold, S.J. 156, 177, 178, 189
Aron, J.L. 280, 281, 286
Axelrod, R. 177, 176, 189

Bachmann, C. 156, 189
Baerends, G.P. 78, 91
Bailey, G.N.A. 39, 57
Baines, G.A. 96, 125
Baker, M.A. 196, 226
Baker, M.C. 229, 230, 263
Baker, R.R. 153, 216
Balen, J.H. van 193
Ball, I.J. 210, 224
Balph, D.F. 95, 104, 124, 168, 189
Balph, M.H. 95, 104, 124, 189
Barlow, G.W. 154, 173, 192
Barnard, C.J. 8, 25, 26, 27, 31, 95, 96, 100-4, 106, 107, 108, 109, 110, 111, 114, 116, 119, 123, 124, 127, 152, 160, 189, 245, 263, 268, 286, 287
Barnett, R.J. 126
Bartholomew, G.A. 190
Bateman, A.J. 179, 189
Beatty, R.A. 185, 189
Bedard, J. 209, 227
Begon, M. 282, 286
Behnke, J.M. 49, 50, 58
Bekoff, M. 248, 263
Bell, G. 141, 152
Bellotti, A.C. 40, 59
Bellrose, F.C. 211, 212, 214, 215, 221, 224
Bender, K. 69, 92
Bennet, F. 197, 224
Benz, G. 65, 87, 92
Bertram, B.C.R. 166, 189, 216, 218, 220, 225, 257, 258, 262, 263, 271, 286
Bethel, W.M. 10, 11, 12, 13, 32
Beuchner, A.K. 185, 190
Bezzel, E. 215, 225
Bignet, J. 58
Birkhead, T.R. 118, 124
Bjorkland, M. 156, 163, 174, 189
Blanchard, G.B. 196, 225
Blanke, R. 84, 91
Bloom, B.R. 14
Boag, P.T. 180, 190
Bohart, G.E. 197, 225
Bolen, E.G. 196, 215, 222, 225, 226
Boray, J.A. 40, 58
Borgia, G. 158, 166, 167, 180, 190
Bradbury, J. 166, 167, 168, 190
Bradley, D.J. 41, 42, 57
Brenner, F.J. 272, 287
Brian, A.D. 64, 91
Brignoli, P.M. 68, 91
Bristowe, W.S. 65, 86, 91
Brockmann, H.J. 8, 9, 31, 88, 91, 95, 96, 104, 106, 123, 124, 127, 130, 152, 196, 225, 286, 287
Brooks, R.J. 14
Brothers, D.H. 258, 265
Brower, L.P. 28, 29, 31
Brown, J.L. 257, 263
Bruning, D. 257, 262, 263
Bucher, T.L. 157, 190
Buckley, F.C. 220, 225
Buckley, P.A. 220, 225
Buckman, S.L. 189
Burger, J.D. 28, 31, 105, 118, 124

Burk, T.E. 25, 26, 31
Burlinghame, P.L. 52, 57
Buskirk, R.E. 85, 91
Butler, R.G. 13, 31

Cade, W.H. 190
Caldwell, G.S. 28, 31
Campanella, P.J. 183, 190
Campbell, C.R.G. 215, 227
Cannon, L.R.G. 53, 58
Capron, A. 45, 47, 49, 58
Caraco, T. 119, 124, 248, 263
Carlson, A. 189
Carney, W.P. 11, 31
Catterall, C.P. 166, 191
Chandler, A.C. 52, 58
Chapin, J.P. 104, 124
Charlesworth, B. 141, 152, 258, 263, 274, 287
Charnov, E.L. 29, 32, 136, 138, 139, 152, 258, 263
Chase, J. 31
Chiszar, D. 162, 190
Christenson, T.E. 84, 91
Christian, J.J. 159, 191
Clawson, E.L. 202, 203, 204, 206, 208, 210, 212-15, 225
Clay, T. 62, 93, 95, 126
Clutton-Brock, T. H. 8, 31, 156, 158, 171, 186, 190, 285, 287
Corkhill, P. 118, 124
Cott, H.B. 18, 31
Cottam, C. 225
Cowardin, L.M. 224
Cox, C.R. 181, 182, 183, 190
Cox, F.E.G. 46, 47, 48, 58
Crofton, H.D. 34, 37, 38, 58
Crompton, D.W.T. 53, 55, 58
Curio, E. 61, 91, 96, 125

Daniel, R. 192
Darwin, C.R. 63, 91, 178, 190
Davies, M.E. 80, 91
Davies, N.B. 6, 26, 31, 118, 125, 154, 156, 158, 159, 174, 175, 181, 184, 190, 261, 263
Davis, D.E. 230, 263
Davison, G.W.H. 166, 190
Dawkins, R. 6, 9, 13, 16, 15, 27, 31, 32, 58, 88, 90, 91, 92, 95, 119-24, 125, 127, 130, 152, 154, 155, 171, 179, 190, 218, 225
Daxboeck, C. 69, 93
Delnicki, D.E. 196, 225
Dense, W.A. 10, 32
Dessaint, J.P. 45, 49, 58
Dewsbury, D.A. 152
Diamond, A.W. 118, 124
Dickey, D.R. 215, 225
Dobben, W.H. van 11, 32
Dow, H. 217, 225
Drager, B. 162, 191
Drobney, R.D. 217, 225
Dunbar, R.I. 138, 152, 154, 159, 191
Dunn, E.K. 118, 125
Dunnet, G.M. 125, 201, 225
Dzink, P.J. 125

Eadie, J. McA. 209, 225
Eberhard, W.G. 70, 86, 92
Edmunds, M. 14, 15, 17, 32, 85, 96, 124
Edney, E.B. 80, 91
Ehmann, M. 216, 225
Eickwort, G.C. 189, 195, 196, 197, 220, 223, 225
Eickwort, K.R. 189, 195
Elgar, M.A. 166, 191
Emlen, S.T. 208, 257, 263
Epple, G. 260, 264
Eriksson, M.O.G. 204, 206, 207, 208, 210, 213, 216, 222, 224, 225
Estep, D.Q. 152
Evans, P.G.H. 197-201, 203, 204, 220, 225
Evans, W.S. 59
Ewald, P.W. 95, 97, 99
Ewer, R.F. 285, 287
Exline, H. 70, 86, 92, 97-100

Fabre, J.H. 197, 225
Falls, J.B. 14
Feare, C.J. 96, 124, 197, 200, 225
Fiscus, C.H. 96, 125
Fisher, R.A. 138, 139, 152, 191
Forel, A. 62, 92
Fox, M.W. 263
Frederickson, L.H. 225
Frederiksson, K.A. 196, 215, 225
Fredga, S. 217, 225
Fretwell, S.D. 129, 143, 148, 153, 193, 268, 287
Fricke, H.W. 171, 191
Fry, C.H. 257, 264

Gerhardt, H.C. 180, 191, 192
Ghiselin, M.T. 139, 152
Gibson, D. 54, 59

Gibson, R.M. 185, 190, 191
Giga, D.P. 283, 287
Gilbert, D.G. 152
Gilbert, L.E. 30, 32
Gillespie, J.H. 210, 225
Gilmer, D.S. 224
Gilpin-Brown, J.B. 69, 92
Gluecksohn-Waelsch, S. 85, 189
Gochfeld, M. 105, 118, 124
Goist, K.C. 84, 91
Goodall, J. 185, 186, 191
Gordon, D.M. 34, 36, 37, 57
Gortmarker, W. 228
Goss-Custard, J.D. 272, 287
Gould, S.J. 207, 225
Gowaty, P.A. 196, 225
Grafen, A. 31, 152, 225
Graham, G.L. 12, 32
Grant, P.R. 95, 109, 116, 117, 118, 124, 125, 180, 190
Greenwood, P.J. 96, 125, 216, 225
Grenquist, P. 203, 226
Grice, D. 203, 208, 212-15, 217, 226
Gromko, M.G. 152
Gross, M.R. 29, 32, 156, 160
Guinness, F. 31, 190, 287
Gustafsson, L. 189

Hahn, C. 31
Hair, J.D. 53, 58
Halliday, T.R. 6, 26, 29, 31, 177, 183, 191
Hamilton, W.D. 154, 172, 176, 177, 189, 191, 196, 203, 225, 258, 264, 272, 287
Hamilton, W.J. III 195, 220, 223, 225
Harcourt, A.H. 186, 191
Harley, C.B. 286, 287
Harrington, F.H. 230, 263
Harris, M. 231, 264
Hartman, G.W. 225
Harvey, P.H. 96, 125, 185, 190, 191
Hassell, M. 96, 125
Hatch, J.J. 117, 118, 123, 125
Hausfater, G. 233, 260, 264
Hawkins, A.S. 224
Hazen, T.C. 58
Hazlett, B.A. 63, 68, 69, 92
Hearn, J.R. 260, 264
Heatwole, H. 104, 125
Hellebrekers, W.Ph.J. 215, 223, 225
Heusmann, H.W. 202, 203, 206, 208, 211, 213, 214, 215, 222, 225
Hildén, O. 215, 226
Hobbs, R.P. 55, 58
Hoekstra, R. 141, 152
Hoffmaster, D.K. 64, 92
Hogan-Warburg, A.J. 8, 32, 165, 166, 168, 185, 191
Holmes, J.C. 10, 11, 12, 13, 32, 52, 53, 55, 58
Holmes, W.G. 209, 236
Holst, D. von 159, 193
Hoogland, J.C. 220, 226
Houston, A.I. 261, 263
Howard, R.D. 8, 32, 157, 179, 191

Ingolffson, A. 119, 125
Innes, G.S. 189
Iwata, K. 62, 63, 89, 92

Jackson, R.R. 86, 92
Janetos, A.C. 85, 92
Jarvis, J.U.M. 230, 257, 261, 264
Jeanne, R.L. 64, 92
Jensen, K. 69, 92
Jewell, P.A. 185, 191
Johnson, K.L. 224
Johnstone, R.C.B. 264
Jones, C.E. 189
Jones, J.W. 184, 191
Jones, R.E. 208, 212, 213, 214, 226
Jonker, L.B. 271, 287
Joste, N. 241, 263
Joyner, D.E. 208, 215, 223, 226

Källander, H. 8, 32, 107, 108, 109, 117, 118, 119, 122, 125
Karlin, A.A. 196, 202, 225
Karlsson, J. 197, 199, 200, 201, 204, 222, 226
Katz, Y. 260, 264
Keenleyside, M.H.A. 184, 191
Kennedy, C.R. 35, 37, 38, 52, 53, 58
Kenward, R.E. 166, 191, 248, 264
Keymer, A.E. 37, 58
King, G.M. 184, 191
King, J.R. 107, 125
Klemper, H.C. 63, 89, 92
Klopfer, P.H. 126
Kluge, A.G. 178, 191
Knowlton, N. 141, 152, 177, 191
Koenig, W.D. 227, 241, 249, 256, 257, 259, 261, 264, 265
Kramer, D.L. 185, 192
Krapu, G.L. 217, 226

Krebs, J.R. 13, 16, 25, 27, 32, 58, 88, 90, 92, 95, 96, 104, 118, 119-24, 125, 155, 171
Krumhansl, C.L. 16, 32
Kruuk, H. 95, 110, 112, 113, 114, 125
Kullmann, E. 83, 92
Kummer, H. 156, 186, 189, 191
Kurt, F. 189, 191
Kushlan, J.A. 109, 125

Lack, D. 166, 167, 168, 191, 197, 226
Lackie, A.M. 46, 47, 58
Lanciani, C.A. 39, 58
Lande, R. 180, 191
Lane, F.W. 154, 192
Lanier, D.L. 144, 152, 220, 226
Larsson, S.G. 191
Lawick, H. van, 95, 113, 114, 119, 125
Lawick-Goodall, J. van 95, 113, 114, 119, 125
Lawlor, L.R. 55, 59, 280, 287
Laws, R.M. 230, 263
Lazarus, J. 166, 192, 287
Le Boeuf, B.J. 158, 171, 181, 182, 183, 190, 192
Lehman, R. 215, 227
Leigh, E.G. 194
Leopold, A.S. 208, 212, 213, 214, 226
Leuthold, W. 14
Levi, H. 70, 86, 92
Lewontin, R.C. 207, 225
Lill, A. 166, 192
Lin, N. 258, 264
Litovitch, E. 227
Lloyd, J.E. 179, 192
Lodge, J.R. 152
Loiselle, P.V. 154, 173, 192
Lokemoen, J.T. 208, 215, 226
Lomas, M.B. 155
Lombardo, M.P. 227
Low, J.R. 224
Lubin, Y.D. 85, 92
Luke, B.M. 153
Lumsden, H.G. 209, 225
Lundberg, A. 189

McArther, R.H. 55, 59
McCamant, R.E. 196, 213, 215, 222, 226
McCleery, R.H. 123, 268, 287
Macdonald, D.W. 257, 260, 264
McKaye, K.R. 195, 209, 226
McKaye, N.M. 209, 226
Mackenzie, K. 54, 59
McLaughlin, C.L. 208, 213, 226
McMillan, R.P. 68, 92
McPhail, J.D. 23, 32
Madden, J.L. 59
Magurran, A.E. 126
Malcolm, J.R. 230, 264
Manwell, C.L.C. 196, 226
Marsh, A.C. 227
Marsh, B.A. 227
Marshall, A.J. 177, 192
Marten, K. 230, 264
Martin, P.A. 144, 152
May, P.G. 39, 58
May, R.M. 34, 37, 43, 57, 96, 125, 280, 281, 286, 287
Maynard Smith, J. 6, 32, 55, 59, 127, 129, 137, 141, 152, 155, 180, 192, 203, 207, 226, 230, 242, 248, 257, 259, 264, 271, 280, 286, 287
Meakins, R.H. 39, 59, 63, 94
Mech, L.D. 114, 125
Meinertzhagen, R. 104, 125
Metcalf, R.A. 239, 241, 265
Meyres, T.U. 224
Michener, C.D. 195, 197, 227, 258, 264, 265
Milinski, M. 6, 32
Miller, R. 52, 59
Mills, M.G.L. 230, 265
Mirick, H. 78, 92
Mitchell, G.F. 45, 46, 47, 48, 59
Miyashita, K. 80, 92
Moehlman, P. 230, 257, 260, 264, 265
Mohler, J.D. 138, 153
Moodie, M.E.E. 24, 32, 156, 159, 192
Moore, P.G. 158, 192
Morse, T.E. 196, 208, 212, 213, 214, 222, 227
Mortimer, M.H. 282, 286
Moss, D. 158, 192
Moynihan, M.H. 27, 32
Mumme, R.L. 212, 227, 260, 264, 265
Munro, J. 209, 227
Murton, R.K. 96, 125

Nagel, W.P. 59

Nakatsuru, K. 185, 192
Nelson, J.B. 121, 125
Nettleship, D.N. 121, 125
Newton, I. 215, 227
Nickol, B.B. 39, 52, 59
Nicolai, J. 90, 92
Niles, D.M. 193
Nolan, V.Jr. 223, 227
Noonan, W.M. 240, 265
Nudds, T.D. 209, 227
Nyffeler, M. 65, 87, 92

Olazarri, J. 70, 71, 78, 83, 93
Orians, G.H. 195, 220, 223
Otte, D. 166, 167, 192
Owen, R.B. 217, 229
Owen, R.W. 11, 31

Packer, C. 176, 177, 192
Paquet, P.C. 230, 263
Parker, G.A. 6, 9, 33, 129, 130, 132, 134, 138, 140, 141, 142, 144, 145, 146, 147, 148, 152, 153, 158, 177, 180, 183, 185, 191, 192, 285, 287
Parker, I.S.C. 264
Partridge, L. 180, 192
Patterson, I.J. 125
Payne, K. 167, 192
Payne, R.B. 167, 192, 195, 200, 210, 223, 227
Peakall, D.B. 94
Perril, S.A. 156, 157, 192
Perrins, C.M. 158, 197, 217, 227
Pettit, R.G. 118, 126
Pimentel, D. 37, 40, 59, 123, 126
Pitcher, T.J. 96, 126
Pitelka, F.A. 227, 257, 264, 265
Platnik, N.I. 69, 92
Podesta, R.B. 48, 59
Power, H.W. 197-8, 227
Price, G.R. 127, 152
Price, P.W. 37, 40, 55, 57, 59, 63, 92
Pringle, J.W.S. 123, 126
Pulliam, H.R. 166, 192

Rand, A.L. 95, 126
Rasmussen, K.L.R. 181, 192
Rau, M.E. 11, 13, 33
Reed, C.H.F. 94
Reimens, T.J. 152
Reinecke, K.J. 217, 227
Reyer, H-U. 248, 265
Rhijn, J.G. van 156, 165, 166, 168, 172, 193
Ribbink, A.J. 209, 227
Richmond, R.C. 152
Ricklefs, R.E. 257, 265
Ridgely, R.S. 126
Ridpath, M.G. 230, 248, 257, 259, 264, 265
Riechman, J.H. 224
Roberts, L.S. 52, 59
Robertson, D.R. 194
Robinson, B. 68, 70, 75, 83, 93
Robinson, M.H. 66, 67, 68, 70, 71, 75, 78, 83, 93
Robinson, R.H. 213, 227
Rogers, J.P. 203, 212-15, 217, 228
Rohde, K. 55, 56, 59
Rohwer, F.C. 26, 33, 97, 126
Rohwer, S. 14, 26, 33, 95, 97, 98-100, 126, 173, 177, 178, 192, 193
Rood, J.P. 320, 260, 265
Rosene, W. 222, 227
Rossem, A.J. van 215, 225
Rothschild, M. 62, 93, 95, 126
Rothstein, S.R. 177, 193, 195, 207, 220, 227
Roughgarden, J. 275, 280, 287
Rowell, R.W. 285, 287
Rowley, I. 243, 244, 248, 257, 265
Rubenstein, D.I. 27, 104, 125, 154, 210
Rutschke, E. 215, 227
Ryan, M.J. 157, 190, 193
Rypstra, A.L. 85, 85, 93

Safriel, U.N. 209, 213, 227
Salisbury, G.W. 146, 147, 153
Salmon, S. 167, 265
Sampson, W.B. 214, 227
Sauer, A. 139, 153
Sayler, A. 257, 265
Schad, G.A. 53, 59
Schaefer, M. 84, 93
Schaller, F. 177, 193
Schaller, G. 95, 113, 114, 126
Scharloo, W. 193
Schloeth, R. 185, 150
Schmidt, G. 87, 93
Schom, C. 40, 59
Semler, D.E. 156, 159, 193
Shadab, M.U. 69, 92
Shaw, R.F. 138, 153
Sheppard, P.M. 22, 23
Sherman, P.W. 209, 220, 226
Short, R.V. 186, 191, 193

Sibly, R.M. 27, 95, 96, 100-4, 110, 123, 124, 127, 152, 168, 189, 245, 263, 268, 286, 287
Sivinski, J. 68, 93, 185, 193
Smith, C.C. 143, 148, 153
Smith, J.N.M. 96, 101, 126
Smith, R.L. 144, 153
Smith, V.G.F. 153
Smithers, S.R. 48, 59
Smith Trail, D.S. 70, 86, 93
Snow, B. 167, 193
Sowls, L.K. 206, 227
Sprent, J.F.A. 47, 59
Stacey, P.B. 185, 192, 241, 257, 255, 261, 265
Stahlbrandt, K. 189
Stanley, S.M. 120, 226
Stearns, S.C. 81, 93
Stephens, H.A. 107, 108, 109, 114, 124
Stoddard, H.L. 222, 227
Stokes, A.W. 168, 194
Stone, F.A. 124, 126
Stowe, M. 68, 93
Sweatman, H. 96, 126

Taft, L.K. 193
Taylor, P.D. 139, 153, 180, 193, 271
Taylor, P.R. 69, 93
Taylor, V.A. 145, 153
Telford, M. 69, 93
Terry, R.I. 49, 59
Thompson, C.F. 223, 277
Thompson, D.B.A. 108, 109, 119, 122, 124, 126
Thornhill, R. 65, 66, 93, 154, 173, 177, 181, 193
Tinbergen, J.M. 198, 227
Tinbergen, L. 101, 126
Tinbergen, N. 78, 93, 183, 193
Trivers, R.L. 173, 176, 179, 193
Tschantz, B. 220, 227
Turnbull, P.L. 84, 85, 87, 93
Tutin, E.G. 181, 186, 193
Tuttle, M.M. 193
Tversky, A. 16, 33

Ulfstrand, S. 189
Uznanski, R.L. 39, 52, 59

Vandemark, N.L. 146, 147, 153
Van den Assem, J. 154, 176, 193
Vehrencamp, S.L. 212, 227, 230-62, 265
Vernes, A. 58
Verrell, P. 156, 173, 180, 193
Vinson, A. 83, 93
Vollrath, F. 8, 32, 65, 66, 67, 68, 70-87, 93, 94, 110, 126
Voous, K.H. 215, 223, 225
Vos, G.J. de 169, 190

Waage, J.K. 154, 193
Wakelin, D. 35, 37, 40, 41, 42, 43, 49, 50, 58, 59
Walker, W.F. 185, 194
Walkey, M. 39, 59, 63, 94
Waltz, E.C. 173, 194
Warner, R.R. 176, 194
Watson, J.R. 285, 287
Watts, C.R. 168, 194
Weller, M.W. 202, 203, 206, 208, 212, 215, 216, 223, 228
Wells, M.C. 248, 263
West-Eberhard, M.J. 233, 237, 258, 265
Westerterp, K. 198, 228
Westman, B. 156, 163, 174, 189
Wheeler, W.M. 62, 63, 94
Whitt, G.S. 239, 240, 265
Wickler, W. 28, 33
Wiehle, H. 71, 94
Wight, H.M. 196, 208, 212, 213, 214, 222, 227
Wijngarden, H. 228
Wiley, R.H. 166, 167, 194
Williams, G.C. 155, 180, 193, 194
Williams, H.H. 53, 59
Willis, E.O. 104, 126
Wilson, E.O. 55, 59, 62, 63, 67, 68, 90, 94, 195, 220, 223, 228, 230, 243, 258, 260, 266
Winfield, I.J. 126
Wirtz, P. 170, 172, 194
Wise, D.H. 70, 86, 94
Witt, C.T. de 5, 281, 286, 287
Witt, P.N. 85, 86, 94
Wittenberger, J.F. 29, 33, 167, 194
Woolfenden, G.E. 230, 257, 266
Wrangham, R.W. 167, 194
Wyllie, I. 122, 126
Wynne-Edwards, V.C. 159, 194

Yom-Tov, Y. 196-202, 203, 216, 220, 221, 227, 228
Young, A.M. 118, 126, 197-201

Zeeman, E.C. 71, 287

Subject Index

acanthocephalan 52, 53
Acanthocephalus clavula 36
Achaeranea 86
adaptation budget 25
'adoption' 209
adornment 178
aggregation 84, 85, 96-7
aggression 29, 63, 64, 89, 90, 97, 98, 112-13, 114, 159, 162, 169, 170, 171, 186
allocreadiid fluke 53
altered host response 11-13
alternative strategy 1, 3, 85, 88, 105, 115, 116, 118, 122, 127-51, 154-88, 203, 248, 251, 252; age-dependent 170
altricial young 4, 197-212, 221
altruism 172, 217; reciprocal 176; worker 258
amensalism 44
amino acid 83
Ammophila 78
amphibian 173
amphipod 69
amplexus 179, 183
Andrena 89, 90
Anelosimus 66, 67, 86
anemone fish (*Amphiprion alkallopisos*) 171
Anguria 30
anisogamy 14, 48; origin of 140-3; ratio 143, 149, 146; stability of 146-50
antbird: *Alethe* 104; *Gymnopithys* 104
antelope 185
anthropomorphism 62
antibody 185
antigen masking 48
anti-predator strategy 1, 16, 64, 96
antlers 26
araneid 80
Araneus 70
Argiope 70-87, 111-12
Argyrodes 111-12; *argentata* 73, 76, 78, 87; *attenuatus* 86; *caudatus* 70-87; *elevatus* 70-87; *gibbosus* 72; *projiciens* 86; *trigonum* 86
Ariamnes 86
armament 178
arms race 2, 18, 20, 23, 25, 26, 28, 45, 48, 88, 95, 155, 171, 172, 185, 188; asymmetric 119; food-scrounging 3, 107-24; interspecific 119; outcome of 119-24; 'period of grace' 48, 123
artificial insemination 151
Ascaris 48
Ascoctyle 11
assassin bug 68
assessment 15, 22, 166, 183; cue 5, 25-6; hierarchy 25
association: stable 172
auk 106, 122
automaton 78
avoidance 109

baboon 181, 186, 233; hamadryas (*Papio hamadryas*) 156, 186, 187; olive (*P. anubis*) 176
bachelor male 170, 171
back-cross 42
background 14, 17, 22
bacteria 46
'battle of the sexes' 3, 155, 179, 188
'beating' relationship 104
beaver 13
behavioural ecology 283
'best of a bad job' strategy 106, 158, 163, 172
bibio-commensal 68
bivalve 119
black-bellied whistling duck (*Dendrocygna autumnalis*) 196, 215
black grouse (*Lyrurus tetrix*) 166, 169
black-headed duck (*Heteronetta atricopilla*) 223
blue-gill sunfish (*Lepomis macrochirus*) 29, 156, 160-3, 175
blue-headed wrasse (*Thalassoma bifasciatum*) 176
blue jay 28
Bombus: *locorum* 64; *terrestris* 64
booby 121
Brachylecithum mosquensis 11
breeding ecology 216
breeding success *see* reproductive success
breeding synchrony 221
broodball 89
brood parasitism 5, 9, 62, 89, 122, 270, 285, 286; evolution of

intraspecific 219-23; interspecific 1, 195, 215, 220, 223; intraspecific 1, 4, 195-224; multiple 230-1; obligate 223; success of 203-4
brood patch 202
brood size 198-224
brown bear (*Ursus arctos*) 113
bryozoa 69

'caller' strategy 155, 156
camouflage 81
canine teeth 186
cannibalism 178
canvasback (*Aythya valisineria*) 217
carnivore 95, 104, 114, 257
carpenter ant (*Camponota* spp.) 11
caste 90; queen 90, 261, 262; soldier 90; worker 90, 261
cattle 151
central nervous system 11
Centris pallida 137, 158
cephalopod 154
Cercocebus 186
cestode 42, 43, 51, 52, 53; tetraphyllidean 53
character convergence 15, 16-25, 27
character displacement 52
cheetah (*Acinonyx jubata*) 107, 114
chimpanzee (*Pan troglodytes*) 181, 186, 187
chorus 156, 168
cichlid 209
cleaner fish 129
cleptobiosis *see* kleptobiosis
cleptoparasitism *see* kleptoparasitism
cloaca 183
clutch size 158, 197-224, 270; indeterminate 206; optimal host 204-6; optimal parasite 211-12
cnidaria 69
coadapted scrounger 124
coalition 246; between dominants 262; between sneaks 174-7, 187; between subordinates 262
coati 257
coercion 180
coevolution 2, 34, 35, 37, 40-5, 54, 56, 63, 120, 123, 186, 280, 285
coexistence 2, 35, 40-50, 129, 160
collembola 177
coloniality 141
colonisation 36, 37
colony 62, 63, 67, 68, 86, 90, 109, 116, 122, 123, 176, 199, 222, 246
commensalism 44, 67, 69
communal brood care 209
communal egg laying 257
communal suckling 230, 257
communication 13
community 1, 7, 53, 55, 105; dynamics 5; stability 55
compensation 3, 107, 113-16
competition 2, 14, 39, 55, 61, 62, 81-3, 84, 91, 105, 116-19, 127, 129, 149, 160, 187, 211, 213, 220, 229, 234, 272; and spacing 120; interference 63; intermale 138, 165, 166, 169, 171, 178, 183, 186; intersexual 167; interspecific 53, 116, 274-86; intraspecific 50, 52, 116; kleptobiotic 63; restriction of 3, 118-19; stable 276
competitive ability 20, 22, 26, 61, 130, 131, 132, 134, 135, 140, 155, 156, 169; asymmetry in 158
competitive advantage 219
compromise 171-88
conditional strategy 130, 157, 187, 273; pure 155
conflict 159, 172, 174, 229; intersexual 178-88; of interest 155, 171-88
contest 61, 76, 88, 119, 156
cooperation 171, 172, 187, 188, 229-62; between males 166, 168, 169; between parent and offspring 243; between siblings 240; between sneaks 175, 176; biased 257; in hunting 260, 261; in suckling 260
'copier' strategy 86, 98, 100, 104, 110
copulation 154, 165, 169, 170, 173, 174, 180, 184, 185, 186, 241; sneak 163, 168, 171
copying 27; area 97, 100-4
cormorant (*Phalocrocorax carbo*) 11
cost of adaptation 121
cost of dominance 260-2
courtship 1, 166, 179, 180; homosexual 177
cowbird 9, 220; *Molothrus ater* 207
'coy' strategy 183, 188
coyote 243
crèche 209
cricket (*Gryllus integer*) 168
cross-fertilization 55

crowding 159; effect 51
crypsis 15, 17, 25, 80, 96
cuckoldry 29, 30, 156, 160-3, 164, 165, 169, 172, 173, 174; opportunistic 156
cuckoo 9, 90, 122, 200, 220; *Coccyzus* 223
Curimagua 87; *bayano* 69
Cyrtophora 70; *citricola* 72, 85
cystacanth 12
cysticercoid 12

defence 1, 8, 66, 90, 98, 99, 106, 114, 118, 119, 154, 175, 245, 260, 261
density-dependence 4, 272
desertion 162, 179, 184, 208, 209, 212
detection cue 14-25
detector/detectee relationship 2, 13-31
development 2; embryonic 201, 220
dietary specialist 123
diet shift 114
digestive enzyme 82
digger wasp *see* golder digger wasp
dingo 260
dioecy 55
Diplura 69, 70
dipso-parasite 68
directional selection 122
dispersal 38, 231, 232, 233, 234, 237, 240, 243, 245, 246, 250, 255, 257, 258
dispersion 35; clumped 100, 167; overdispersion 35, 36, 37, 40, 50; random 36; underdispersion 36; uniform 100
display 1, 27, 166, 167, 168, 169; communal 166-8; group 168, 169; 'hula' 180; 'rookooing' 166
disruptive selection 140
distastefulness 20
division of labour 243, 246
dominance relationship 1, 13, 14, 25, 81, 84, 97-100, 119, 131, 168, 170, 171, 174-5, 176, 177, 180-2, 183, 186, 187, 188, 229-62
dragonfly 67, 154, 183; *Mecistogaster* 65; *Megalopresprus* 65
dungfly (*Scatophaga stercoraria*) 158, 180, 183
dunnock (*Prunella modularis*) 174, 181, 183

earthworm 107, 108, 109, 114, 115, 116, 118, 119, 122, 267-9
eastern bluebird (*Sialis sialis*) 202
ecosystem 34
effective immune response 36
effective productivity 245
egg: cost of laying 217; dumping 4, 215; laying hypersynchrony 201; laying season 221; laying sequence 201, 208, 223; laying synchrony 201, 210, 221. laying tactic 224; laying time 217; ownership 235, 250; recognition 216; rejection 220, 235, 241, 256, 262; removal 200, 210, 211, 216; 'scatter laying' 210, 211; stealing 177
egret 105, 109; cattle *(Bubuleus ibis)* 104, 170; snowy (*Egretta thula*) 28
eider (*Somateria mullissima*) 106, 119, 209; raft 119
elephant 230
elephant seal (*Mirounga angustirostris*) 158, 170, 171, 181, 182
encapsulation 46
energy budget 141; for gamete production 146, 147
energy expenditure 38, 107
energy requirement: daily 117, 122
epidemiology 5
equal payoff line (*EPOL*) 5, 272-86
equal payoff point 272
Eriophora fuliginea 87
Ero fulcata 86
Euglossa cordata 197
eusociality 230, 258, 261; communal route to 230; familial route to 230; semi-social route to 230
evasion 3, 46, 47, 48, 114, 173-4
evolutionarily stable flock composition 268
evolutionary stable point 268, 269
evolutionary stable state 127-51
evolutionary stable strategy (*ESS*) 6, 158, 176, 203; mixed 130, 132, 134, 155, 156, 157, 165, 172, 187
evolutionary equilibrium 5
exclusion 52
experience 78, 113, 166, 177, 203
exploratory activity 11
extinction 45, 210

facilitation 166

Fasciola hepatica 40
fecundity 38, 39, 51
female choice 24, 166, 179, 185
fertilization 162, 175, 181; external 185; internal 140, 144, 185
fighting 26, 119, 165, 178, 197; ability 85, 262, 269
fish 173
fitness 177, 268, 282; biasing 4, 229-62; coefficient of variation in individual 240; frequency-dependent 127; lifetime 236, 240; lifetime inclusive 242; inclusive 216, 219, 233-62; individual 132, 233, 234, 237; of producers and scroungers 128-51; parental 243; *per capita* 248; surface 280
flocking 27, 99-104, 110, 111, 116, 119, 166, 267-9, 283; equilibrium flock size 115; local information advantage 27, 28, 97-104, 108; mixed species 108; 'species flock' 53; unstable equilibrium in 101
flounder (*Platichthys flesus*) 54
flour beetle 276; *Tribolium confusum* 12, 40
following 100
food-scrounging 95-124; interspecific 104-24; intraspecific 96-104
foraging efficiency 201
frequency-dependence 4, 29, 88, 99, 100, 112-13, 116, 131, 134, 137, 156, 162, 165, 187, 203, 267-86
frigate bird (*Fregata* spp.) 118, 121
frog 155, 173, 179; bullfrog (*Rana catesbeiana*) 8, 157, 179; gladiator (*Hyla rosenbergi*) 178; green tree (*H. cinerea*) 156, 157; *Physaleamus pustulosus* 168; prepollical spines in 178
fruit fly (*Drosophila* spp.) 180
fulvous tree duck (*Dendrogyna bicolor*) 215

games theory 155, 245; and producer/scrounger relationships 271-4; hawk/dove game 271; n-persons game 129; zero-sum game 232
gamete 3, 138, 140, 141; disassortative fusion between 142; division of resources between 141; expenditure on 147; investment in 9-10, 138, 145; megagamete 140; microgamete 140
Gammarus lacustris 12
gannet (*Sula bassana*) 118
gastric ulcer 159
gene fixation 178
generalist scrounger 121, 122
generation time 120
genetic monomorphism 129, 157
genetic polymorphism 43, 44, 129, 156, 157, 162, 165, 193
genetic quality 180
geotaxis 12
golden digger wasp (*Sphex ichneumoneus*) 8, 9, 88, 106, 134
golden plover (*Pluvialis apricaria*) 107, 108, 109, 114, 115, 119
great tit (*Parus major*) 158
groove-billed ani (*Crotophaga sulcirostris*) 5, 212, 229, 236, 241, 250, 255, 256, 257
group living 4, 17; and defence 260; and thermal regulation 272; anti-predator effect 248, 271, 272; conspicuousness 248; cooperation within 257; effect on foraging 248, 271; effect on reproduction 248, 256, 258; natal group 251, 255; size 123, 169, 234-62
group selection 262
guarding 81, 84, 176, 257, 274, 285
gull 104, 105, 106, 109, 117, 118, 127, 123; black-headed (*Larus ridibundus*) 107-9, 114-19, 122, 267-9, 284, 285, 286; glaucous (*L. hyperboreus*) 119; great blackbacked (*L. marinus*) 109; herring (*L. argentatus*) 105, 109, 268; laughing (*L. atricilla*) 117, 123; ring-billed (*L. delawarensis*) 105-8

habitat: marginal 257; saturation 257, 260
handling cost 27, 38, 106, 108, 109, 110, 111, 116
hanging fly (*Bittacus apicalis*) 87, 177
harem 8, 156, 160, 171, 248, 268, 285
Harris sparrow (*Zonotrichia querula*) 14, 97-100, 119
hatching success 213
heliconiid 30
helminth 10, 46
helper 243, 256; at the nest 230, 243,

245, 255, 258, 260; sneaks as 165-71
Hemiptera: *Arachnocoris* 66; Phylinae 66
Heriades 196
heritability 180
hermaphrodite 55, 139; sequential 139; simultaneous 140
hermit crab 63, 69; *Pagurus bernhardus* 69
homeostasis 57
homoiotherm 57
hooded merganser (*Mergus cucullatus*) 209
'hopeful reproductive' 171, 261
hormonal suppression 260
host 1-3, 34-57, 61-91, 105, 107-24, 129, 195-224; accidental 35; additional 35; alternative 73; auxiliary 35; defence 2, 34; definitive 43; immune system (*see also* immune response) 14, 44; intermediate 10-13, 43; principal 35; required 35; resistance 41, 49; response 206-9
house sparrow (*Passer domesticus*) 97, 100-4, 111, 117, 268, 286
'hoverer' strategy 156
humming bird 67, 118, 120; *Phaethornis longirostris* 65
hunting dog (*Lycaon pictus*) 112-13, 114, 119, 230, 250
hunting efficiency 114
hyaena (*Crocuta* spp.) 104, 105, 114, 117, 119, 230, 257; spotted (*C. crocuta*) 112-13
Hymenolepis: *citelli* 40; *diminuta* 39, 51
Hymenoptera 63, 220, 223, 243, 257, 258, 259; aculeate 62

ideal free distribution 129, 268
imitation 95
immune privileged site 47
immune response 45, 47, 49, 52; non-specific 45
immunodepression 47, 49, 50
immunopathology 46
inbreeding 256; cost 261
incubation 203, 212, 216, 217, 234; period 208, 256
individual recognition 26, 172, 176
individual selection 179
individual strategy 118
infection: level of 40; patency 49
infectious disease 5, 267, 280-1, 285
ingestion rate 114
inquiline 63
interference 272
internal state 286
interspecific brood care 209
Ischnothele 69
isogamy 140

jackal (*Canis mesomelas*) 114, 230, 243
jackdaw (*Corvus monedula*) 118

'kidnapping' 209
kin 202, 216-19, 223, 231; group 237; recognition 209; selection 4, 168, 209, 216-19, 221, 222, 258
kittiwake (*Rissa tridactyla*) 118
kleptobiosis 2, 61-91; generalist 70, 80, 81; specialist 70, 81
kleptogamy 1, 8
kleptoparasitism 1, 2, 3, 7, 8, 27, 61-91, 95-124; age-related 105, 118; ecto 68; generalist 115, 123; secondary 117; specialist 121
K-selection 38, 81, 120

lactation 50, 257, 260, 261, 262
lapwing (*Vanellus vanellus*) 106, 107, 108, 109, 114, 115, 116, 117, 118, 119, 122, 267-9, 284, 285, 286
'law of battle' 178
learning 78, 80, 105, 286; ability 123-4
Leishmania 42, 48, *donovani* 42
lek 8, 165-88
leopard (*Panthera pardus*) 110
life-cycle 34, 37, 48, 285
life-dinner principle 121-2
life history strategy 120
Ligula intestinalis 10, 11
lion (*Panthera leo*) 104, 112-13, 114, 119, 230
little blue heron (*Egretta caerulea*) 28
longear sunfish (*Lepomis megalotus*) 184
Lymnea tomentosa 40
lymphocyte 47, 48, 49

Macaca 186
macrophage 47, 48

maladaptation 155, 212
malaria 280, 281, 285
mallard (*Anas platyrhynchos*) 12
manakin (*Manacus manacus*) 166
manipulation 1, 10-13, 27, 90, 91, 177, 229, 231, 243, 250, 251, 258, 260
marmoset 14, 243
mate choice *see* female choice, mate selection
male searching 154
mate selection 14, 167
mating strain 141
mating strategy 137
mating success 166
mating system 84, 230, 242, 258; monogamy 163, 164, 173, 230, 234, 242, 243, 257; polyandry, sequential 185; polygamy 173, 178, 248; polygyny 163, 164, 179, 185, 188, 200; promiscuity 181, 183-7, 242; serial polygamy 197
mating type 140
mealworm (*Tenebrio molitor*) 100-4, 111
megachilid bee 196-7; *Chalcidoma* 196; *Dianthidium* 196; *Hoplitis* 196; *Osmia* 196; *Stelis bilineolata* 197
membrane synthesis 48
metacercaria 11, 36
microhabitat 52, 69
migration 257
milkweed (*Asclepias curassavica*) 28
millipede 177
mimicry 2, 30-1, 48, 96; age-dependent 28; aggressive 28; automimicry 28, 29; Batesian 7, 22, 28, 29; egg 122; molecular 47, 50; social 27-8, 30; transsexual 29-30, 160, 173
Mischocyttarus drewseni 64
mite 68, 154; orobatid 177
mixed strategy 129, 134, 139
mobbing 112
model; cost-increasing 2, 7, 13, 19-22; frequency-dependent 4; games theory 271-4; negative binomial 36
mole rat 230, 243, 257
monarch (*Danaus plexippus*) 28, 29
mongoose 230; banded 257; dwarf 257, 260
monogenea 56
monophyletic group 87
mosquito: anopheline 278, 280-1
motivation 65
multiple paternity 163
muskrat 13
mussel 106
mustelid 113
mutualism 1, 67, 258; in breeding 168, 195, 275
Mysmena 87
Mysmenidae 69, 70
Mysmenopsis 80; *dipluramigo* 69; *gamboa* 69; *ischnamigo* 69

nectar 64
negotiation 188
nematode 42, 50, 53, 55
Neopalthis madates 67
Nephila clavipes 68, 70-87, 111-12
Nereis fucata 69
nest 178; building 1, 195, 203, 233; clay 178; communal 212; competition for sites 222; dump 196, 212; 'kolchoz' 196; predation 220
newt 156; red-spotted (*Notophthalmus viridescens*) 180
niche 52, 53, 56, 129; diversification 53; expansion 55; foraging 104; fundamental 53; overlap 52; realized 53; scrounger 285; segregation 53, 55; space 134
Nomeda 89, 90
Nuctenea 87
nuptial gift: in birds 183; in insects 154, 177, 183

oestrus 144, 170, 176, 186
offspring recognition 219
offspring survival 197, 206, 209, 212
opportunism 2, 35, 91, 104, 105, 106, 117, 122, 123, 158, 268, 273, 286
optimal brood size 257
optimal diet 52
optimal division of labour 246
optimal fitness-biasing 229, 245
optimal foraging 6
optimal group size 117, 229, 232, 239; composition 250-8
optimality model 206
optimal social strategy 252-4
ostrich (*Struthio ramelus*) 216, 220, 229, 262

ovarian cycle 260; olfactory suppression 260
overlap of generations 258
oviposition 67, 89, 179, 180, 183
ovum 143, 160, 206
owl 222
ownership convention 156

pack 112-13
pair formation 212
parasitism 2, 4, 7, 8, 10-13, 14, 15, 34-57, 63, 68, 82, 83, 89, 90, 95, 96, 111, 127, 129, 134, 149, 150, 167, 195-224, 271; energetics of 37-40; generalist 35; intersexual 137-51; intrasexual 138; obligate 35, 197; opportunist 35; pathogenicity 43, 44; site selection 38; social 63; specialist 35; transmission 40, 43, 44, 46, 49, 50; virulence 43
parentage: biochemical analyses of 196, 198, 199, 202, 239
parental care 1, 3, 4, 145, 149, 160-3, 183, 194, 196, 212, 217, 219, 221, 223, 229, 234, 255; male 242
parental cost 136-7
parental investment 9, 63, 87-91, 138, 179
pareto optimum 262
Passiflora 30
paternity 184
'patroller' strategy 156
'paying for permission' 187
payoff asymmetry 172, 217
payoff surface 5, 284, 285
peculation 2, 61, 62, 63, 68, 90
perch (*Perca fluviatilis*) 36; yellow (*P. flavescens*) 10
'persuasion' 183
Phasianidae
phenotype: age and 136; continuous 130-4; discrete 134-6; extended 13; flexible 203, 214; material determination of 137; parental determination of 127-8, 136-7
phenotype – limited strategy 3, 156, 158, 165, 266, 270, 286; in producers and scroungers 130-51
philanderer 163
Philanthus 78
Philomyza 68
philopatry 216
Pholcus 86
photophobia 12
phototaxis 12
pied flycatcher (*Ficedula hypoleuca*) 156, 163-5, 173, 179, 181
pied kingfisher 248
pied wagtail 261
pika: *Ochotona collaris* 55; *O. princeps* 55
pilfering 2, 61-91
piracy *see* kleptoparasitism
pituitary gland 159
Plasmodium 41, 48, 280
'playing the field' 129
plerocercoid 10
pocellid crab 69
pochard (*Aythya ferina*) 215; red-crested (*Netta rufina*) 215
Polistes fuscatus 240; *metricus* 221, 239, 259
pollinator 64
polyclonal stimulation 49
Polymorphus paradoxus 12
population 1, 5, 7, 34-57, 196; cycle 81; density 129, 213; dispersal pattern 250; dynamics 5, 81, 268, 274-86; intrinsic rate of increase 39, 274, 282; regulation 50, 222; size 120; stability 5, 50, 257; structure 216, 250
Portia fimbriata 86
preadaptation 66
precocial young 4, 204, 209, 216, 221
predation risk 157, 159, 166, 167, 216-17
predator 16, 96, 118, 121-2, 123, 212, 222; generalist 25; specialist 25
pregnancy 50
prey availability 119
prey rejection 109
prey selection 115
primary producer 64
Prionyx atratus 89
probing 20, 21, 26
procellariiform 123
proximate explanation 180
pseudoscorpion 177
Ptinella aptera 145
puffin (*Fratercula arctica*) 106, 109, 110, 116, 117, 118
puma (*Panthera concolor*) 113, 114

pup-killing 260
pure strategy 157

quail 196, 216; bobwhite (*Colinus virginianus*) 222

rabbit 280
raccoon 212
'raffle principle' 144, 145
Raillietina cesticillus 12
rank *see* dominance relationships
rate of evolution 120-1; asymmetry in 120-1
rate-limitation 267, 269
ratite 196
ray 53
red deer (*Cervus elephus*) 8, 156, 158, 170, 171, 285
redhead (*Aythya americana*) 202, 203, 208, 215, 223
reference cue 26
relatedness 4, 216, 247; coefficient of (*r*) 218-22, 236-62; 'double-accounting' of 236, 242
replication rate 127
replicator 127, 138
reproductive success: and age 139; and brood parasitism 201; annual 234, 244, 256; breeding season 268; fledging 243; lifetime 159, 234, 241, 256; offspring production 229; of wasp nests 239; *per capita* 243, 256
reptile 173
resident advantage 158
resource accrual 26, 122
resource gradient 53
resource partitioning 2, 35, 50-6, 57, 245; interspecific 52-6; intraspecific 50-2; of kleptoparasite hosts 116, 119; of scrounger hosts 120; spatial 53, 54; temporal 52
respiration 80; lung 80; tracheal 80
resting stage 57
retaliation 1, 3, 106, 107, 110-13; aggressive 114
ritualization 61
roach (*Rutilus rutilus*) 11
robber fly 68
r selection 38, 39, 57, 81, 120
ruddy duck (*Oxyura jamaicensis*) 215
ruff (*Philomachus pugnax*) 8, 156, 165, 169, 172, 173
'rule of thumb' 157

safe cue space 16-25
sage grouse (*Centrocercus urophasianus*) 166
salamander 156, 178, 196; *Ambyostoma maculatum*) 177; *A. tigrinum* 177; *Ensatina* 177; *Plethodon* 177
salmon (*Salmo salar*) 184
satellite strategy 1, 8, 83, 165-77, 261; coercion in 172; in mating 7, 29, 155, 157
satin bower bird (*Ptilinorhynchus violaceus*) 177, 178
Scarabeidae 63, 88, 89; *Aphodius rufipes* 89; *Geotrupes spiniger* 89
scavenger 95, 104
Schistocephalus solidus 10, 39
Schistosoma 14, 42, 48, 49; *mansoni* 48
scorpion fly (*Panorpa* spp.) 65, 66, 67, 87, 154, 177, 180, 181
scrounger specialisation 121
scrounging by deception 2, 13-31
scrub jay 230
'searcher' strategy 100-4, 286
secondary sexual characteristic 15
selection pressure 121-3
sequestering 28, 29
sex 3, 127-51; evolution of the sexes 5, 127-51; sexes as producers and scroungers 137-51
sex ratio 128, 134, 138-40; autosomal 138; equal expenditure in 139; functional 140; skewed 248
sexual defence 178
sexual dimorphism 173, 186
sexual interference 156, 177
sexual selection 10, 30, 183, 185, 186, 188
sexual swelling 186, 187
sheep 185
sheepshead minnow (*Cyprinodon variegatus*) 11
'shepherding' effect 97, 119
shiner (Notropis cornutus) 10; spottail (*N. hudsonius*) 10
sickle cell haemoglobin 43
'side payment' strategy 156, 159, 105, 170
signal 13, 71, 81, 168, 184; auditory 168; economy 27-8; form and function 28; olfactory 168; range 167; visual 168

similarity 16
skua 118, 123; arctic (*Stercorarius parasiticus*) 62, 104, 106, 109, 110, 116, 117; great (*S. skua*) 118
slave-taking 90
snatching 100
'sneaky' strategy 3, 4, 8, 29, 154-88, 274
society 63, 254; despotic 257; egalitarian 229-62
solitary bee 196
solitary wasp 196
song 179
spawning 144, 176, 184
speciation 40
species divergence 55
species recognition 180
speckled wood butterfly (*Pararge aegeria*) 156, 158
sperm 9-10, 138, 160, 177, 185, 188; competition 138, 143-6, 184, 185; contribution to zygote 138; displacement 144; ejection 183; pelagic 183; quality 186; storage 144; transfer 87
spermatophore 177, 178, 180
spite 177-8
splendid wren (*Malurus splendens*) 243, 244, 258
stag 171
starling (*Sturnus vulgaris*) 105, 197-201, 202, 203, 222
status *see* dominance relationships
status-limited cue 26
stereotypy 78, 83; in foraging behaviour 123
stick insect 14
Stizoides unicinctus 89
superstimulus 166
sapraoesophageal ganglion 11
survivability 180
survivorship: adult 239
swan 106
switching strategy 116-18, 130, 131, 132, 134, 135, 139, 157, 160-3, 268, 273, 286
symbiosis 67, 87, 129
symphile 67, 68

tamarin 243
Tasmanian native hen 229, 248, 259
termite 243
tern (*Sterna* spp.) 118, 170, 123
territoriality 8, 30, 62, 119, 165-88, 232, 238, 242, 256, 261; long-term benefits of 119; mobile 119; natal territory 237, 243; parental territory 237, 259; polyterritoriality 179
testis 185, 186
testosterone 97
theridiid 80, 86; *Theridion* 86
Thomphaea 86
three-spined stickleback (*Gasterosteus aculeatus*) 10, 22, 156, 158, 177
tick 68
toad 169, 173, 179; natterjack (*Bufo calamita*) 156, 157, 158, 173, 183
toleration 3, 113-16, 165, 168
tortoise 53
total cue space 16
trade-off 169
'transvestism' 29, 173
travel cost 232
trematode 11
Trichinella spiralis 11, 13
Trichuris muris 41, 49
trophallaxis 68
trout 23
'trust' 176
Trypanosoma 48; *brucei* 48, 49; *cruzi* 48
Trypoxylon politum 82, 196
tufted duck (*Aythya fuligula*) 215
tumour 159
Tylodelphis podicipina 36

uganda kob (*Adenota kob thomasi*) 14
uropygial gland 121
usurpation 1, 2, 3, 4, 7-13, 29, 30, 85, 87, 88, 89, 90, 96, 100, 243, 271, 273; interspecific 197; obligatory 86

virus: *Myxoma* 280; transmission 281, virulence 280
Volvocidae 141

waterbuck (*Kobus ellipsiprymnus*) 170-1, 172
weaponry 15
web 2, 64-87, 104, 110; barrier 84; communal 67
weevil: *Callosobruchus maculatus* 283; *C. rhodesianus* 283

white-crowned sparrow (*Zonotrichia leucophrys*) 14
white-throated sparrow 14
widowbird (*Vidua*) 90
wild turkey (*Meleagris gallopavo*) 168
wolf (*Canis lupus*) 114, 230, 243, 257
wolverine (*Gulo gulo*) 113
wood duck (*Aix sponsa*) 202, 203, 204, 208, 210, 212-15, 223
woodpecker 222; acorn (*Melanerpes formicivorus*) 212, 240, 257, 259

Zygote 3, 138, 141-51